Advances in Disease Vector Research

Advances in Disease Vector Research

Edited by

Kerry F. Harris
Virus Vector Laboratory, Department of Entomology, Texas A&M University, College Station, Texas 77843, USA

Editorial Board

Advances in Disease Vector Research

Volume 6

Edited by Kerry F. Harris

With Contributions by
S. Barbagallo M.R. Brown W. Burgdorfer
F.H. Collins V. Finnerty D.A. Golino
S.F. Hayes H. Hibino A.O. Lea G.P. Martelli
J. Nedelman G.N. Oldfield S.M. Paskewitz
R.T. Plumb A.H. Purcell B. Raccah
C.N. Roistacher C.E. Taylor R.D. Ward

With 36 Illustrations

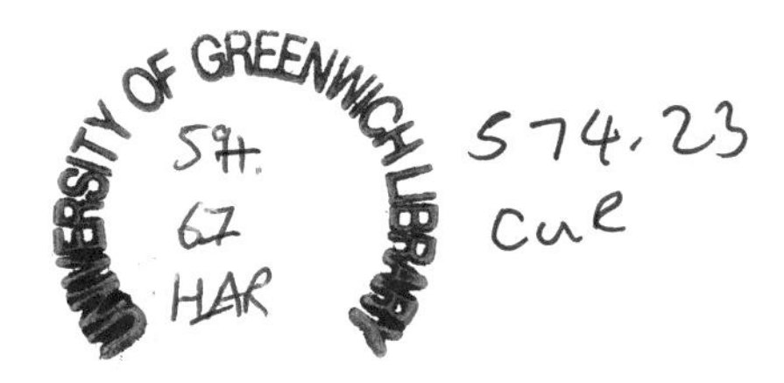

Springer-Verlag
New York Berlin Heidelberg
London Paris Tokyo Hong Kong

Kerry F. Harris
Virus Vector Laboratory
Department of Entomology
Texas A&M University
College Station, Texas 77843
USA

Volumes 1 and 2 of *Current Topics in Vector Research* were published by Praeger Publishers, New York, New York.

ISSN: 0934–6112

Typeset by TCSystems, Inc., Shippensburg, Pennsylvania.
Printed and bound by Quinn-Woodbine, Woodbine, New Jersey.
Printed in the United States of America.

9 8 7 6 5 4 3 2 1

ISBN 0-387-97080-0 Springer-Verlag New York Berlin Heidelberg
ISBN 3-540-97080-0 Springer-Verlag Berlin Heidelberg New York

A Letter from the Editor

Dear Readers and Authors,

Starting with Volume 5, this series will be entitled

Advances in Disease Vector Research

We feel that this new title more accurately defines the scope of the series and will help prevent any confusion about the scientific disciplines covered in it. The subject matter and editorial policy remain otherwise unchanged.

Kerry F. Harris, Editor
Springer-Verlag, Publisher

Preface

I think the reader will agree that we have attained a good balance in Volume 6 between human- or animal-host and plant-host–related topics from outstanding research scientists.

In Chapter 1, Frank Collins, Susan Paskewitz, and Victoria Finnerty explore the potential of recombinant DNA technology to distinguish individual species and to establish phylogenetic relationships among member species in the *Anopheles gambiae* species complex, which includes the principal malaria vectors. Currently, relatively little is known about these morphologically identical species that are sympatric over most of their range but are not always equally involved in malaria transmission. With respect to individual species identification, the researchers have thus far described two DNA fragments, derived from the ribosomal DNA intergenic spacer region, that reliably distinguish five species in the complex by means of an RFLP visualized on a Southern blot. They have also described other species-specific fragments derived from a ribosomal DNA intron that could form the basis for a rapid dot blot assay.

With respect to the phylogenetic relationships among member species in the complex, Collins, Paskewitz, and Finnerty focus on a comparison at the level of restriction site mapping and Southern analysis of the rDNA intergenic spacer regions. As expected, the two spacer regions near the coding region junctions are well conserved among the species, whereas the central regions tend to be highly variable among member species in the complex. This variability appears, at least partially, to be due to differences in the reiteration number of a small repeated sequence. Considered together, these recombinant DNA-based data suggest a phylogenetic relationship that differs from the one based on the cytogenetic data. However, the authors point out that their analysis is in its infancy and much more data, especially from the study of other genetic regions, are needed in order to draw firm conclusions concerning the evolution of this complex.

Mark Brown and Arden Lea, in Chapter 2, advance the concept that the neuroendocrine and midgut endocrine systems form a regulatory axis in the adult mosquito. Brain neurosecretory cells with axons to the corpora cardiaca are known to be the source of several different types of peptide hormones. The midgut only recently has been recognized as an endocrine organ in insects. Within the midgut epithelium of the mosquito are several hundred cells that resemble the endocrine cells that are dispersed in the vertebrate gut. Morphological features of cells from both systems in the mosquito are compared, and results from immunocytochemical surveys are summarized to demonstrate that at least one type of peptide messenger occurs in both systems. The authors examine unique and shared characteristics of the neuroendocrine and midgut endocrine systems and discuss regulatory roles that may affect physiological coordination and vector–pathogen interaction.

In Chapter 3, Jerry Nedelman reviews and investigates patterns of gametocytemia and the relationship between gametocytemia and infectiousness in *Plasmodium falciparum* malaria. The point of view taken is that of a biometrician who wants to know what insights mathematical models might provide, what biological principles such models are built on, and how strong the data that support those principles are. New analyses of data from the Garki project, conducted by the World Health Organization in Nigeria during the 1970s, are presented.

In his chapter on the biology of phlebotomine sandflies, Chapter 4, Richard Ward has selected topics that are either new (pheromones) or that reflect significant advances made during the last decade. He has ignored most aspects of the sandfly/*Leishmania* relationship, since this was reviewed extensively in a recent two-volume publication edited by Peters and Killick-Kendrich (Reference 76 in bibliography). One exception is the exciting new observations on sugar feeding by sandflies that were made in long-term investigations in France and Israel. Almost every aspect of research into sandfly biology has benefited recently from improved colonization techniques, and it is this subject that introduces the topics discussed. In this respect, studies on sandfly viral transmission, age determination of adults, behavior, and, in particular, systematics have received increased attention, which has led to new insights.

Willie Burgdorfer and Stanley Hayes provide a comprehensive overview of the arthropod-borne borrelioses in Chapter 5; they compare vector–spirochete relationships in louse- and tick-borne borrelioses with those recorded so far by Lyme disease. Accordingly, the development of the recently discovered spirochete *Borrelia burgdorferi,* the causative agent of Lyme disease, is unique in that it primarily takes place in the midgut diverticula of its tick vectors, which are members of the *Ixodes ricinus* complex. Penetration of the gut epithelium and subsequent invasion of tissues may occur during feeding and may lead to tissue infections, including infections of the salivary glands. The spirochetes are thought to be transmitted by infectious saliva and possibly by regurgitated infectious midgut fluids.

The authors review associations of *B. burgdorferi* with other species of ticks and with such hematophagous insects as biting flies and mosquitoes. They have also found that spirochetal cycles in nature differ from those established during the past few years.

In a final section, Burgdorfer and Hayes address the potential significance of vesicles, often filled with membrane-bound chromatic bodies or granules, associated with *B. burgdorferi.* Whereas molecular analyses revealed the presence of genetic material, there is, as yet, no evidence to suggest that vesicles or their granular contents play a part in the complex developmental cycle to give rise to new generations of spirochetes.

In Chapter 6, Giovanni Martelli and Charles Taylor discuss the origins and geographical distribution of longidorid and trichodorid nematodes relative to their association with nepoviruses and tobraviruses. This highly developed vector specificity, which is especially apparent among nepoviruses, is examined relative to the ecology of the viruses, with particular reference to recent investigations that have shown that the efficiency of transmission of a particular virus may differ among geographically separated populations of its nematode vector. Mechanisms of virus transmission by nematodes are reviewed and the role played by the vectors in virus dissemination is outlined. Evidence of the adaptability of nepoviruses and tobraviruses to a complex ecological situation—in which selection for nematode transmission, host plant compatibility, and transmission through seed is a continuous process—has been obtained.

Methods for detecting plant viruses in their vectors are presented and discussed by Roger Plumb in Chapter 7. Bioassays for the presence of viruses in vectors using test plants have a long history, but only recently have methods based on fine structure, immunology, or nucleic acid hybridization become available. Each of these methods has advantages and disadvantages. The principal advantage in epidermiological studies is speed, as the disadvantage is interpretation relative to the infectivity of the vectors. Although such methods are relatively new, they are already being used to speed up decisions on the need to apply pesticide control against vectors. This use will increase where it can be demonstrated that such indirect methods give an accurate assessment of infectivity. In the longer term, visualization and detection of viruses in vectors will add much to our knowledge of virus–vector relationships and thus will aid virus control. Plumb considers some of these aspects and reviews the history of detection to date for the most important plant virus vector groups.

In Chapter 8, Hiroyuki Hibino presents a comprehensive coverage of insect-borne viruses affecting rice. Of fifteen rice viruses worldwide, thirteen are insect-borne. Many of these insect-borne viruses occur in Asia. Of the thirteen viruses, six are borne by leaf-hoppers, five by planthoppers, one by aphids, and one by beetles; eleven persist in their vector insects and two are semipersistent. Three of the persistent viruses are also known to be vertically transmitted through the vectors' eggs. The vector–virus interactions of rice viruses are generally specific. All the viruses so far analyzed have RNA either in the single- or double-stranded form. Each insect-borne rice virus is fully described with respect to its geographical distribution, properties, host range, vector relations, plant-host relations, epidemiology, and control.

Alex Purcell, author of Chapter 9, reviews the research on the transmission of xylem-limited bacteria by suctorial insects. The bacterium *Xulella fastidiosa* was first isolated in culture a decade ago, although it was classified only a few years ago. Since then, various strains of this bacterium have been isolated from an expanding list of plant species. The plant diseases caused by this group of bacteria thus may have been largely underestimated. There are numerous interesting questions concerning its vector transmission, many of which remain largely unanswered. What is the basis for vector group-specificity? Why is *X. fastidiosa* limited to the Americas? Or is it? Purcell concludes the chapter with the prediction that the xylem habitat will not prove to be so sterile as many have previously assumed, and that its microbiology should in fact provide a fertile field for future research and possibly for applications of new biotechnology.

In Chapter 10, Deborah Golino and George Oldfield discuss the available research data on *Spiroplasma citri, Spiroplasma kunkelii,* and *Spiroplasma phoeniceum,* their leafhopper vectors, and the diseases they cause. Our understanding of the transmission of the plant pathogenic spiroplasmas by their leafhopper vectors is limited by the complexity of the events that disseminate disease in nature. The identity and ecology of the individual leafhopper species that transmit these spiroplasmas in large part determines the economic impact of the diseases they cause. Because of the technical difficulties posed by detailed field studies of vector relationships, only limited data are available on the natural vectors of each species of spiroplasma. The authors define the differences between natural, experimental, and artificial vectors of spiroplasmas and suggest that these differences might be important criteria in evaluating the information available on transmission. Much of the information on spiroplasma–leafhopper relationships has been obtained under experimental conditions; the limitations of those conditions are discussed.

Benny Raccah, Chet Roistacher, and Sebastiano Barbagallo conclude the volume, in Chapter 11, on the semipersistent mode of transmission of plant viruses. Descriptions of the viruses, and the vectors involved, are included, with special emphases being given to virus–vector interactions, vector–virus-specificity, and the dependence of one virus on another. The authors also speculate on the mechanism of transmission in light of our present knowledge of virus retention sites in the vector.

Among the semipersistent viruses, two are known to be of outstanding economic importance: citrus tristeza virus (CTV) and, to a lesser extent, beet yellows virus (BYV). The role of aphid vectors in CTV epidemics is discussed in detail, as is the biology of important vector species. A helpful key for the recognition of common aphid species in the field is in a separate appendix.

In reporting on the epidemics of CTV in South and North America and the Mediterranean, the authors discuss at length the lessons to be learned from former pandemics and precautions to be adopted to prevent future disasters for the citrus industry. Methods for diagnosis, eradication, and suppression of disease spread are discussed in detail. Various control measures that have been used or that show a potential for use are discussed, including quarantine, cross-protection, and the use of resistant rootstocks.

I thank each of the authors to Volume 6 of *Advances in Disease Vector Research* for their outstanding contributions. I thank them too for their patience and understanding as we worked together, during sometimes trying times, to meet a demanding publication schedule. The support, understanding, and technical assistance of the members of the Editorial Board and the staff of Springer-Verlag are humbly acknowledged and sincerely appreciated.

Kerry F. Harris

Contents

A Letter from the Editor v
Preface vii
Contributors xvii

1 Ribosomal RNA Genes of the *Anopheles gambiae* Species Complex
Frank H. Collins, Susan M. Paskewitz, and Victoria Finnerty 1

Introduction 1
The *Anopheles gambiae* Complex 2
General Characteristics of Dipteran rDNA 4
Anopheles gambiae rDNA: General Structure 6
Intergenic Spacer Structure 8
Intervening rDNA Sequences 12
Applied Uses of rDNA 15
Evolution of rDNA Within the Complex 21
Summary 23
References 23

2 Neuroendocrine and Midgut Endocrine Systems in the Adult Mosquito
Mark R. Brown and Arden O. Lea 29

Introduction 29
The Nervous System 31
The Neuroendocrine System 32
The Midgut Endocrine System 35
Neuropeptide-Like Immunoreactivity in the Nervous System and Midgut Endocrine System ... 39
Comparison of the Neuroendocrine and Midgut Endocrine Systems 46
References 50

3 Gametocytemia and Infectiousness in Falciparum Malaria: Observations and Models
Jerry Nedelman 59

Introduction 59
Gametocytogenesis 60
Patterns of Gametocytemia 61
Circadian Rhythm 65
Infectiousness Versus Gametocytemia 65
Reservoir of Infection 67
Immune Response 71
Gametocytemia and Infectiousness in Mathematical Models of Malaria 73
Conclusions 86
References 87

4 Some Aspects of the Biology of Phlebotomine Sandfly Vectors
Richard D. Ward 91

Introduction 91
Colonization 92
Sandflies as Reservoirs and Vectors of Viruses 96
Age Determination 102
Sugar Feeding 104
Species Groups and Complexes 107
Chemically Mediated Behavior 116
References 119

5 Vector-Spirochete Relationships in Louse-Borne and Tick-Borne Borrelioses with Emphasis on Lyme Disease
Willy Burgdorfer and Stanley F. Hayes 127

Introduction 127
The Behavior of Louse-Borne and Tick-Borne Spirochetes in Their Vectors 130
The Behavior of *Borrelia theileri* in Its Tick Vectors 133
The Behavior of the Lyme Disease Spirochete *Borrelia burgdorferi* in Its Tick Vectors 134
The Relationship of *Borrelia burgdorferi* to Nonspecific Tick Vectors and Other Hematophagous Arthropods 139
Borrelia burgdorferi: Subject of a Complex Development Cycle? 141
Conclusion 144
References 147

6 Distribution of Viruses and Their Nematode Vectors
Giovanni P. Martelli and Charles E. Taylor 151

Introduction 151
Nematode-Transmitted Viruses as Plant Pathogens 152
Causal Agents 156
Vectors 168
Virus–Vector Associations 172
Discussion 177
References 180

7 Detecting Plant Viruses in Their Vectors
Roger T. Plumb 191

Introduction 191
Aphids 192
Plant and Leafhoppers 197
Nematodes 199
Mites 200
Fungi 201
Beetles 201
Other Vector Groups 202
Concluding Remarks 202
References 203

8 Insect-Borne Viruses of Rice
Hiroyuki Hibino 209

Introduction 209
Geographical Distribution 211
Vector Species 212
Vector-Virus Interactions 213
Nucleic Acids and Proteins 215
References 229

9 Homopteran Transmission of Xylem-Inhabiting Bacteria
Alexander H. Purcell 243

Introduction 243
Historical Background 243
Pathogens 245
The Xylem Environment 247
Vectors 250
Epidemiology 257
Summary 259
References 260

10 Plant Pathogenic Spiroplasmas and Their Leafhopper Vectors
Deborah A. Golino and George N. Oldfield 267

Introduction 267
Spiroplasmas and Their Habitats 268
Leafhoppers That Transmit Spiroplasmas 269
Transmission Terminology 272
Spiroplasma citri 274
Spiroplasma kunkelii 280
Spiroplasma phoeniceum 283
Spiroplasma–Leafhopper Relationships 283
Concluding Remarks 288
Summary 288
References 289

11 Semipersistent Transmission of Viruses by Vectors with Special Emphasis on Citrus Tristeza Virus
Benjamin Raccah, Chester N. Roistacher, and Sebastiano Barbagallo 301

Introduction 301
Viruses 303
Vectors 305
Virus–Vector Relationships 312
Epidemiology and Control 316
Principles of the Approaches to Control 322
Conclusion and Summary 328
Appendix 330
References 332

Index 341

Contributors

Sebastiano Barbagallo
Department of Entomology, University of Catania, Via Valdisavoia 4, Catania, Italy

Mark R. Brown
Department of Entomology, University of Georgia, Athens, Georgia 30602, USA

Willy Burgdorfer
Department of Health and Human Services, Public Health Service, National Institutes of Health, National Institute of Allergy and Infectious Diseases, Laboratory of Pathobiology, Rocky Mountain Laboratories, Hamilton, Montana 59840, USA

Frank H. Collins
Malaria Branch, Division of Parasitic Diseases, Center for Infectious Diseases, Centers for Disease Control, Atlanta, Georgia 30333, USA

Victoria Finnerty
Department of Biology, Emory University, Atlanta, Georgia 30322, USA

Deborah A. Golino
USDA-ARS, Department of Plant Pathology, University of California, Davis, California 95616, USA

Stanley F. Hayes
Department of Health and Human Services, Public Health Service, National Institutes of Health, National Institute of Allergy and Infectious Diseases, Laboratory of Pathobiology, Rocky Mountain Laboratories, Hamilton, Montana 59840, USA

Hiroyuki Hibino
National Agriculture Research Center, Kannondai, Tsukuba 305 Japan

Arden O. Lea
Department of Entomology, University of Georgia, Athens, Georgia 30602, USA

Giovanni P. Martelli
Dipartimento di Patologia vegetale, University of Bari and Centro di Studio del CNR sui Virus e le Virosi delle Colture Mediterranee, 70126 Bari, Italy

Jerry Nedelman
Department of Mathematical Sciences, Clemson University, Clemson, South Carolina 29634-1907, USA

George N. Oldfield
Department of Plant Pathology, University of California, Riverside, California 92521, USA

Susan M. Paskewitz
Malaria Branch, Division of Parasitic Diseases, Center for Infectious Diseases, Centers for Disease Control, Atlanta, Georgia 30333, USA

Roger T. Plumb
Institute of Arable Crops Research, Rothamsted Experimental Station, Harpenden, Herts, AL5 2JQ, United Kingdom

Alexander H. Purcell
Department of Entomological Sciences, University of California, Berkeley, California 94720, USA

Benjamin Raccah
Department of Virology, The Volcani Center, Bet Dagan, Israel

Chester N. Roistacher
Department of Plant Pathology, University of California, Riverside, California 92521, USA

Charles E. Taylor
Scottish Crops Research Institute, Invergowrie, Dundee, DD2 5DA, Scotland

Richard D. Ward
Department of Medical Entomology, Liverpool School of Tropical Medicine, Pembroke Place, Liverpool L 35QA, United Kingdom

1
Ribosomal RNA Genes of the *Anopheles gambiae* Species Complex

Frank H. Collins, Susan M. Paskewitz, and Victoria Finnerty

Introduction

The *Anopheles gambiae* complex is a group of six sibling species of Afro-tropical anophelines. The two most widely distributed species *Anopheles gambiae* and *An. arabiensis* are among the principal vectors of malaria in Africa. *An. melas, An. merus,* and *An. bwambae,* are vectors of limited regional importance, whereas the highly zoophilic *An. quadriannulatus* is not believed to contribute significantly to the transmission of malaria (25, 63, 71, 72). Although some species-specific differences have been described in the distributions of morphometric criteria (15, 20), no morphological taxonomic characters have been observed that permit the unequivocal identification of the species of individual adult or larval specimens. Because the habitats of these six species overlap considerably, both on geographical and ecological scales, two or more members of the complex are often present in the same adult or larval collections.

Definition of the taxonomic entities in the complex was initially achieved by demonstrating reproductive isolation between laboratory colonies (28, 29, 31). Subsequent study has revealed that species identity can be determined, with varying degrees of reliability, by analyzing allozymes, cuticular hydrocarbons, morphometric characters, larval saline tolerance, mitotic chromosome structure, and the distribution of paracentric inversions in the polytene chromosome complement (10, 24, 43, 50, 54, 55). Among these methods, only the latter is absolutely reliable for determining the species of an individual specimen (21, 47). Unfortunately, chromosomes with the appropriate degree of polyteny are present only in the salivary glands of late fourth instar larvae and the

---

Frank H. Collins, Malaria Branch, Division of Parasitic Diseases, Center for Infectious Diseases, Centers for Disease Control, Atlanta, Georgia 30333, USA.
Susan M. Paskewitz, Malaria Branch, Division of Parasitic Diseases, Center for Infectious Diseases, Centers for Disease Control, Atlanta, Georgia, 30333, USA.
Victoria Finnerty, Department of Biology, Emory University, Atlanta, Georgia 30322, USA.

ovarian nurse cells of half gravid females. Moreover, the technical skills and specimen processing requirements of this method constrain its use in routine vector surveillance programs. Nevertheless, it has been an extremely powerful and informative method for studying population, ecological, and evolutionary relationships among the member species in the complex (8, 23).

Our interest in the ribosomal RNA genes (with their associated intervening DNA sequences, collectively known as rDNA) of the *An. gambiae* complex derives from two sources. The most immediate objective has been to develop molecular tools that would permit malariologists and vector control specialists to deal with the problem of species identification. A second, more basic interest relates to the wider range of evolutionary questions which molecular analysis of the rDNA of a group of closely related species can address.

As a target for dealing with the basic question of species identification, rDNA has a number of useful attributes. The rDNA of several groups of closely related Diptera has been studied in detail, and gene copy number, although varying somewhat between taxa, is generally in the range of several hundred to a thousand copies per diploid genome (3). Such repeated DNA sequences can form the basis of very sensitive DNA probe-based assays that are not stage- or sex-specific. Additionally, because DNA is a highly stable molecule, which is relatively undegraded in desiccated or alcohol-preserved samples, sample collection and handling can be greatly simplified.

As a target for evolutionary studies, rDNA has several particularly useful characteristics. This large gene family has a basic architecture and function that has been conserved across virtually all multicellular organisms; thus, there is sequence homology even between rather distantly related organisms (46, 65, 75). Specific regions within this architectural framework, however, appear to evolve extremely rapidly, so that significant differences between very closely related taxa can be detected. For example, rRNA transcription is carried out by its own specific polymerase, RNA polymerase I, which appears to be a more rapidly evolving enzyme than the polymerases responsible for transcription of other genomic coding sequences (49, 68). Thus, even such important signal regions such as promotors can yield valuable information when closely related species are compared. Therefore, structural features and DNA sequences can be identified within the rDNA to permit evolutionary comparisons over a broad range of taxonomic hierarchies.

The *Anopheles gambiae* Complex

The taxa in the *An. gambiae* complex clearly reflect very recent speciation events. All combinations of crosses involving the six species produce viable progeny; although the progeny males are generally sterile, hybrid females are always fertile (30). Field studies in areas where two or more species are sympatric have recorded natural hybrids, on occasion at frequencies approaching 0.1% (7, 69), a level sufficient to suggest that interspecies gene flow may affect the gene pool of some of these six species. Three species are highly tolerant of saline. *Anopheles melas* and *An. merus* are allopatric species that breed primarily in the coastal marshes of western and eastern Africa, respectively. *Anopheles bwambae* has a limited range defined by its breeding sites in the brackish water mineral springs in the Semliki forest of Uganda (70). The other species are freshwater breeders, and, where their ranges overlap, their larvae often share breeding habitats. With the exception of *An. quadriannulatus,* all members of the complex feed upon man to a significant extent (25).

The two most widespread and important malaria vector species *An. gambiae* and *An. arabiensis* are only encountered in association with man, although *An. arabiensis* consistently exhibits a somewhat greater propensity than *An. gambiae* to feed on domestic animals also (25, 71). These two species are chromosomally the most polytypic members of the complex, possibly a consequence of within-species specialization associated with adaptation to the varied ecological niches occupied by man (22, 25). Indeed, detailed studies of *An. gambiae* in West Africa have revealed the presence within this taxon of sympatric populations of chromosomally defined forms exhibiting levels of reproductive isolation that suggest incipient speciation (8, 23).

Phylogenetic relationships among the six defined species have been inferred from the distribution of fixed paracentric inversions in the polytene chromosome complement, as illustrated in Figure 1.1. *Anopheles quadriannulatus* has been proposed as the ancestral form on the basis of ecological and behavioral characteristics, a patchy (presumably relict) distribution in eastern and southern Africa, and a somewhat central polytene chromosome arrangement. All paracentric inversion terminology is defined relative to the standard *An. quadriannulatus* complement. *Anopheles melas* and *An. bwambae* share a fixed inversion on the left arm of chromosome 3, whereas *An. merus* and *An. gambiae*

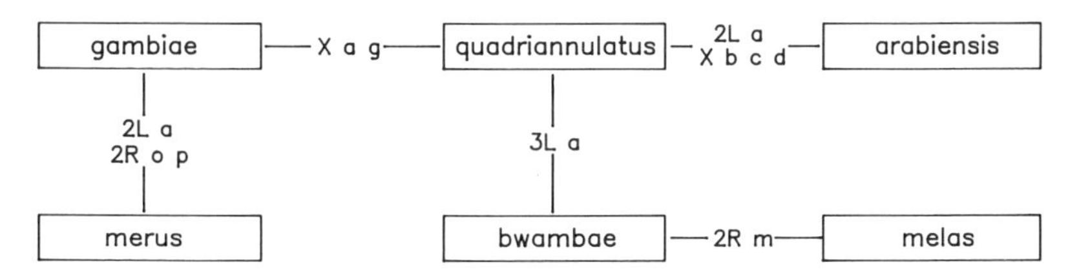

FIGURE 1.1. Phylogenetic relationships in the *An. gambiae* complex as inferred from the distribution of fixed paracentric chromosomal inversions. Adapted from Coluzzi et al. (1).

are grouped because of two shared fixed inversions on the X chromosome. The large number of chromosome 2 polymorphic inversions shared between *An. gambiae* and *An. arabiensis* is presumed to reflect genetic introgression between these two species (reviewed in ref. 25).

Biological and morphological characters, such as habitat, behavior, and egg and adult morphology, would suggest a different phylogeny, with *An. merus–An. melas* and *An. gambiae–An. arabiensis* being the most similar pairs (25). As we shall show below, analysis of rDNA structure of these species lends support to this latter grouping.

General Characteristics of Dipteran rDNA

Ribosomal DNA of species in at least six dipteran families has been cloned and characterized. Among the better studied examples are several *Chironomus* species (Chironomidae), *Sciara coprophila* (Sciaridae), several *Drosophila* species (Drosophilidae), *Calliphora erythrocephala* (Calliphoridae), *Sarcophaga bullata* (Sarcophagidae), and several *Glossina* species (Muscidae) (26, 27, 33; reviewed in ref. 3).

These studies have revealed a number of characteristics common to dipteran rDNA. All have ribosomal RNA genes arranged in a small number of chromosomal locations (often only one), with diploid gene copy number ranging from approximately 100 in *S. coprophila* to more than 1000 in *C. erythrocephala* and several *Drosophila* species (3). In each location there is a long array of ribosomal RNA genes, each separated by an intergenic spacer (IGS). Transcription of the IGS is rare, and transcripts that do occur are rapidly degraded (66, 67). The large (approximately 7- to 8-kb) primary transcript of the gene itself is processed to produce several different rRNA and regulatory components (see Figure 1.2). These include (starting at the 5′ end) the external transcribed spacer (ETS), which varies in length from about 100 bp in *S. coprophila* (60) to nearly 1 kb in *Drosophila* and *Glossina* species (3, 26, 68), the approximately 2-kb, 18S coding sequence, a 1- to 2-kb,

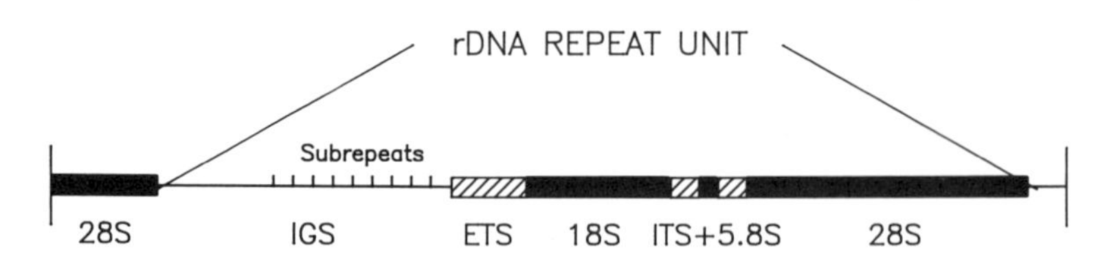

FIGURE 1.2. Typical arrangement of dipteran rDNA genes. The length of the noncoding regions is variable both within and between species. A common characteristic of some genes (not shown) is the presence of non-rDNA sequences inserted at the 3′ end of the 28S.

internal transcribed spacer which contains the 5.8S coding sequence, and an approximately 4-kb 28S coding sequence, which is cleaved into two nearly equal-sized fragments, the 28Sα and 28Sβ sequences (39).

The DNA sequences in much of the transcribed region, particularly the coding regions, are highly conserved among different species of Diptera. The spacers, in contrast—particularly the IGS—diverge rapidly, even between closely related species. Furthermore, considerable IGS length variation is typically observed within individuals (see, e.g., ref. 13, 14). Much of the intraindividual length variation appears to be due to variable numbers of small (< 500-bp), repeated sequences that are typically grouped in a block immediately upstream of the ETS. Some repeats show significant sequence homology with the true promotor in the ETS; thus, they are probably RNA polymerase I loading sites (11, 44, 66). Another source of IGS length variation may be a large region without significant promotor sequence homology located 5′ of the subrepeat array (see, e.g., ref. 68).

A slightly different IGS structure has been reported in some chironomids. Many species have short cistrons of a single invariant size class, which suggests little or no IGS length variation (33). However, *Chironomus thummi thummi* spacer architecture is similar to the other Diptera, with varying numbers of a 120-bp subrepeat in the region immediately proximal to the ETS (62). Unlike the other Diptera, in which the subrepeats are unique to the rDNA, this 120-bp repeat exhibits more than a 90% sequence homology with a satellite DNA family present in approximately 70,000 copies per genome in centromeric heterochromatin (48, 62).

All Diptera examined, except Chironomidae, have noncoding rDNA sequences, variously called introns, intervening sequences, or insertion sequences (hereafter called ISs). The principal insertion region in all species examined is within the 3′ end of the 28 Sβ sequence (3, 36). At least two IS classes have been recorded from some species, and genomic hybridization studies show between-species homologies of portions of certain ISs (2, 36, 64). Each species typically has several IS-size variants, some of which may represent copies deleted in the 5′ end. Quite large insertions are present in some Diptera [a 10-kb IS is common in *D. virilis* (2, 3)], but others show smaller IS classes, such as *S. coprophila,* in which 900- and 1400-bp classes predominate (3). No ISs are reported from the *Chironomidae,* although the presence of ISs of 500 bp or less cannot be excluded.

Most studies suggest that IS containing (IS+) rDNA genes are transcriptionally silent or produce only low levels of aborted transcripts, and some IS+ genes may be clustered in tandem blocks separate from the transcriptionally active, IS free (IS−) genes (4, 5, 44). The IS+ genes can be quite abundant. In *D. virilis,* for example, as many as 75% of the rDNA genes appear to be IS+ (3). Some insect IS DNA contains sequences that have the characteristics of mobile elements, such as open reading frames with reverse transcriptase homology, flanking target site duplications, and 3′ poly A sequences (32, 61, 76). Retroposons lacking long terminal repeats have been isolated from the 28S region of the rDNA of a number of insect species (see, e.g., refs. 9. 76). Thus, IS+ rDNA genes are probably typical pseudogenes, and their high frequency in certain species may reflect a combination of factors, including IS mobility and the molecular amplification characteristically observed in frequently repeated, tandem DNA sequences (34, 35).

Anopheles gambiae rDNA: General Structure

Anopheles gambiae complex rDNA studies have been based on a large number of colonies from different geographical isolates, including twelve of *An. gambiae,* six of *An. arabiensis,* two each of *An. merus* and *An. melas,* and one of *An. quadriannulatus. Anopheles bwambae* is not currently available as a laboratory colony. Although many of these colonies have been in laboratories long enough to compromise the geographical integrity of their gene pools, at least one colony of each species is a new isolate established from wild specimens specifically for these studies. In addition, field collected mosquitoes from Kenya, Zimbabwe, and The Gambia, which represent all five of the species studied, have been examined.

Ribosomal DNA has been cloned from four members of the complex: *An. gambiae, An. arabiensis, An. merus,* and *An. quadriannulatus.* In all cases, the same general cloning strategy was followed. Genomic DNA was extracted from 1 to 2 g of adult mosquitoes containing approximately equal numbers of males and females derived from a colony representing a single geographical isolation. The DNA was purified either on a cesium chloride gradient or with Elutip-d columns (Schleicher & Schuell)[1], partially restricted with the enzyme Sau3A, size fractioned on an agarose gel, and eluted and cloned into the BamHI site of the phage vector EMBL-3 (38). In the first library (*An. gambiae* G3 colony), rDNA insert-containing clones were detected with heterologous rDNA from *S. coprophila* (18). For identification of rDNA-containing clones in other *An. gambiae* complex libraries, the G3 rDNA clones were used.

From an estimate of the *An. gambiae* diploid genome as 0.48 pg of DNA (N. Besansky, personal communication), an average recombinant phage insert of 14 kb, and a typical rDNA unit length of 12 kb, the frequency of rDNA-containing clones in the G3 library suggests that each diploid genome contains approximately 500 to 600 rDNA repeat units, a

[1] Use of trade names is for identification only and does not imply endorsement by the Public Health Service, the U.S. Department of Health and Human Services, or the U.S. Department of Defense.

figure within the range reported for other Diptera. Studies of the assortment of species-specific rDNA restriction fragments in interspecies hybrids and sex-specific differences in the intensity of genomic Southern blots probed with cloned rDNA indicate that all *An. gambiae* rDNA is located on the X chromosome (18). Thus, in *An. gambiae*, the rDNA copy number in males is approximately 350 versus 700 per genome in females. Genomic Southern blots loaded with equal amounts of DNA of the five species and probed with conserved *An. gambiae* rDNA sequences give virtually the same signal intensity, suggesting that all species in the complex have a similar number of copies per genome. *Anopheles arabiensis* is like *An. gambiae* in having X-linked rDNA. However, the rDNA of *An. quadriannulatus, An. merus,* and *An. melas* is not located exclusively on the X chromosome (S.M.P., unpublished data).

Four *An. gambiae* rDNA phage clones (and some plasmid subclones) have been examined at the level of restriction site mapping (Figure 1.3). The approximate limits of the 18S and 28S coding regions were inferred by Southern blot analysis using *C. erythrocephala,* rDNA-containing plasmids (6). Two different classes of clones are represented in this group. Although λAGr12 and λAGr19 contain fairly typical coding regions and IGSs, λAGr23 and λAGr24 contain coding regions and ISs. Analysis of these clones has shown that the basic architecture of *An. gambiae* rDNA is similar to that observed in other Diptera (51). A 2- to 3-kb region of XhoI-defined subrepeats is present in the spacer regions of λAGr12 and λAGr19. Downstream from the subrepeat region is a 1.5-kb HindIII-bounded region that probably contains most of the ETS. Between the

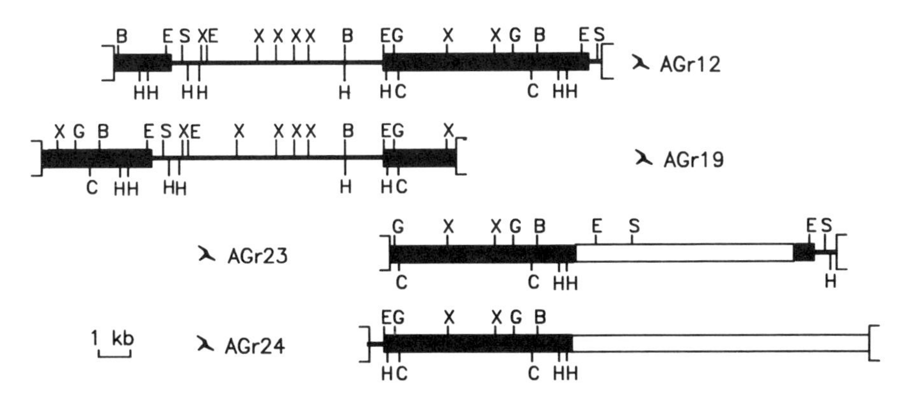

FIGURE 1.3. Restriction map of four *An. gambiae* rDNA clones. The restriction sites are (B) BamHI; (C) SstI; (E) EcoRI; (G) BglII; (H) HindIII; (P) PstI; (S) SalI; (X) XhoI. The λAGr12 and λAGr19 represent typical dipteran rDNA genes at the level of restriction site mapping. The λAGr23 and λAGr24 represent different IS containing clones, which are also typical of dipteran rDNA. Filled rectangles represent the transcribed regions, open rectangles represent ISs, narrow lines represent intergenic spacer regions. Boundaries between regions are approximate. Figure modified from McLain and Collins (51).

upstream of these two HindIII sites and the EcoRI site downstream from the 28S region, all *An. gambiae* IS− genes appear to be identical at the level of restriction site mapping involving the enzymes shown.

The two IS+ clones λAGr23 and λAGr24 contain large insertions in virtually the same location, a region a few hundred base-pairs upstream of the 28S EcoRI site. Although only the IS of λAGr23 has been mapped in some detail (below), the length and internal structure of the two sequences are quite different.

The basic architecture of *An. arabiensis* and *An. quadriannulatus* rDNA is similar to that of *An. gambiae* throughout the coding region. *Anopheles merus* and *An. melas,* however, show some significant differences, especially in the 18S and ITS regions.

Intergenic Spacer Structure

Anopheles gambiae IGS Structure

Different restriction fragments of λAGr12 have been subcloned into plasmid vectors and used as probes to examine variation in the distribution of restriction sites flanking these sequences, as shown in Figure 1.4. All the variation in IGS length is delineated by the two HindIII sites defining the limits of subclone 12H4. The IS− *An. gambiae* genes are invariant in the two regions flanking 12H4, as determined by probing various genomic digests with 12H3 and 12H5. These two fragments probably define regions within the ETS (12H5) and at or just beyond the end of the 28Sβ coding region (12H3).

By using a series of λAGr12 spacer sequences as probes, we have examined the IGS length variation within individuals from various *An. gambiae* colonies and field isolates. Figure 1.5 shows a HindIII digest of *An. gambiae* individuals probed with 12H4, the sequence spanning the entire IGS of λAGr12. First, most of the large number of spacer length variants within an individual specimen differ in an ordered manner,

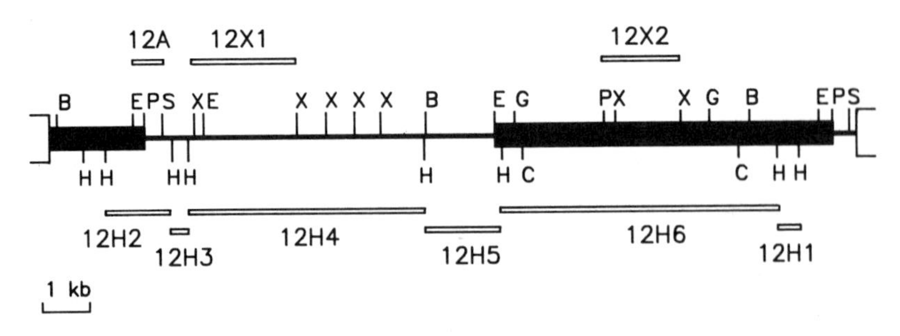

FIGURE 1.4. Restriction map of λAGr12. The restriction fragments used to examine rDNA gene variation throughout the complex are shown above and below the map.

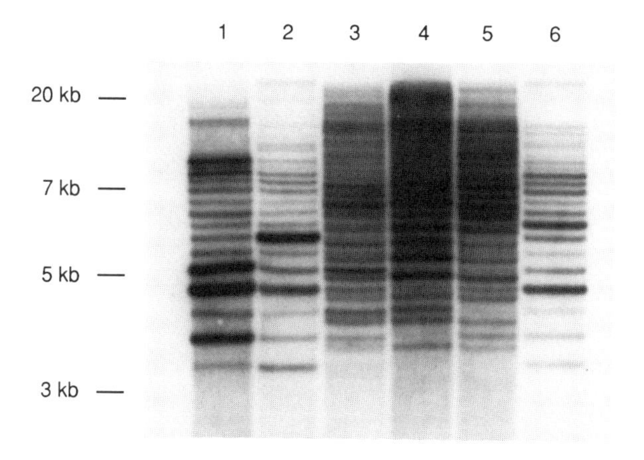

FIGURE 1.5. HindIII digest of individuals from an *An. gambiae* colony probed with 12H4. Each lane contains one-half of the DNA extracted from a single specimen. Origin of colonies are (1) Nigeria, (2) Kenya, (3) Zanzibar, (4) The Gambia, (5) Burkina Faso, (6) Kenya.

suggestive of variation in the frequency of a monomeric unit, probably a 400-bp subrepeat defined by XhoI in λAGr12. Second, certain repeat length variants are more abundant than are others. Studies involving colony specimens show little difference between individuals of a given colony in the relative frequencies of their length variants. Differences between colonies, however, are often quite pronounced, as can be seen in Figure 1.5. Third, some length variants, particularly in certain colonies, are of a size class that indicates that a factor other than the standard monomer is contributing to length variation. Some arrays of spacer length variants detected by the probe 12H4 suggest two or more superimposed ladders, with the rungs of each ladder differing by the same monomeric length added to invariant regions of slightly different lengths. This is seen when the simple monomeric ladders in lanes 1, 2, and 6 of Figure 1.5 are compared with the more complex ladders in lanes 3 through 5. A similar phenomenon has been noted in *Glossina* and other species (see, e.g., ref. 27). This basic spacer architecture is also evident in other members of the complex.

The internal structure of the *An. gambiae* IGS was investigated by probing HindIII/XhoI double digests of mosquitoes from several colonies with two fragments derived from AGr12: 12X1, the 2.0-kb XhoI fragment at the 5′ end of the spacer, and 12H4. The DNA is from the same individuals examined in Figure 1.5, one-half of the DNA from each individual being used in each of the two digests. These results are shown in the panels of Figure 1.6. As is evident in panel A, the 2.0-kb XhoI fragment is a principal component of most or all of the IGS sequences of the specimens in lanes 3 through 5, which also possess what appear to be

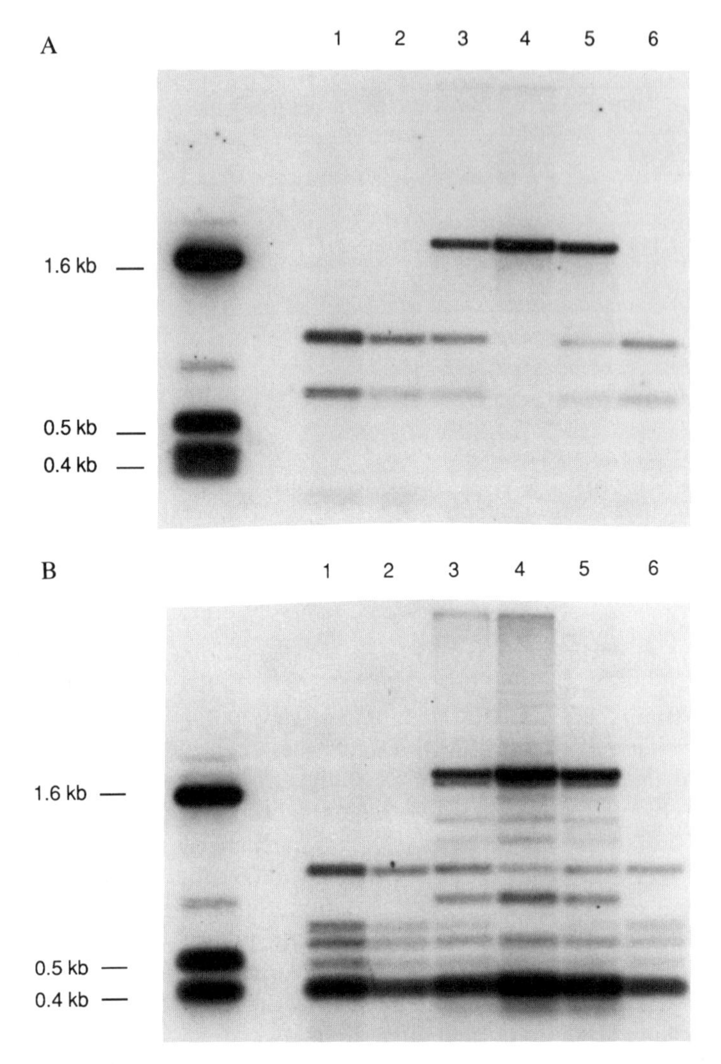

FIGURE 1.6. HindIII-XhoI double digest of individuals from an *An. gambiae* colony. Each lane contains one-half of the DNA extracted from the same single specimens shown in Figure 1.5. Panel A shows the filter probed with 12X1; Panel B, the same filter stripped and reprobed with 12H4.

two spacer ladders with different invariant region lengths (see Fig. 1.5). It is totally absent from lanes 1, 2, and 6, which have IGS ladders built on only one invariant region of the same length. When these filters are stripped and reprobed with 12H4 (panel B), the principal new bands are the expected small subrepeats. However, certain individuals have longer sequences, which probably represent adjacent copies of the subrepeat where nucleotide changes have occurred in the XhoI-recognition sequence. These results suggest that the IGS regions delineated by the HindIII sites contain only two classes of sequence, the small monomeric repeat units and the longer, 5′ blocks of DNA homologous with the 2.0-kb XhoI fragment, which itself may be composed of several copies of an as yet unidentified subrepeat.

Variation Between Members of the Complex

In contrast to the basic similarity between *An. gambiae* complex species in the architecture of the transcribed rDNA regions, IGS structures differ significantly between species. Figure 1.7 shows HindIII-digested DNA from individuals of the different species probed with the *An. gambiae* spacer 12H4. *Anopheles gambiae* and *An. quadriannulatus* each have IGSs that vary in length from approximately 4 to 20 kb (clustering around 6 to 9 kb); *An. arabiensis* IGSs are shorter, typically between 3 and 8 kb. The ladder-like array of IGS fragments in these three species suggests the presence in each of an internal repeat structure based on subrepeats of very similar sizes, approximately 400 bp. The IGSs of *An. merus* and

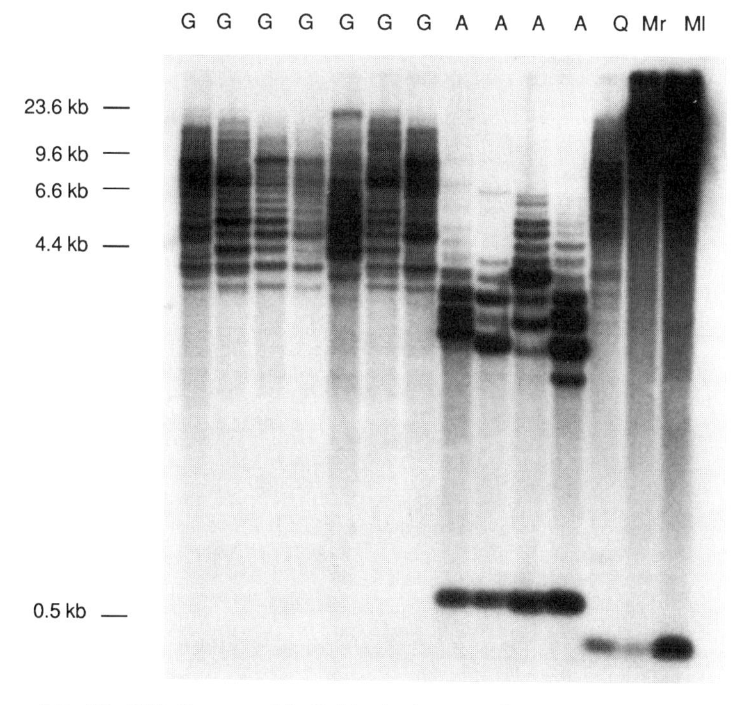

FIGURE 1.7. HindIII digests of individuals from various *An. gambiae* complex colonies probed with the *An. gambiae* spacer fragment 12H4. Lanes are labeled as follows: (G) *An. gambiae;* (A) *An. arabiensis;* (Q) *An. quadriannulatus;* (Mr) *An. merus;* (Ml) *An. melas.*

An. melas are considerably longer, as indicated in Figure 1.7 by high molecular weight fragments, probably exceeding 20 kb. The intensity of hybridization in these lanes probably is the result of the presence of a high copy number of an internal repeat element that is homologous with sequences in 12H4. Analysis of cloned DNA and other genomic digests has revealed ladders of highly repeated IGS-homologous fragments in these two species, which distribute as EcoRI ladders, with about a 175-bp subrepeat (Figure 1.8). The frequency and size of these subrepeats produces the same ladders in *An. merus* and *An. melas*. Preliminary evidence suggests that these 175-bp repeats may also be present in non-rDNA genomic locations.

Intervening rDNA Sequences

Members of the *An. gambiae* complex are like most other Diptera in that a portion of the rDNA coding regions is interrupted by intervening sequences. A preliminary study indicated that ISs may be found in two

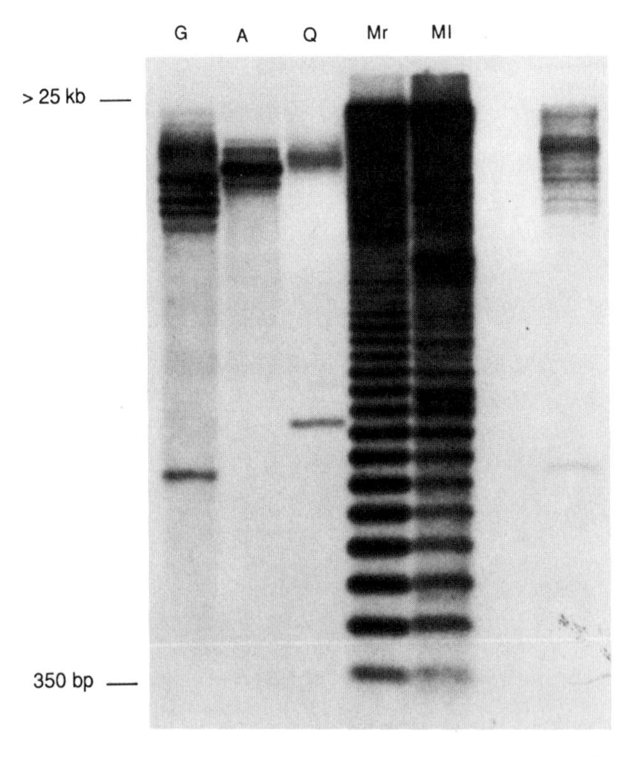

FIGURE 1.8. EcoRI digest of individuals from various *An. gambiae* complex colonies probed with the *An. gambiae* spacer fragment 12H4: (G) *An. gambiae;* (A) *An. arabiensis;* (Q) *An. quadriannulatus;* (Mr) *An. merus;* (Ml) *An. melas.*

locations in the 28S coding region. Figure 1.9 shows HindIII-digested genomic DNA from members of each of the five species probed with 12A. In an uninterrupted 28S region, this probe should hybridize with an 800-bp HindIII fragment. But in addition to the principal 800-bp band, a series of other fragments is evident. This pattern of secondary bands, which probably represents between 5 and 10% of the rDNA units—depending on the species and colony—is caused by insertions within this 800-bp region.

Two IS+ *An. gambiae* rDNA clones have been examined, and both are interrupted in this region. Two subclones of the IS+ rDNA clone λAGr23 (23A and 23B) have been used to test for homology with λAGr24 (Figure 1.10). Neither subclone hybridizes to sequences from λAGr24. Where IS variation has been studied in other Diptera, conservation of restriction sites in members of an IS class has generally been observed at the 3′ end. Since one end of the λAGr24 IS is not contained in this clone, we cannot

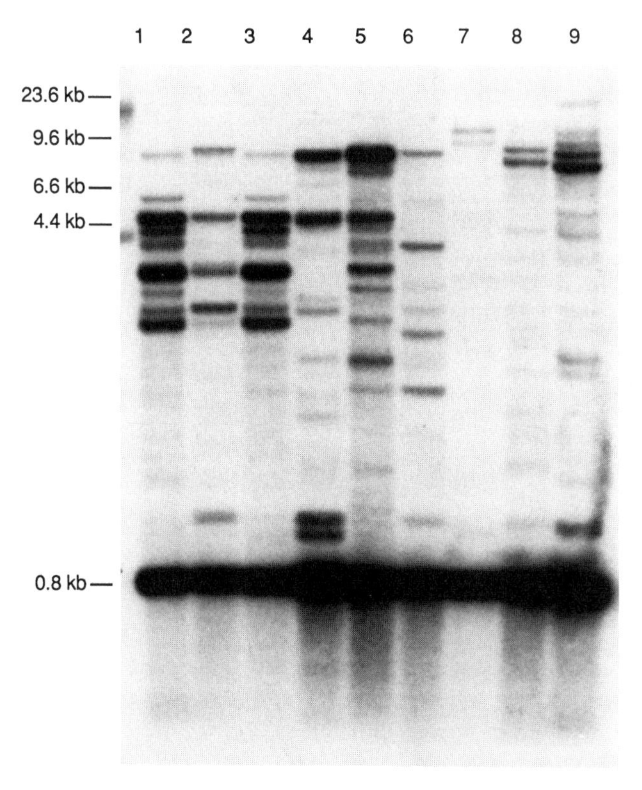

FIGURE 1.9. HindIII digest of individuals from an *An. gambiae* complex of different colonies probed with the fragment 12A. Lanes 1–3, *An. gambiae;* lanes 4–5, *An. arabiensis;* lane 6, *An. quadriannulatus;* lane 7, *An. merus;* lanes 8–9, *An. melas.*

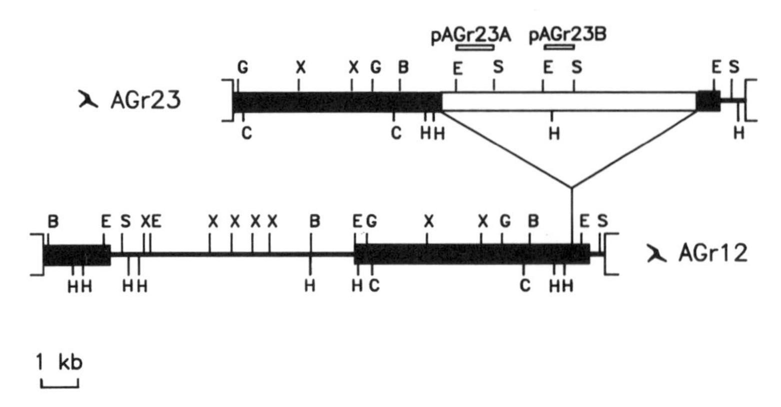

FIGURE 1.10. Restriction map of an *An. gambiae* IS+ clone showing approximate insertion site with respect to the 28S coding region of an IS− clone.

exclude the possibility that the missing portion would show homology with the λAGr23 sequence. When used to probe genomic Southern blots, these subclones hybridize primarily with 4.0- and 5.0-kb HindIII genomic fragments in the *An. gambiae* lanes, fragments that correspond to genomic copies of the 4.0- and 5.0-kb HindIII fragments in λAGr23 from which the subclones were derived. These two HindIII bands represent only a subset of the bands visible in Figure 1.10; thus other IS classes may be found in this region of the rDNA.

DNA homologous with the subclones 23A and 23B is not detected in other species of the complex, which suggests that the entire IS is absent from or highly diverged in the other species. By probing several hundred *An. gambiae* individuals from different colonies and field sites, we have observed a considerable variation in the genomic abundance of this sequence. However, DNA homologous to clones 23A and 23B appears to be present in all *An. gambiae* individuals from all parts of Africa (37).

A second 28S insertion site apparently exits 5′ to the one previously discussed (S.M.P., unpublished data). All individuals of all species have a 5-kb IS in approximately 15% of the 28S coding sequences. No variation has been observed—either within or between species—in the size or abundance of this IS, which suggests that it is an evolutionarily much older and more stable sequence. No example of this has yet been cloned. A similar phenomenon has been seen in *An. albimanus* rDNA (F.H.C., unpublished data), in which 5 to 10% of the 28S sequences in male mosquitoes have a large IS, which does not vary among specimens in either size or abundance. Possibly such sequences are associated with only one of several nucleolar organizer regions (see, e.g., refs. 45, 74).

Applied Uses of rDNA

Species-Diagnostic DNA Probes

Because the member species of the *An. gambiae* complex are indistinguishable on the basis of external morphology, considerable research effort has been directed toward the development of species-diagnostic criteria that would be of use in the context of malaria control or epidemiological studies. Furthermore, many other important vectors of malarial parasites and other insect-borne disease agents have been shown to be members of such cryptic species complexes (16, 56). The only completely reliable method currently in use for identifying members of cryptic species, an identification based on fixed inversions in the polytene chromosome complement, has two serious limitations. First, it is applicable only to certain life stages, fourth instar larvae or half gravid females in the case of *Anopheles* mosquitoes, and second, the method requires a very demanding level of technical expertise that is only acquired through long and careful training by an expert.

Such limitations have prompted a number of recent investigations of the potential for use of DNA probes to identify the species of individual specimens from cryptic complexes (17, 37, 40–42, 57–59). Two different approaches have been used to isolate appropriate DNA sequences. One has involved identifying species-specific sequences, often by sequential hybridization of radiolabeled total genomic DNA of the two or more species being compared to cloned DNA fragments of one species (see, e.g., refs. 40, 58). Individual clones that show strong hybridization to the homologous species and little or no hybridization to the heterologous species are used. This approach typically identifies frequently repeated, often heterochromatic DNA sequences. An alternative approach involves cloning repeated gene families, such as the ribosomal RNA, 5S RNA, or histone genes, which are then examined for species-specific differences in such variable portions of these sequences as intergenic spacers and insertion sequences (see, e.g., refs. 1, 18, 37).

Although the former approach is relatively simple and direct, the DNA sequences identified may not be species-specific. Because most such probes are selected on the basis of a difference in signal, they may reflect differences in copy number rather than absolute species-specificity. In this case, species identification would depend upon relative differences in hybridization signals involving two or more different probes, a technically difficult comparison. Furthermore, repeated DNA sequences may vary in copy number over several orders of magnitude, not only between species but also between different geographical populations or even individuals of the same species (ref. 53; Figure 2 of ref. 58). Therefore, DNA probes that are tested by comparing a limited number of colonies may not be consistently species-diagnostic over the entire geographical range of the species of interest.

The ribosomal RNA genes have proved to be extremely useful diagnostic characters for identifying members of the *An. gambiae* complex species (17–19). The five species examined thus far differ significantly in IGS and ETS architecture; thus, virtually any rDNA probe that detects restriction fragments spanning these regions can be used as a species-diagnostic probe. Because of the high level of length variation in the interior of the IGSs, however, the simplest patterns of species-specific restriction fragment differences are found at the 5′ and 3′ ends of the spacers. These regions are relatively invariant within a species but differ markedly between species.

Two different probes have been employed to detect these species-diagnostic IGS differences (16). Figure 1.11 presents a simplified diagram of the basic rDNA repeat unit of the five *An. gambiae* complex species, along with a map of λAGr12 showing two subclones that reveal species-diagnostic, restriction fragment length polymorphisms (RFLPs). The

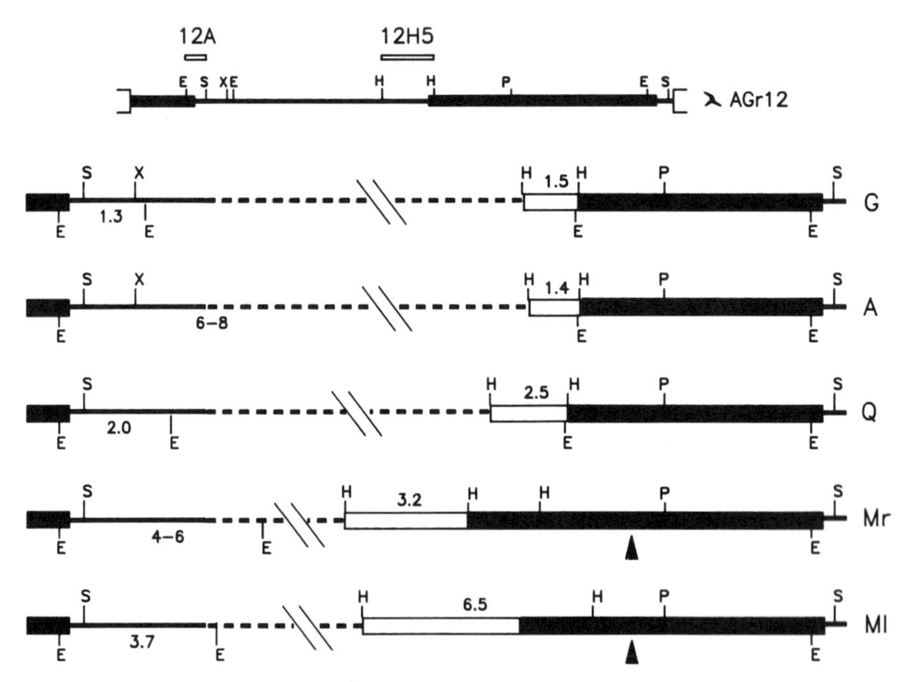

FIGURE 1.11. Diagram of a typical rDNA repeat unit found in the five *An. gambiae* complex species. A simplified map of λAGr12 is shown above to illustrate the two fragments that reveal species-diagnostic RFLPs. The 12H5 fragment shows the HindIII RFLP at the left of the spacer (probably the ETS region), and the 12A fragment shows the EcoRI RFLP at the right end of the spacer, representing the 3′ end of 28S and adjacent IGS regions.

RFLPs shown in panel A of Figure 1.12 are revealed by probing EcoRI-digested mosquito DNA with the subclone 12A, derived from the 3′ end of the 28S coding region. The different restriction fragments detected probably reflect a combination of factors. The fragments in the *An. gambiae* and *An. quadriannulatus* lanes are believed to be 5′ IGS regions that are relatively conserved within each of these species. The 1.3-kb EcoRI fragment detected in the *An. gambiae* lane is present in the spacers of all *An. gambiae* specimens thus far examined, including field and laboratory specimens from geographically different parts of Africa. The 2.0- and 3.7-kb RFLP in the *An. quadriannulatus* and *An. melas* lanes probably also represent relatively conserved 5′ IGS regions in these species, but other minor variants are observed, which indicates that these fragments may extend into the variable region of the spacer. The array of fragments in the *An. merus* lane is probably due to some heterogeneity in the location of the most upstream IGS EcoRI site. The middle IGS regions of both *An. merus* and *An. melas* appear to have long arrays of small subrepeats characterized by the presence of EcoRI sites. The greater variability in the EcoRI-bounded 5′ IGS region of *An. merus* relative to *An. melas* may be due to a higher level of sequence variability in the region of the subrepeat of *An. merus* that contains the EcoRI-recognition sequence. The cluster of bands in the *An. arabiensis* lane is the result of restriction fragments that span the entire IGS to an EcoRI site at the 5′ end of the conserved coding region.

The second species-diagnostic probe (Figure 1.12, panel B) detects differences in the 3′ end of the IGS–ETS region. The 1.4-, 1.5-, and 2.5-kb HindIII fragments of *An. arabiensis, An. gambiae,* and *An. quadriannulatus* detected by this probe are highly conserved, probably ETS sequences. The longer sequences in the *An. merus* and *An. melas* lanes clearly extend into regions that vary somewhat between rDNA units in an individual specimen. Nevertheless, the principal bands associated with these two species, a 3.1- to 3.2-kb doublet in *An. merus* and a 6.5-kb band in *An. melas,* are invariant within the species.

Both these probes have been used to identify large numbers of field-collected mosquitoes as to species. The rDNA RFLP method has been validated with field specimens from sites in Kenya, Zimbabwe, and The Gambia, which included *An. gambiae, An. arabiensis, An. quadriannulatus, An. merus,* and *An. melas* (19). Because of the high copy number of rDNA and the stability of DNA in desiccated specimens, we routinely identify specimens with DNA extracted from the mosquito abdomen, whereas the head and thorax portion is used for the analysis of *Plasmodium* sporozoite antigen.

Although this RFLP is a clearly valid method, it is technically cumbersome in spite of its advantages. A far simpler species-diagnostic approach is currently being developed. Preliminary data indicate that the IGS subrepeat sequences differ sufficiently between *An. gambiae* com-

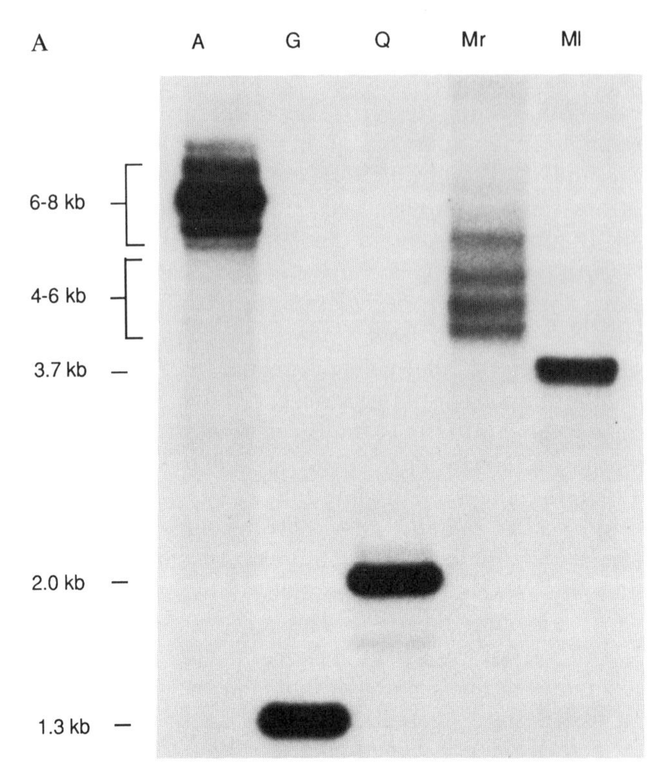

FIGURE 1.12. Species-diagnostic RFLPs revealed with rDNA spacer probes. Panel A shows an autoradiogram of a Southern blot containing one-half of a mosquito per lane digested with EcoRI and probed with 12A: (G) *An. gambiae;* (A) *An. arabiensis;* (Q) *An. quadriannulatus;* (Mr) *An. merus;* (Ml) *An. melas.* Panel B shows an autoradiogram of a Southern blot containing HindIII digests of the remainder of the DNA prepared for panel A, probed with 12H5.

plex species to permit the use of synthetic oligonucleotide probes targeted at this region. As with the other regions of the IGS, these sequences are conserved within a species but diverge significantly between species. In addition, they are frequently repeated. The average subrepeat copy number probably varies from approximately half a dozen copies per IGS in *An. gambiae* to several dozen in the spacers of *An. merus* and *An. melas.* Synthetic probes developed to detect species-specific regions of these rDNA IGS subrepeats will clearly be sensitive to the order of several thousand copies per genome. Moreover, species-specific probes can be used on dot blots of crude, undigested DNA preparations.

Not only will dot blot, species-diagnostic assays based on such sequences have all the advantages of simplicity and sensitivity afforded by any frequently repeated, species-specific sequence, they will also be free of the uncertainty associated with using a probe of unknown function

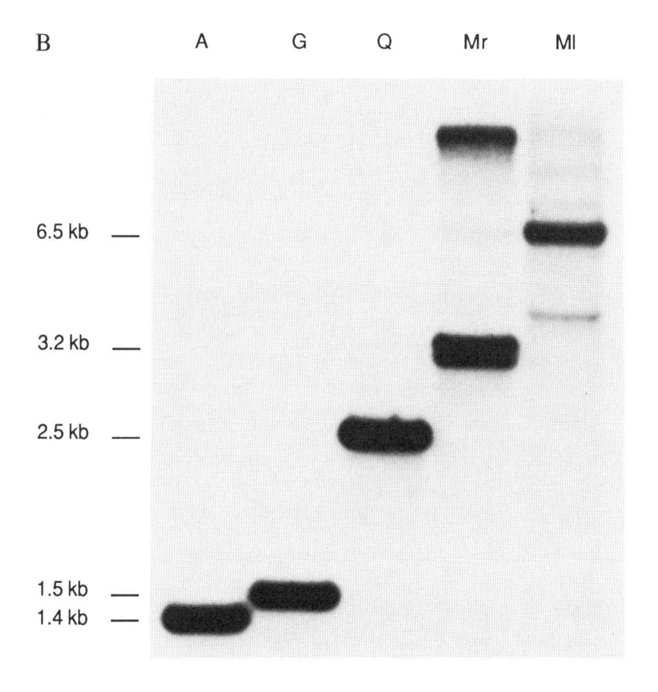

and distribution within the group of interest. Since the architecture of rDNA has been thoroughly studied in a number of insect taxa, species-diagnostic methods based on the common features of this architecture offer considerable advantages. Furthermore, because of the fundamental similarities in insect rDNA, this basic strategy for developing simple methods to identify individual members of morphologically identical complexes of closely related species should be broadly applicable (16).

Measures of Population Structure

Ribosomal DNA is also a particularly useful tool for the genetic analysis of within-species population structure (see, e.g., ref. 73). The architecture of the IGSs, coupled with a function that appears to be somewhat independent of nucleotide sequence, permits the evolutionarily rapid accumulation of new variants (68). Unequal crossing-over and perhaps, to a lesser extent, gene conversion can result in a very rapid gain or loss of subrepeat and IGS variants. Even when selection has not taken place, homogenization of a sexually reproducing population for a given sequence variant will proceed more or less in unison throughout the population. Thus, the IGS region can be a very sensitive indicator of population subdivision.

The arrays of *An. gambiae* and *An. arabiensis* IGSs shown in Figures

1.4 and 1.7 illustrate the interindividual differences that are typical, with certain IGS length variants being distinctly more abundant than others. Analysis of the frequencies of IGS variants in different colonies has clearly revealed that individuals in a colony are often nearly identical in their IGS profiles.

Examination of specimens from field populations, however, has revealed considerably more variation than has been observed in colonized mosquitoes. The distribution of this interindividual variation is clearly correlated with genetic similarity. Interindividual variation in IGS profiles increases as comparison group distances are increased—from family, through progressively larger and more geographically separated populations.

The large number of IGS variants present in an individual, as is generally observed in *An. gambiae* or *An. quadriannulatus,* combined with the clustering of variants in length classes that are difficult to resolve on gels, as in *An. merus* and *An. melas,* can greatly complicate interpretation of these data. A more effective approach in analyzing IGS variation is to focus on limited regions of the spacers. For example, in our study of population structure in *An. gambiae* in East Africa, we probed DNA from single mosquitoes restriction-digested with the enzyme XhoI, which cleaves the DNA between each of the subrepeats as well as releasing a block of DNA at the 5′ end of the spacer. These genomic Southern blots were then probed with a subcloned fragment from λAGr12 that included both the 5′ block and one of the subrepeats (52). The net effect of this strategy is that most variations in IGS length could be categorized as one of two basic types of fragments: a cluster of subrepeat monomers and monomer multiples resulting from degeneracy of XhoI sites and a comparable array of fragments homologous to the 5′ block of IGS DNA. Compare, for example, the variation revealed between specimens by the digest in Figure 1.5 with the same mosquitoes examined in this latter manner in both panels of Figure 1.6.

Using this approach to study gene flow among *An. gambiae* populations in seven different villages in western Kenya, we were able to identify seven different rDNA spacer genotypes based on the presence or absence of XhoI bands of variable length. Eight of twenty-one possible comparisons revealed significant differences between populations in the frequency distribution of these genotypes. In some cases, restricted gene flow was apparent between populations separated by less than 10 km (52). Thus, with an understanding of the underlying architecture, rDNA IGS probes can be extremely useful in detecting subdivided population structure. If such studies are extended to comparisons involving behavioral or physiological characters that relate to malaria transmission or susceptibility, it may even be possible to identify genetic subdivisions within a vector species that could bear on malaria control.

Evolution of rDNA Within the Complex

Of all the criteria thus far applied to establish molecular phylogenies, rRNA has been the most widely used. This is primarily the result of the high degree of sequence conservation, even between distant taxa, that allows placement of species into a broadly verifiable context. Moreover, because the function of rRNA is constant in all organisms, it is a relatively good molecular clock. Finally, because the rRNA genes also contain more labile regions, data on the structure of rDNA can be used to assess the phylogenetic relationships between more closely related species, such as the members of the *An. gambiae* complex.

Morphological and ecological data suggest that the two saltwater species (*An. melas* and *An. merus*) form one sister group and that the two species adapted to man's environment (*An. arabiensis* and *An. gambiae*) form another. However, when the fixed inversions of the polytene chromosomes are used to determine the phylogeny of the complex, quite different relationships are seen (Figure 1.1).

The rDNA of five of the member species in the *An. gambiae* complex has been examined by combining two related criteria. Cloned rDNA of *An. gambiae, An. arabiensis, An. quadriannulatus,* and *An. merus* has been examined at the level of restriction enzyme analysis and, in some limited instances, at the level of nucleotide sequence analysis. But because of high levels of intraindividual and intraspecies variation in even the conserved regions of the rDNA repeat unit, all conclusions about the consensus rDNA structure have been verified by Southern analysis using cloned DNA.

The consensus-transcribed regions within these five species have been mapped with eleven restriction enzymes (S.M.P., unpublished data). The basic interspecies similarities and differences determined by a limited group of these enzymes is illustrated in Figure 1.11. The transcribed regions of all five species are essentially identical from a landmark PstI site, which represents a region in the 5.8S coding region, downstream through the 28S region to approximately 200 bp beyond the SalI site shown in Figure 1.11. *Anopheles gambiae* and *An. arabiensis* remain identical on the level of restriction site analysis through to the XhoI site illustrated in Figure 1.11, but they differ with respect to the EcoRI site less than 100 bp beyond.

From the PstI site upstream through the 18S sequence, *An. gambiae, An. arabiensis,* and *An. quadriannulatus* show considerable similarity. The *An. gambiae* and *An. arabiensis* coding regions are identical to the 3′ HindIII site that flanks the sequence homologous with 12H5; *An. quadriannulatus* shows the curious feature of apparently having two different-size classes in this region, with one variant identical to the above and the alternate—as shown in Figure 1.11—having a slightly longer

PstI/HindIII fragment. All three of these species also have an identical group of restriction sites in this region.

Although the *An. merus* and *An. melas* coding regions have not been mapped as carefully as the regions of the other three species, they do share several unique characters. Both have a variable length region of DNA between the 18S and 28S coding sequences, possibly in the ITS. And both have extremely long, relatively conserved sequences homologous with 12H5, the presumed ETS sequence from *An. gambiae*.

These two species differ from the others in their IGS architecture. *Anopheles merus* and *An. melas* have very long (> 20-kb) spacers with 175-bp EcoRI subrepeats (Figure 1.8). The IGS length is much shorter in *An. gambiae, An. arabiensis,* and *An. quadriannulatus,* and the major subrepeat length, as inferred from the distance between IGS ladder rungs, is approximately the same in these three species—roughly 400 bp.

Anopheles quadriannulatus, An. merus, and *An. melas* share one structural feature that is absent from *An. gambiae* and *An. arabiensis,* an approximately 500-bp HindIII fragment located at the 5′ end of the IGS that shows homology with the *An. gambiae* sequence 12X1 (see Figure 1.7). Additionally, these three species have rDNA in locations other than the X chromosome, in contrast to *An. gambiae* and *An. arabiensis,* in which rDNA is exclusively X-linked.

These various characteristics suggest a very close relationship between the rDNA of *An. gambiae* and *An. arabiensis; An. merus* and *An. melas* similarly show strong affinities; and *An. quadriannulatus* holds a central position. The phylogeny indicated by rDNA (Figure 1.13) is distinctly different from that proposed on the basis of chromosomal inversions (Figure 1.1), the only point of similarity being the central position of *An. quadriannulatus*. Clearly, the disparity between these two phylogenies may reflect such complications as rDNA or inversion introgression between species and possibly some adaptive convergence (see, e.g., ref. 12). Resolution of this difference will probably require the examination of a completely independent group of characteristics.

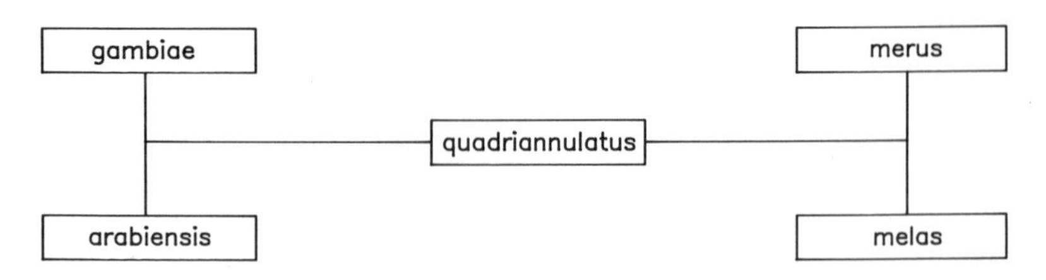

FIGURE 1.13. *Anopheles gambiae* complex phylogeny based upon the molecular analysis of rDNA.

Summary

Current interest in mosquitoes of the *An. gambiae* complex is prompted by their involvement in the transmission of malaria in Africa. The morphologically identical member species have been reliably distinguished only by cytogenetic criteria. This factor has hindered both epidemiological and ecological studies, and basic studies of population structure and evolution were extremely difficult. Both the evolutionary and diagnostic questions were approached by exploring the utility of rDNA sequences, which have numerous advantages. These include a high copy number, as well as highly conserved and highly variable regions that lend themselves to evolutionary studies. Coupled with this is the almost certain probability of finding species-diagnostic sequences within the highly variable intergenic spacer regions.

We find that the rDNA genes of this complex are generally similar in architecture to those of other Diptera. The intergenic spacer sites flanking the coding regions are highly conserved within a species, and these reveal RFLPs that are perfectly valid for species diagnosis. As expected, the medial regions of the spacers tend to vary both within and between species. Preliminary results indicate that small, frequently repeated sequences within these medial spacer regions may prove useful as species-specific, species-diagnostic DNA probes. As with many other eukaryotes, many of the rDNA genes have intervening sequences, some of which appear to be completely species-specific.

Our studies thus far suggest a phylogenetic relationship that differs from the relationship inferred from cytogenetic data. However, much more data, preferably drawn from the study of other regions of the genome, are necessary to draw firm conclusions concerning evolution of this complex.

Acknowledgments. The authors thank Diane Mills, Melissa Rasmussen, and Theresa Blue for technical assistance. This research was supported by U.S. Agency for International Development PASA No. BST-0453-P-HC-2086-02 to the C.D.C. and U.S. Army Contract DAMD 17-85-C-5184 to V.F.

References

1. Arnold, M.L., Shaw, D.D., and Contreras, N., 1987, Ribosomal RNA-encoding DNA introgression across a narrow hybrid zone between two subspecies of grasshopper, *Proc. Natl. Acad. Sci. USA* **84:**3946–3950.
2. Barnett, T., and Rae, P.M.M., 1979, A 9.6-kb intervening sequence in *D. virilis* rDNA, and sequence homology in rDNA interruptions of diverse species of *Drosophila* and other Diptera, *Cell* **16:**763–775.

3. Beckingham, K., 1982, Insect rDNA, in Busch, H., and Rothblum, L. (eds.): The Cell Nucleus, X, New York: Academic Press, pp. 205–269.
4. Beckingham, K., and Rubacha, A., 1984, Different chromatin states of the intron$^-$ and type 1 intron$^+$ rRNA genes of *Calliphora erythrocephala, Chromosoma* **90:**311–316.
5. Beckingham, K., and Thompson, N., 1982, Under-replication of intron$^+$ rDNA cistrons in polyploid nurse cell nuclei of *Calliphora erythrocephala, Chromosoma* **87:**177–196.
6. Beckingham, K., and White, R., 1980, The ribosomal DNA of *Calliphora erythrocephala:* An analysis of hybrid plasmids containing ribosomal DNA, *J. Molec. Biol.* **137:**349–373.
7. Bryan, J., 1979, Observations on the member species of the *Anopheles gambiae* complex in The Gambia, West Africa, *Trans. Royal Soc. Trop. Med. Hyg.* **73:**463–466.
8. Bryan, J.H., DiDeco, M.A., Petrarca, V., and Coluzzi, M., 1982, Inversion polymorphism and incipient speciation in *Anopheles gambiae* s.str. in The Gambia, West Africa, *Genetica* **59:**167–176.
9. Burke, W.D., Calalang, C.C., and Eickbush, T.H., 1987, The site-specific ribosomal insertion element Type II of *Bombyx mori* (R2Bm) contains the coding sequence for a reverse transcriptase-like enzyme, *Mol. Cell Biol.* **7:**2221–2230.
10. Carlson, D.A., and Service, M.W., 1980, Identification of mosquitoes of *Anopheles gambiae* species complex A and B by analysis of cuticular components, *Science* **207:**1089–1091.
11. Chooi, W.Y., and Lieby, K.R., 1984, Electron microscopic evidence for RNA polymerase loading at repeated sequences in non-transcribed spacers of *D. virilis, Exp. Cell Res.* **154:**181–190.
12. Cluster, P.D., Marinkovic, D., Allard, R.W., and Ayala, F.J., 1987, Correlations between development rates, enzyme activities, ribosomal DNA spacer-length phenotypes, and adaptation in *Drosophila melanogaster, Proc. Natl. Acad. Sci. USA* **84:**610–614.
13. Coen, E.S., and Dover, G.A., 1983, Unequal exchanges and the coevolution of X and Y rDNA arrays in *Drosophila melanogaster, Cell* **33:**849–855.
14. Coen, E.S., Thoday, J.M., and Dover, G.A., 1982, Rate of turnover of structural variants in the rDNA gene family of *Drosophila melanogaster, Nature* **295:**564–568.
15. Coetzee, M., Newberry, K., and Durand, D., 1982, A preliminary report on a morphological character distinguishing important malaria vectors in the *Anopheles gambiae* complex in southern Africa, *Mosq. Syst.* **14:**88–93.
16. Collins, F.H., Finnerty, V., and Petrarca, V., 1989, Ribosomal DNA-probes differentiate five cryptic species in the *Anopheles gambiae* complex, *Parassitologia:* in press.
17. Collins, F.H., Mehaffey, P.C., Rasmussen, M.O., Brandling-Bennett, A.D., and Finnerty, V., 1988, Comparison of DNA probe and isoenzyme methods for differentiating *Anopheles gambiae* and *Anopheles arabiensis, J. Med. Entomol.* **25:**116–120.
18. Collins, F.H., Mendez, M.A., Rasmussen, M.O., Mehaffey, P.C., Besansky, N.J., and Finnerty, V., 1987, A ribosomal RNA gene probe differentiates member species of the *Anopheles gambiae* complex, *Am. J. Trop. Med. Hyg.* **37:**37–41.

19. Collins, F.H., Petrarca, V., Mpofu, S., Brandling-Bennett, A.D., Were, J.B.O., Rasmussen, M.O., and Finnerty, V., 1988, Comparison of DNA probe and cytogenetic methods for identifying field collected *Anopheles gambiae* complex mosquitoes, *Am. J. Trop. Med. Hyg.* **39:**545–550.
20. Coluzzi, M., 1964, Morphological divergences in the *Anopheles gambiae* complex, *Riv. di Malariol.* **43:**197–232.
21. Coluzzi, M., 1968, Chromosomi politenici delle cellule nutrici ovariche nel complesso *gambiae* del genera *Anopheles, Parassitologia* **10:**179–183.
22. Coluzzi, M., 1984, Heterogeneities of the malaria vectorial system in tropical Africa and their significance in malaria epidemiology and control, *WHO Bull.* **62 (Suppl.):**107–113.
23. Coluzzi, M., Petrarca, V., and DiDeco, M.A., 1985, Chromosomal inversion intergradation and incipient speciation in *Anopheles gambiae, Boll. Zool.* **52:**45–63.
24. Coluzzi, M., and Sabatini, A., 1967, Cytogenetic observations on species A and B of the *Anopheles gambiae* complex, *Parassitologia* **9:**73–88.
25. Coluzzi, M., Sabatini, A., Petrarca, V., and DiDeco, M.A., 1979, Chromosomal differentiation and adaptation to human environments in the *Anopheles gambiae* complex, *Trans. Royal Soc. Trop. Med. Hyg.* **73:**483–497.
26. Cross, N.C.P., and Dover, G.A., 1987a, Tsetse fly rDNA: An analysis of structure and sequence, *Nucl. Acids Res.* **15:**15–30.
27. Cross, N.C.P., and Dover, G.A., 1987b, A novel arrangement of sequence elements surrounding the rDNA promoter and its spacer duplications in tsetse species, *J. Mol. Biol.* **195:**63–74.
28. Davidson, G., 1962, *Anopheles gambiae* complex, *Nature* **196:**907.
29. Davidson, G., 1964, The five mating types in the *Anopheles gambiae* complex, *Riv. di Malariol.* **43:**167–183.
30. Davidson, G., Paterson, H.E., Coluzzi, M., Mason, G.F., and Micks, D.W., 1967, The *Anopheles gambiae* complex, in Wright, J.W., and Pal, R. (eds): Genetics of Insect Vectors of Disease, Elsevier, Amsterdam, New York, pp. 211–250.
31. Davidson, G., and White, G.B., 1972, The crossing characteristics of a new, sixth species in the *Anopheles gambiae* complex, *Trans. Royal Soc. Trop. Med. Hyg.* **66:**531–532.
32. Dawid, I.B., DiNocera, P.P., and Mandal, R.K., 1984, Ribosomal RNA genes and transposable sequences in *Drosophila melanogaster,* Genetics: New Frontiers, pp. 235–245.
33. Degelmann, A., Royer, H.-D., Hollenberg, C.P., 1979, The organization of the ribosomal genes of *Chironomus tentans* and some closely related species, *Chromosoma* **71:**263–281.
34. Dover, G.A., 1982, Molecular drive: A cohesive mode of species evolution, *Nature* **299:**111–117.
35. Dover, G., and Coen, E., 1981, Spring cleaning ribosomal DNA: A model for multigene evolution, *Nature* **290:**731–732.
36. Eickbush, T.H., and Robins, B., 1985, *Bombyx mori* 28S ribosomal genes contain insertion elements similar to the Type I and II elements of *Drosophila melanogaster, EMBO J.* **4:**2281–2285.
37. Finnerty, V., Mendez, M.A., Rasmussen, M.O., Berrios, S.I., and Collins, F.H., Intervening sequences in the 28S ribosomal DNA of member species of the *Anopheles gambiae* mosquito complex, manuscript.

38. Frischauf, A.M., Lehrach, H., Poustka, A., and Murray, N., 1983, Lambda replacement vectors carrying polylinker sequences, *J. Molec. Biol.* **170:** 827–842.
39. Fujiwara, H., Ishikawa, H., 1986, Molecular mechanisms of introduction of the hidden break into the 28S rRNA of insects: Implication based on structural studies, *Nucl. Acids Res.* **14:**6393–6401.
40. Gale, K.R., and Crampton, J.M., 1987a, DNA probes for species identification of mosquitoes in the *Anopheles gambiae* complex, *Med. Vet. Entomol.* **1:**127–236.
41. Gale, K.R., and Crampton, J.M., 1987b, A DNA probe to distinguish the species *Anopheles quadriannulatus* from other species of the *Anopheles gambiae* complex, *Trans. Royal Soc. Trop. Med. Hyg.* **81:**842–846.
42. Gale, K.R., and Crampton, J.M., 1988, Use of a male-specific DNA probe to distinguish female mosquitoes of the *Anopheles gambiae* species complex, *Med. Vet. Entomol.* **2:**77–79.
43. Gatti, M., Bonaccorsi, S., Pimpinelli, S., and Coluzzi, M., 1982, Polymorphism of sex chromosome heterochromatin in the *Anopheles gambiae* complex, in Steiner, W.W.M., Tabachnick, W.J., Rai, K.S., and Narang, S. (eds): Recent Developments in the Genetics of Insect Disease Vectors, Stipes, Champaign, Illinois, pp. 32–48.
44. Gerbi, S.A., 1985, Evolution of ribosomal DNA, MacIntyre, R. (ed): in Molecular Evolutionary Genetics, Plenum Press, New York, pp. 419–517.
45. Gillings, M.R., Frankham, R., Speirs, J., and Whalley, M., 1987, X–Y exchange and the coevolution of the X and Y rDNA arrays in *Drosophila melanogaster, Genetics* **116:**241–251.
46. Gray, M.W., Sankoff, D., and Cedergren, R.J., 1984. On the evolutionary descent of organisms and organelles: A global phylogeny based on a highly conserved structural core in small subunit ribosomal RNA, *Nucl. Acids Res.* **12:**5837–5852.
47. Hunt, R.H., and Coetzee, M., 1986, Chromosomal and electrophoretic identification of a sample of *Anopheles gambiae* group (Diptera: Culicidae) from the island of Grand Comoros, Indian Ocean, *J. Med. Entomol.* **23:** 655–660.
48. Israelewski, N., and Schmidt, E.R., 1982, Spacer size heterogeneity in ribosomal DNA of *Chironomus thummi* is due to a 120-bp repeat homologous to a predominantly centromeric repeated sequence, *Nucl. Acids Res.* **10:** 7689–7700.
49. Kohorn, B.D., and Rae, P.M.M., 1983, Localization of DNA sequences promoting RNA polymerase I activity in *Drosophila, Proc. Natl. Acad. Sci. USA* **80:**3265–3268.
50. Marchand, R.P., and Mnzava, A.E.P., 1985, A field test of a biochemical key to identify members of the *An. gambiae* group of species in northeast Tanzania, *J. Trop. Med. Hyg.* **88:**205–210.
51. McLain, D.K., and Collins, F.H., 1989, Structure of the rDNA cistron in the mosquito *Anopheles gambiae* and rDNA sequence variation within and between species of the *An. gambiae* complex, *Heredity* **62:**233–242.
52. McLain, D.K., Collins, F.H., Brandling-Bennett, A.D., and Were, J.B.O., 1989, Microgeographic variation in rDNA intergenic spacers of *Anopheles gambiae* in western Kenya, *Heredity* **62:**257–264.

53. McLain, D.K., Rai, K.S., and Fraser, M.J., 1987, Intraspecific and interspecific variation in the sequence and abundance of highly repeated DNA in mosquitoes of the *Aides albopictus* subgroup, *Heredity* **58:**373–381.
54. Miles, S.J., 1979, A biochemical key to adult members of the *Anopheles gambiae* group of species (Diptera: Culicidae), *J. Med. Entomol.* **15:**297–299.
55. Mnzava, A.E.P., and Kilama, W.L., 1986, Observations on the distribution of the *Anopheles gambiae* complex in Tanzania, *Acta Trop.* **43:**277–282.
56. Onori, E., and Muir, D., 1984, Malaria Vector Species Complexes and Intraspecific Variations: Relevance for Malaria Control and Orientation for Future Research, UNDP/World Bank/WHO, Geneva, 102 p.
57. Panyim, S., Yasophornsrikul, S., Tungpradabkul, S., Baimai, V., Rosenberg, R., Andre, R.G., and Green, C.A., 1988, Identification of isomorphic malaria vectors using a DNA probe, *Am. J. Trop. Med. Hyg.* **38:**47–49.
58. Post, R.J., 1985, DNA probes for vector identification, *Parasitol. Today* **1:**89–90.
59. Post, R.J., and Crampton, J.M., 1987, Probing the unknown. *Parasitol. Today* **3:**380–383.
60. Renkawitz, R., Gerbi, S.A., and Glatzer, K.H., 1979, Ribosomal DNA of the fly *Sciara coprophila* has a very small and homogeneous repeat unit, *J. Mol. Gen. Genet.* **173:**1–13.
61. Roiha, H., Miller, J.R., Woods, L.C., and Glover, D.M., 1981, Arrangements and rearrangements of sequences flanking the two types of rDNA insertion in *D. melanogaster, Nature* **290:**749–753.
62. Schmidt, E.R., Godwin, E.A., Keyl, H-G., and Israelewski, N., 1982, Cloning and analysis of ribosomal DNA of *Chironomus thummi piger* and *Chironomus thummi thummi, Chromosoma* **87:**389–407.
63. Service, M.W., 1985, *Anopheles gambiae:* Africa's principal malaria vector, 1902–1984, *ESA Bull.* (Fall):8–12.
64. Smith, V.L., and Beckingham, K., 1984, The intron boundaries and flanking rRNA coding sequences of *Calliphora erythrocephala* rDNA, *Nucl. Acids Res.* **12:**1707–1724.
65. Sogin, M.L., Elwood, H.J., and Gunderson, J.H., 1986, Evolutionary diversity of eukaryotic small subunit rRND genes, *Proc. Natl. Acad. Sci. USA* **83:**1383–1287.
66. Sollner-Webb, B., and Tower, J., 1986, Transcription of cloned eukaryotic ribosomal RNA genes, *Ann. Rev. Biochem.* **55:**801–830.
67. Tautz, D., and Dover, G.A., 1986, Transcription of the tandem array of ribosomal DNA in *Drosophila melanogaster* does not terminate at any fixed point, *EMBO J.* **5:**1267–1273.
68. Tautz, D., Tautz, C., Webb, D., and Dover, G.A., 1987, Evolutionary divergence of promoters and spacers in the rDNA family of four *Drosophila* species: Implications for molecular coevolution in multigene families, *Mol. Biol.* **195:**525–542.
69. White, G.B., 1970, Chromosomal evidence for natural interspecific hybridization by mosquitoes of the *Anopheles gambiae* complex, *Nature* **231:**184–185.
70. White, G.B., 1973, Comparative studies on sibling species of the *Anopheles gambiae* Giles complex (Diptera: Culicidae). III. The distribution, ecology, behavior and vectorial importance of species D in Bwamba County, Uganda,

with an analysis of biological, ecological, morphological and cytogenetical relationships of Ugandan species D, *Bull. Entomol. Res.* **63:**65–97.
71. White, G.B., 1974, *Anopheles gambiae* complex and disease and transmission in Africa, *Trans. Royal Soc. Trop. Med. Hyg.* **68:**278–301.
72. White, G.B., 1985, *Anopheles bwambae* sp.n., a malaria vector in the Semliki Valley, Uganda, and its relationships with other sibling species of the *An. gambiae* complex (Diptera: Culicidae), *Syst. Entomol.* **10:**501–522.
73. Williams, S.M., DeSalle, R., and Strobeck, C., 1985, Homogenization of geographical variants at the nontranscribed spacer of rDNA in *Drosophila mercatorum, Mol. Biol. Evol.* **2:**338–346.
74. Williams, S.M., Furnier, G.R., Fuog, E., and Strobeck, C., 1987, Evolution of the ribosomal DNA spacers of *Drosophila melanogaster:* Different patterns of variation on X and Y chromosomes, *Genetics* **116:**225–232.
75. Woese, C.R., 1987, Bacterial evolution, *Microbiol. Rev.* **51:**221–271.
76. Xiong, Y., and Eickbush, T.H., 1988, The site-specific ribosomal DNA insertion element R1Bm belongs to a class of non-long-terminal-repeat retrotransposons, *Mol. Cell Biol.* **8:**114–123.

2 Neuroendocrine and Midgut Endocrine Systems in the Adult Mosquito

Mark R. Brown and Arden O. Lea

Introduction

In mosquitoes, as well as in other insects, three regulatory systems, the neural, the neuroendocrine, and the endocrine, interact to control physiological processes and behavior on a daily basis. These systems also integrate general physiological processes with the episodic events of feeding, ecdysis, diapause, metamorphosis, and reproduction. All three systems accomplish this complex, coordinated regulation by the same basic process—their component cells secrete diverse chemical messengers that affect specific target cells.

The neural and neuroendocrine systems in insects exist as distinct, though interrelated elements within the nervous system; when we refer to both regulatory systems, we will use the general term *nervous system*. The neural system consists of conventional neurons that are highly organized and integrated to transmit electrical signals to other neurons or muscle fibers by releasing chemical messengers into a specialized area, the synaptic cleft at the axon terminal. Four types of messengers are secreted by neurons (82): acetylcholine, amino acids (e.g., γ-aminobutyric acid and glutamate), biogenic amines (e.g., octopamine and serotonin), and peptides. The first three types of messengers are synthesized in the axon terminal and stored in the synaptic vesicles. The vesicles fuse with the cell membrane, open, and release the messengers into the cleft. The role of peptides as neurotransmitters or neuromodulators has only recently been recognized and is best exemplified by proctolin (79).

The neuroendocrine system is composed of neurosecretory cells that are intrinsic in the ganglia of the nervous system or associated with other

Mark R. Brown, Department of Entomology, University of Georgia, Athens, Georgia 30602, USA.
Arden O. Lea, Department of Entomology, University of Georgia, Athens, Georgia 30602, USA.

tissues in the insect body. Biogenic amines and peptides, ranging in size from 10 to more than 200 amino acids, are the only known messengers secreted by the neuroendocrine system. In contrast to neurons, these messengers are exclusively synthesized and packaged in secretory granules in the cell body. The granules are larger and more electron-dense than the neuronal synaptic vesicles. Secretory granules move down the axons, and the neuroendocrine messengers are released into the open circulatory system for transport to distant target cells or into the extracellular space from which they diffuse to adjacent cells.

The endocrine system in insects has both glandular and "diffuse" components that secrete different types of chemical messengers into the circulatory system. The glandular corpora allata, which secrete juvenile hormones, are found in all developmental stages. The prothoracic glands in larvae and the gonads in adults secrete ecdysteroids (43, 88). Juvenile hormones and ecdysteroids affect behavioral, developmental, and reproductive processes in adult mosquitoes (7, 38, 67, 97) and will not be considered in this review.

The midgut endocrine system is the "diffuse" component of the endocrine system. The term *diffuse* describes the distribution of solitary endocrine-type cells of the midgut epithelium. Most of the endocrine cells extend into the midgut lumen and, presumably, are receptive to chemical signals there. Secretory granules, which resemble those in neurosecretory cells, are packed in the basolateral regions of midgut endocrine cells. The contents of these granules appear to be released into the hemolymph and extracellular space between cells, although biochemical evidence for this release has yet to be substantiated. Nevertheless, hormone-like peptides have been detected by immunocytochemistry in the secretory granules of such cells (12). At present, application of the term *endocrine* to these cells is based on criteria obtained from immunocytochemical studies and observed morphological similarities to those described in the extensive literature on the gut endocrine system in vertebrates (102).

The primary purpose of this review is to examine the neuroendocrine system of the female mosquito relative to the midgut endocrine system. The neural system is discussed in the context of the broadly defined nervous system. Since knowledge about these systems in mosquitoes is limited, unambiguous definitions are elusive. However, by emphasizing the known morphological features of the neuroendocrine and midgut endocrine systems, we can establish morphological criteria for delineating the two systems that may later be defined by specific chemical messengers and functions. *Aedes aegypti* is almost exclusively the model for endocrine research on dipteran vectors of pathogens; therefore, we will refer principally to this species.

In the first part of this review, we will briefly describe the nervous system. Then we will focus on the anatomy of the neuroendocrine system in the female mosquito and the physiological activity of the several factors and peptides originating in this system. Third, we will describe the ultrastructure and distribution of endocrine cells in the midgut and compare aspects of the midgut endocrine system in the mosquito with similar systems in other insects. We will summarize our immunocytochemical survey of the neural, neuroendocrine, and midgut endocrine systems for putative neurotransmitters and hormones, and conclude with a brief description of the structural characterizations of peptides from mosquito heads that are antigenically similar to other invertebrate neuropeptides.

Finally, we will examine both the unique and the shared features of the neuroendocrine system and the midgut endocrine system and consider the regulatory roles specific to each system. A comparison of the two systems supports the concept of a neuroendocrine–midgut endocrine regulatory axis, applicable not only to the endocrinology of mosquitoes but also to insects in general. Implications for future research in vector–pathogen interactions and mosquito endocrinology will be discussed.

The Nervous System

The nervous system is ectodermal in origin; in the adult mosquito, it has three divisions: the central, visceral, and peripheral (19). The central division includes: (a) the supraesophageal ganglion and optic lobes, which we will refer to as the *brain;* (b) the subesophageal ganglion; (c) four fused ganglia in the thorax (prothoracic, mesathoracic, metathoracic, and first abdominal ganglia); (d) the ganglia in the second to sixth abdominal segments; (e) the fused terminal ganglia in the seventh abdominal segment; and (f) the connections between the ganglia (Figure 2.1).

The brain and other ganglia consist of an outer layer of neuronal cell bodies (perikaryal rind) and inner masses of axons (neuropil), which together are enclosed within a sheath, the neurolemma. The brain and subesophageal ganglion, joined by two thick circumesophageal connections, form a globular mass, with a narrow passage between them for the esophagus. The supraesophageal ganglion is divided into the protocerebrum, deutocerebrum, and tritocerebrum. The deutocerebrum and associated sensory nerve tracts are the only extensively mapped cerebral regions in the female *A. aegypti* (18), but there is a detailed morphological description of the male brain and subesophageal ganglion (77a).

The visceral division consists of the stomatogastric, ventral visceral, and caudal visceral nervous systems (80). The stomatogastric system is made up of the frontal ganglion, the recurrent nerve, the hypocerebral ganglion, the paired ventricular ganglia, and nerves to the mouthparts and digestive tract. The ventral and caudal visceral nervous systems innervate the respiratory, reproductive, and digestive organs.

The peripheral division consists of motor nerves, originating in the central ganglia, and sensory nerves associated with the cuticle, muscles, viscera, and sense organs.

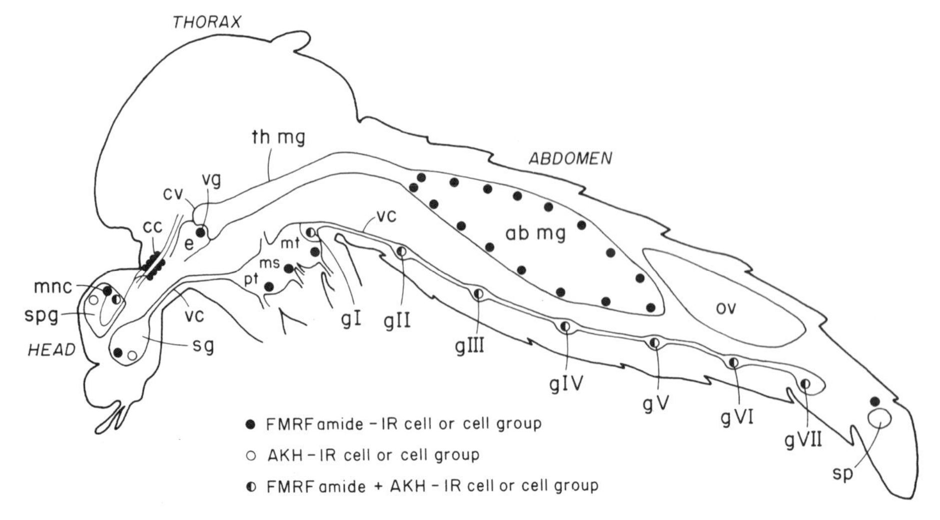

FIGURE 2.1. Summary of FMRFamide- and AKH-like immunoreactivity in the nervous system and midgut of the mosquito *A. aegypti; ab mg* abdominal midgut, *cc* corpus cardiacum, *cv* cardiac valve, *e* esophagus, *gI-VII* abdominal ganglia, *mnc* medial neurosecretory cells, *ms* mesathoracic ganglion, *mt* metathoracic ganglion, *ov* ovary, *pt* prothoracic ganglion, *sg* subesophageal ganglion, *sp* spermatheca, *spg* supraesophageal ganglion, *th mg* thoracic midgut, *vc* ventral nerve cord, *vg* ventral ganglia.

Neurotransmitters in the mosquito nervous system have not been identified. The existence of receptors for two putative neurotransmitters, dopamine and octopamine, is indicated by increased levels of cyclic AMP in head homogenate of adult *Culex pipiens,* resulting from stimulation of adenylate cyclase (83).

The Neuroendocrine System

Anatomy

The neuroendocrine system is composed of neurosecretory cells in the protocerebrum and cardiacal cells associated with the corpora allata (14, 20, 66, 76). The cardical cells are equivalent to the intrinsic cells of the corpus cardiacum of other insects. Other neurosecretory cells or systems, as yet unidentified in the mosquito, may be dispersed in the body or incorporated within other ganglia. Perisympathetic organs associated with the ventral nerve cord are examples of such organs; they have been described for a few species of Nematocera (41) and many other insects (85).

The neurosecretory cells in the supraesophageal ganglion, which are clustered into four groups, were first identified histochemically (paraldehyde-fuchsin staining) by light microscopy (14, 66). The medial neurosecretory group has approximately 12 cells on each side of the median furrow in the posterior protocerebrum. There are five lateral neurosecretory cells in a group located in each posteriolateral side of the protocerebrum. The other two groups of neurosecretory cells are bilaterally located in the ventral regions of the anterior and posterior protocerebrum.

The organization and ultrastructure of the neuroendocrine system have been extensively investigated by electron microscopy (20, 66). Axons from the medial and lateral neurosecretory cells emerge from the posterioventral protocerebrum as the nervi corporis cardiaci I and II, respectively; they enter the aortal wall and form a neurohemal organ, the corpus cardiacum. This organ is located in the head above the dorsal plate of the pharyngeal pump. Two neurohemal regions form the corpus cardiacum: (1) an anterior aortic neurohemal site where axons terminate in the lumen of the aorta and (2) a posterior lacunar site where axons terminate in extracellular channels open to the hemocoel (66). Some axons from the neurosecretory cells may continue posteriorly to the corpus allata or to the ventricular ganglia on each side of the thoracic midgut.

Each corpus allatum in the anterior prothorax is associated with five to six cells, the cardicacal neurosecretory cells (20, 66). Their axons, which join the nervi corpora allati, have been traced for only a short distance; therefore, release sites for cellular products are as yet unknown.

Neuroendocrine Regulatory Factors

Neuroendocrine factors, which are presumed to be peptides, affect more than 20 physiological processes in a variety of insects (54), although the origin and molecular structure, of only a few of these factors is known. Several were discovered when the neuroendocrine control of aedine reproduction was studied. Ablation of the medial neurosecretory cells (MNC) at eclosion blocked oocyte maturation in several mosquito species when they were subsequently fed blood (58). Implantation of the same cells restored the maturation process and demonstrated that a brain factor, egg development neurosecretory hormone (EDNH), was required for oogenesis. Later parabiosis experiments, in which decapitated blood-fed females and their intact sugar-fed partners both produced eggs, provided convincing evidence that EDNH release is stimulated by a humoral rather than a neural signal (59). More recent experiments have demonstrated that a releasing factor, secreted by the ovaries, stimulates EDNH release (61).

Peptides that can stimulate ecdysteroid secretion by isolated ovaries or yolk deposition in decapitated blood-fed females have been partially purified from extracts of heads or whole bodies (8, 44, 62, 118, 119).

Active peptides isolated by these different groups vary in estimated molecular weight from 6500 to 24,000. Although these ecdysteroidogenic factors are believed to be EDNH, their secretion by the MNC has not been demonstrated.

Recently, Matsumoto et al. (65) purified to homogeneity several neuropeptides with both gonadotropic and ecdysteriodogenic activity. The peptides, extracted from approximately 800,000 female heads, were isolated by conventional chromatography followed by ion exchange and reversed-phase, high performance liquid chromatography (HPLC). The chromatographic fractions were tested for stimulation of yolk deposition in autogenous *Aedes atropalpus* decapitated at eclosion and in *A. aegypti* decapitated after blood feeding. The purified gonadotropic peptides also stimulated isolated ovaries of *A. aegypti* to secrete ecdysteroids, as determined by radioimmunoassay (RIA). The amino acid composition of one of the peptides (approximate molecular weight, 11,000) confirmed the basic nature of the peptide observed during purification. Attempts to sequence the peptides were unsuccessful, in part, because of presumed amino-terminal blocks.

Other physiological experiments indicate that additional regulatory factors are secreted by the mosquito MNC. Synthesis of glycogen from sugar is limited by a factor from the MNC. When the MNC are removed, abnormally high levels of glycogen accumulate in the fat body after sugar feeding; normal glycogen accumulation is restored when MNC are implanted in MNC-ablated females before feeding (60). It has also been shown that secretion of proteolytic enzymes that aid in blood digestion is reduced 50% when the MNC are removed from females before blood feeding (10).

Diuretic factors have been isolated from mosquito heads. A factor extracted from the heads or thoracic ganglia of female *Anopheles freeborni* induces rapid excretion of urine by isolated Malpighian tubules (73). Two peptides (1900 to 2700 MW) with diuretic activity were isolated by HPLC from extracts of *A. aegypti* adult heads (3, 81). Another purified peptide factor depolarizes transepithelial voltage of isolated tubules without stimulating diuresis (3, 81). Recently, it has been demonstrated that the head is the source of diuretic factors which are released into the hemolymph shortly after a blood meal (119a). Hemolymph from blood fed mosquitoes stimulated fluid secretion by isolated Malpighian tubules, and hemolymph from unfed and blood fed decapitated mosquitoes did not.

Several biogenic amines, including serotonin, noradrenaline, adrenalin, octopamine, and dopamine, and leucokinins, myotropic peptides isolated from cockroaches, have been tested for effects on the transepithelial voltage and fluid secretion of isolated Malpighian tubules from *A. aegypti* (46a, 111a). Serotonin and the leucokinins depolarized the epithelium and increased fluid secretion; the other biogenic amines had no activity.

The Midgut Endocrine System

Anatomy

The alimentary tract forms a sensory and functional interface between the mosquito and its food. It consists of a foregut, midgut, and hindgut.

The foregut differentiates from the embryonic ectoderm as a layer of cells, which are covered by a cuticule on the luminal side. In the foregut of the adult mosquito, nectar or blood passes over a variety of sensory elements (sensilla) en route to the midgut. Nectar is diverted into the diverticula (part of the foregut) for storage and slow transfer to the midgut for digestion and absorption.

In the embryo, the midgut is formed by bipolar foregut and hindgut invaginations (87), presumably of endodermal origin. However, there is some controversy on this point because there is evidence of an ectodermal origin in other insects (25). In both sexes of *A. aegypti,* the midgut has two regions. The tubular thoracic region begins in the prothorax at the cardiac sphincter and extends through the thorax and first abdominal segments to enter the bulbous abdominal region that terminates at the Malpighian tubules and pyloric sphincter (Figure 2.1). The blood meal is stored entirely in the abdominal region during digestion.

The midgut epithelium is a monolayer of polarized cells, with numerous digestive cells and scattered endocrine and regenerative cells (48, 51). The digestive cells secrete enzymes into the lumen. The peritrophic membrane lining the midgut is secreted either by midgut cells as a whole (mosquito adults) or by a specialized group of cells in the midgut (mosquito larvae). Water, ions, and nutrients released from the digested food pass through the peritrophic membrane and are transported through the midgut epithelium to the hemolymph (25, 96). Tracheoles, circular and longitudinal muscles, and nerve fibers, including neurosecretory axons (prevalent in the thoracic region and at the base of the Malpighian tubules), are separated from the cells of the midgut epithelium by a basal lamina.

After digestion, the undigested food passes to the hindgut for excretion. The hindgut (of ectodermal origin) has a luminal cuticular layer covering the epithelium and is often differentiated into structures that enhance resorption of water. The Malpighian tubules (of ectodermal origin) extend from the junction of the midgut and hindgut into the body cavity. The polarized cells of the tubule epithelium transfer water and waste products from the hemolymph to the tubule lumen and on to the hindgut (9).

Midgut Endocrine Cells

Midgut endocrine cells have been described in detail for representative species from 10 insect orders (11, 16, 21, 32–34, 48, 89). Hecker (46–48) was the first to observe "clear" or "granular" endocrine-like cells in male and female *A. aegypti* and in other culicids. Similar cells have been found in the midgut of several species of such blood-feeding insects as female sandflies (92), adult fleas (89), and *Rhodnius* (4, 16).

In all the insects examined, including mosquitoes (11) (Figure 2.2), endocrine cells, which are conical or flask-like in shape, are dispersed as solitary cells in the midgut epithelium. These cells have numerous basal secretory granules, but they do not have a basal membrane labyrinth. In contrast, the larger columnar digestive cells have relatively few apical secretory granules and an extensive basal membrane labyrinth. Endocrine cells often have apical or basolateral extensions or both. Corresponding features are characteristic of endocrine cells in the vertebrate gut (102). Such properties define a unique class of midgut cells that are likely present in the midgut of all insects.

In female *A. aegypti,* there are more endocrine cells in the abdominal region of the midgut than in the thoracic region (11). To further define the distribution of midgut endocrine cells, Brown et al. (11) sampled the number of such cells in the anterior, central, and posterior areas of the abdominal region. The posterior area contained almost two times as many cells as the anterior area and four times as many cells as the central area. Endocrine cells were also observed in the midguts of *A. aegypti* larvae and *A. atropalpus* and *Toxorhynchites* spp. adults.

The most surprising result of our study was that the midgut of an adult mosquito may contain more peptidergic endocrine cells than the nervous system contains. There are an estimated 500 endocrine cells in the mosquito midgut (11), whereas the dipteran nervous system has about 200 neurosecretory cells (91). Neither this number nor this distribution of midgut endocrine cells has been described for any other insect.

Types of Endocrine Cells

Brown et al. (11) described the ultrastructure and types of endocrine cells distributed throughout the midgut of adult *A. aegypti.* Some endocrine cells extend to the lumen and are capped with microvilli (open cells), but others do not reach the lumen (closed cells). These observations were confirmed by serially sectioning both types of cells. Another variable trait of midgut endocrine cells is that of cytoplasmic density (or electron density). Both open and closed cells can exhibit either clear or dark cytoplasm.

In comparison, regenerative cells, thought to differentiate into digestive cells, also have dark cytoplasm but lack secretory granules and apical extensions. Both regenerative cells and endocrine cells are scattered basally in the epithelium. Of the endocrine cells observed in a mosquito midgut, approximately one-fourth were paired with a regenerative cell.

The association of both types of cells has been observed in the midguts

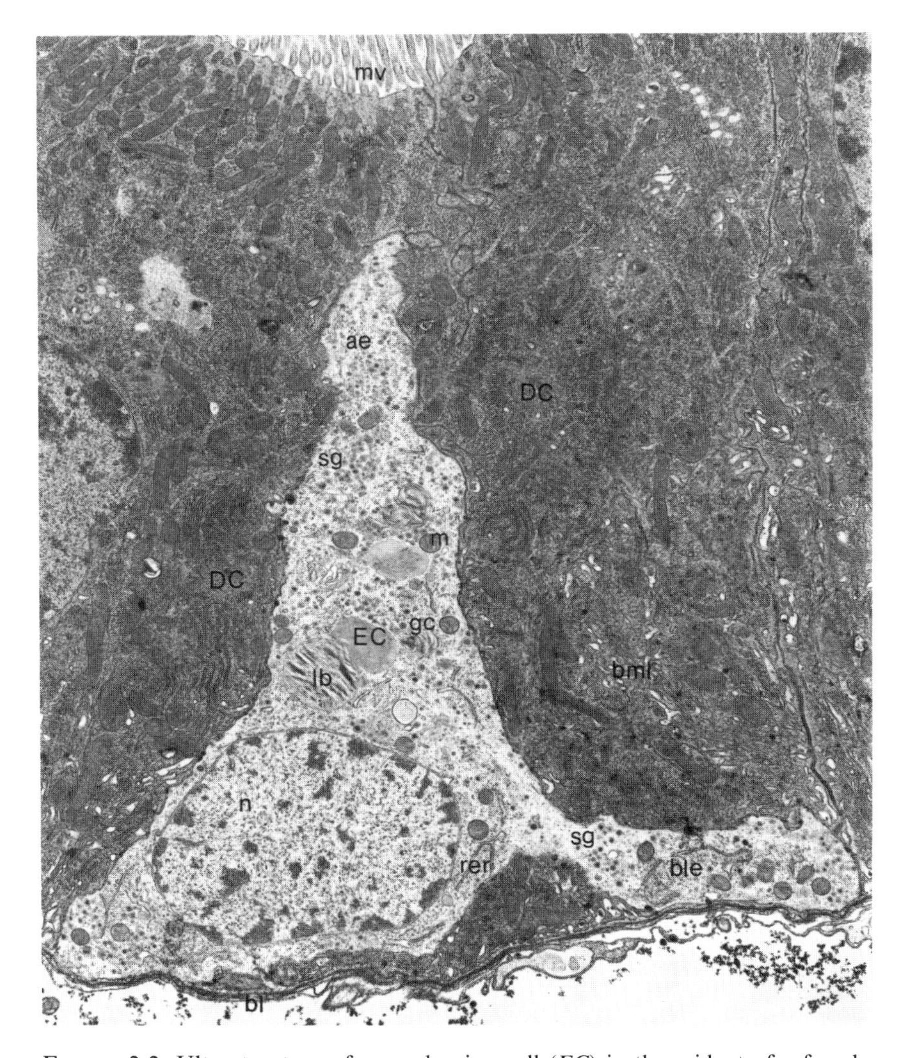

FIGURE 2.2. Ultrastructure of an endocrine cell (*EC*) in the midgut of a female mosquito. Note the presence of secretory granules (*sg*) in the "clear" cytoplasm of the endocrine cell and the lack of a basal membrane labyrinth (*bml*), a characteristic of neighboring digestive cells (*DC*); *ae* apical extension, *bl* basal lamina bordering the hemocoel, *ble* basolateral extension from endocrine cell, *gc* Golgi complex, *lb* lamellar body, *m* mitochondrin, *mv* microvilli into the midgut lumen, *n* nucleus, *rer* rough endoplasmic reticulum.

of other insects (16, 21, 34, 37), which indicates that endocrine cells may differentiate from regenerative cells. The first evidence of differentiation of a regenerative cell into an endocrine cell may be the presence of secretory granules. Such cells would exhibit the features described for closed midgut endocrine cells with dark cytoplasm. During further development, these cells may extend to the lumen and become open cells with either clear or dark cytoplasm.

Subcellular Organization

Formation and Release of Secretory Granules

The most definitive characteristic of midgut endocrine cells of *A. aegypti* is the presence of round, membrane-bounded secretory granules (60 to 120 nm in diameter) in the basal cytoplasm. Among the cells observed in a mosquito midgut, the number of granules range from a few dispersed granules to many hundreds of granules. There are two types of granules, solid or haloed, but any one cell will contain only one type. Solid granules are completely filled with a dense matrix; haloed granules have a clear space between the membrane and the dense matrix core.

The ultrastructure of midgut endocrine cells is typical of cells that secrete extracellular proteins. These proteins or prohormones, for example, are translated on the rough endoplasmic reticulum and transported to Golgi complexes by transition vesicles. These vesicles fuse with the *cis* (forming) face of the complexes and empty their contents into the cisternae. At the *trans* (maturing) face of the complexes, secretory granules containing these products are observed to bud from the cisternae into the cytoplasm.

Secretory granules clustered along the basal and basolateral cell membrane release their contents into the extracellular space by exocytosis (11, 35). The exocytic process appears to begin with a fusing of the granule and plasma membranes. The fused membranes open and release the contents of the granule into the extracellular space, as indicated by the empty omega-shaped figures along the basal cell membrane.

Several investigators have based a classification of endocrine cells in the midgut of a particular insect on certain characteristics of the secretory granules, as follows: size (up to 800 nm in diameter in cockroach endocrine cells), shape (irregular to rod-shaped), and density of the inner matrix (16, 32, 75). Brown et al. (11) classified approximately 100 endocrine cells in a mosquito midgut by the size (60 to 120 nm) and type (solid or haloed) of secretory granules in each cell (clear or dark). Particular cell types had a specific distribution within the midgut.

Lamellar Bodies

Large (up to 1 μm in diameter) membrane-bounded structures in the apical and lateral regions of endocrine cells resemble lamellar bodies and exhibit considerable structural diversity from one endocrine cell to another. The presence of granule-like structures inside the bodies suggests that the bodies may be involved in endocytosis or recycling of secretory granule membranes after exocytosis. Comparable lamellar bodies in neuroendocrine cells of the land snail *Lymnaea stagnalis* are thought to have such a role during periods of active secretion (90). The absence of acid phosphatase in lamellar bodies of midgut endocrine cells in the mosquito, contrasted to its presence in lysosomes in adjacent digestive cells, indicates that lamellar bodies are functionally different from lysosomes, which are degradative organelles (11).

Apical Extensions

Open endocrine cells, which have apical processes capped with microvilli, may be specialized for the reception of molecular signals from the food ingested by the mosquito. Secretogogues could bind to receptors on the microvilli or be taken into the vesicular–tubular system of the apical extension by endocytosis or pinocytosis. Membrane-bounded vesicles appear to bud inwardly from the cell membrane between microvilli and fuse with an extensive system of membrane-bounded tubules that intertwines among microtubule bundles along the length of the apical extension. Through such interactions, diet components may directly affect the release of chemical messengers from the cells.

Neuropeptide-Like Immunoreactivity in the Nervous System and Midgut Endocrine System

During the last two decades, comparative endocrinologists have noted two significant trends in the evolution of peptidergic nervous and endocrine systems. First, the structure and function of peptide hormones is highly conserved among diverse vertebrate groups. And second, the distribution of peptidergic cells in the nervous system, alimentary tract, and other tissues is similarly conserved. These discoveries led investigators to examine invertebrates for antecedent peptides and peptidergic cell systems by immunoassay.

Vertebrate and Invertebrate Neuropeptide-Like Immunoreactivity in Insects

The immunochemical detection of putative regulatory peptides in insects is based on the assumption that antigenic amino acid sequences are similar in the peptide hormones from phylogenetically different animals. The "small" prothoraciocotropic hormone from the silkworm *Bombyx mori,* which has steroidogenic activity similar to the mosquito ecdysteroidogenic peptides, is an example of an insect hormone that is structurally similar to a vertebrate hormone, although it differs functionally. The silkworm peptide has an amino acid sequence 40% homologous to human insulin (72). Other examples are the melanization and reddish coloration hormones from *B. mori* that have N-terminal sequence similarities to insulin-related peptides (63). Peptides in the blowfly (28) and the tobacco hornworm (56) have insulin-like activities in immunoassays and in both insect and vertebrate bioassays. The amino acid composition of the blowfly and hornworm insulin-like peptides is similar to that of vertebrate insulins.

Immunocytochemical detection of regulatory peptides is a simple and precise method with which to localize the cell or tissue source of such peptides. In one of the first immunocytochemical studies of insect tissues, insulin, somatostatin, glucagon, and other peptides were detected in the ganglia of hoverfly larvae (29). Since then, numerous vertebrate-like peptides have been found in the brain, corpora cardiaca, and other ganglia of several different species of insects (28, 30, 45, 55, 112, 114).

Immunochemical methods have been used exclusively to identify putative midgut hormones. With immunocytochemical methods for light microscopy, vertebrate hormone-like immunoreactivity, such as pancreatic polypeptide and enkephalin, has been localized in midgut endocrine cells of several insects, including the blowfly *Calliphora* (1, 27, 31, 33, 52). Insulin- and glucagon-like factors have been detected in midgut extracts with RIA (106, 108).

These reports substantially support the assertion that the neural, neuroendocrine, and midgut endocrine regulatory systems of insects secrete peptides related to neurotransmitters or neurohormones found in vertebrates and in other invertebrate groups (55, 95). The identification and localization of peptide hormone immunoreactivity in an insect can be the first step in the extraction, purification, and, ultimately, the structural and functional characterization of the native peptide.

Before we describe research on FMRFamide/pancreatic polypeptide-like and adipokinetic hormone-like immunoreactivity in mosquito tissues, we will briefly review the available information on these different peptides.

Peptides in the FMRFamide (Phe-Met-Arg-Phe-amide) family have cardioexcitatory and myotropic activity in molluscs (84). Related peptides have been identified in the nervous system and gut of invertebrates and vertebrates (23, 24, 40). Boer et al. (5) were the first to detect FMRFamide-like immunoreactivity in the insect nervous system by immunocytochemistry. The only dipteran species examined, in addition to mosquitoes, has been the fruitfly *Drosophila melanogaster*. In the adult *D. melanogaster,* FMRFamide-like immunoreactivity was detected in MNC and in several cell groups symmetrically distributed in the supraesophageal, subesophageal, and thoracic ganglia (121). For tobacco hornworm larvae, *Manduca sexta,* FMRFamide-like immunoreactivity was detected in the neuroendocrine system; it could be released from isolated corpora cardiaca–corpora allata when incubated in potassium-rich media (15).

The FMRFamide peptide family has limited sequence similarities (i.e., the last few amino acids at the amidated carboxyl terminal) to the pancreatic polypeptide (PP) family in vertebrates. Pancreatic polypeptide was first identified as one of the pancreatic islet hormones in mammals, and related peptides were also found in the nervous system (78). This peptide released from pancreatic islets inhibits secretion of enzymes by the pancreas (117), and another peptide in the family, neuropeptide Y, is a powerful stimulant of feeding behavior upon injection into the brain of satiated rats (103).

Pancreatic polypeptide-like immunoreactivity has been localized in the nervous system of two dipterans (26, 27, 29) and insects from two other orders (30, 36, 122). Although the MNC in the adult blowfly *Calliphora erthrocephala* were not reactive to the antiserum, most of the cells in the hypocerebral ganglion and a neurohemal area along the dorsal sheath of the thoracic ganglia contained immunoreactive material (26, 27). In addition, PP- and FMRFamide-immunoreactivity is co-localized in neurons of three other adult insects (68, 69, 111, 113). The structural relatedness of PP- and FMRFamide-like peptides in these insects can only be determined when their amino acid sequences are known.

Peptides structurally related to adipokinetic hormone (AKH) currently constitute the only recognized family of peptide hormones in insects (93). The peptides may function as neurohormones by elevating lipid or carbohydrate levels in the hemolymph (53, 101) or as myotropic neurotransmitters (93). It is interesting to note that a portion of the sequence shared by these peptides is similar to that of glucagon, a vertebrate peptide, which elevates carbohydrate levels in blood.

In a series of immunocytochemical studies with sequence-specific antisera, Schooneveld et al. (98–100) examined the nervous systems of insects from several orders for AKH-like immunoreactivity. There are six or more AKH-immunoreactive neurons in the brain of the adult housefly *Musca domestica* and of *D. melanogaster* (98). Neurons in the medial neurosecretory area, cells and axons in the corpus cardiacum, and axons to the corpora allata were not reactive. In contrast, among the hemimetabolous insects examined, there are more than 100 immunoreactive cells in the corpus cardiacum.

Immunocytochemical Survey of the Nervous System and Midgut Endocrine System in the Mosquito

Brown et al. (12, 13) used immunocytochemical methods to examine tissues from female mosquitoes (*A. aegypti*) for factors antigenically related to peptide hormones from vertebrates and other invertebrates. The types of immunoreactive cells and their distribution in the nervous system and midgut were described in detail.

FMRFamide-Like Immunoreactivity

Approximately 120 cells that are immunoreactive to a FMRFamide antiserum are distributed throughout the different ganglia of the nervous system (summarized in Figure 2.1) (13). In the brain and subesophageal ganglion alone, there are 75 to 100 immunoreactive cells (Figure 2.3A). Several females were examined for such cells, and although their location in the ganglia was the same for each mosquito, their number varied slightly.

All the cells exhibited cross-reactivity to a bovine PP antiserum; however, preabsorption experiments with the FMRFamide and PP antisera and their antigenic peptides indicated that the immunoreactive peptide in the cells was probably more closely related to FMRFamide than to PP. For this reason, the immunoreactivity is considered to be FMRFamide-like.

Brain and Subesophageal Ganglion

FMRFamide-like immunoreactivity is evident in paired clusters of five to eight MNC, the nervi corporis cardiaci, and the corpus cardiacum. As observed by electron microscopy, immunoreactivity in these cells is limited to the secretory granules (Figure 2.3B). Some immunoreactivity is also present in axons extending to the corpora allata in the anterior prothorax; however, the corpora allata and adjacent cardiacal cells are not reactive.

There are also 12 to 18 immunoreactive cells in each side of the posterior lateral regions of the protocerebrum, although some of these cells may be lateral neurosecretory cells. Several posterior neuropil masses in the protocerebrum, especially those proximal to the MNC, contain many immunoreactive axons. In the anterior portion of the brain, four to ten cells on each side appear to be neurons. Axons in the nerve tract from the tritocerebrum to the frontal ganglia are immunoreactive. Each optic lobe contains three to six cells, and stained axons surround the external medulla.

In the subesophageal ganglion, there are paired clusters of 10 to 14 cells at the anterior edge. Three to eight cells are scattered throughout each posterior side. Positive axons from the cells are interspersed in the neuropil masses of the ganglion.

Thoracic and Abdominal Ganglia

In each of the thoracic ganglia, immunopositive cells are found ventrally on each side of the midline. Stained axons from these cells are evident in the neuropil masses of the fused ganglia. The first abdominal ganglion, which is fused to the metathoracic ganglion, contains a single immunoreactive cell, as does each ganglion in the abdomen.

The cells of the ventricular ganglia, which lie on each side of the cardiac

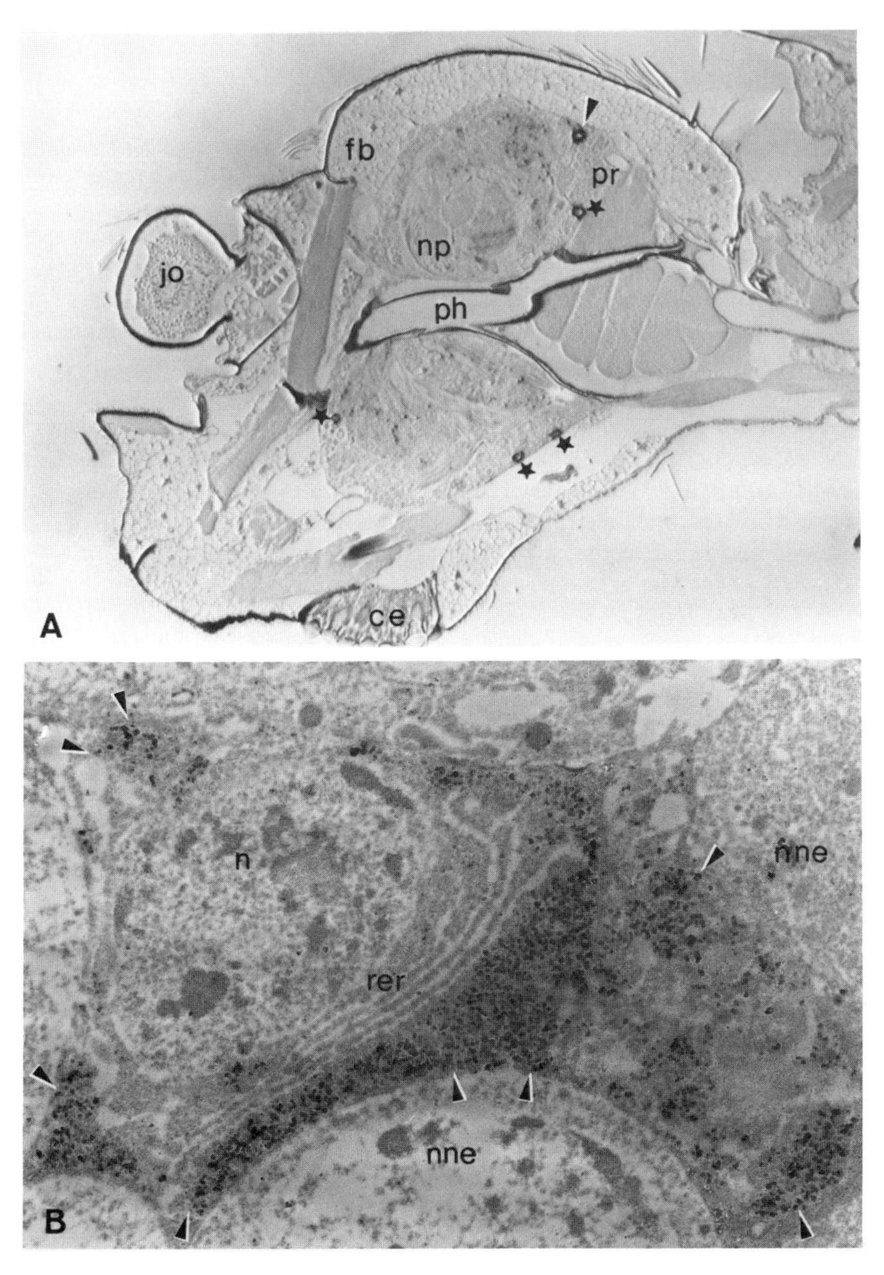

FIGURE 2.3. (A) In a sagittal section (anterior, left) through the head of a female mosquito, a medial neurosecretory cell (arrow) and four other cells (stars) contain FMRFamide-like immunoreactivity; *ce* compound eye, *fb* fat body, *jo* Johnston's organ, *np* neuropil, *pr* perikaryal rind, *ph* pharynx. (B) FMRFamide-like immunoreactivity localized in numerous secretory granules (arrows) of a neurosecretory cell in the supraesophageal ganglion; *n* nucleus, *rer* rough endoplasmic reticulum, *nne* nonreactive neurons.

valve of the thoracic midgut, are all immunoreactive. Stained axons from these cells extend to the cardiac valve and along the thoracic midgut. Finally there is a single immunopositive cell adjacent to the anterior-most spermatheca.

Midgut Endocrine System

Approximately 250 of the estimated 500 endocrine cells in a female mosquito midgut are immunoreactive to the FMRFamide antiserum (12). By light microscopy, these cells have the characteristics of endocrine cells, rather than digestive cells. At the ultrastructural level, the immunoreactive substance is found only in the secretory granules of such cells. Therefore, the cells are considered to be peptidergic endocrine cells.

FMRFamide-like and PP-like immunoreactivity co-localize in the same midgut endocrine cells, but the immunoreactivity appears to be more FMRFamide-like, based on preabsorption of the sera with the antigenic peptides. Glattli et al. (39) also localized PP-like immunoreactivity in midgut endocrine cells of two mosquitoes, *A. aegypti* and *Anopheles stephensi,* and the tsetse fly, *Glossina morsitans morsitans,* but did not test for FMRFamide cross-reactivity. Immunocytochemistry with antisera to several other insect and vertebrate peptides, such as insulin, glucagon, and AKH, did not reveal other types of immunoreactive cells in the mosquito midgut (12, 39).

Functional Significance

In the nervous system of the female mosquito, the majority of the cells immunoreactive to the FMRFamide antiserum appear to be conventional neurons with axons directed to intrinsic neuropil masses. In these cells, the immunoreactive peptide may function as a neurotransmitter.

In contrast, the immunoreactive cells located in the MNC region of the brain are presumably neurosecretory, since their axons extend to the corpus cardiacum. The specific localization of immunoreactivity in the secretory granules of these cells is strong circumstantial evidence for a neurohormone that has some sequences in common with FMRFamide.

Endocrine cells, which have FMRFamide immunoreactivity, are distributed exclusively in the abdominal midgut region where the blood meal is digested (Figure 2.1). Ingestion of vertebrate blood reduces the apparent number of such cells and the intensity of the reaction in others. These two observations suggest that a blood meal either stimulates release of the immunoreactive substance from the cells or reduces its synthesis.

ADIOPOKINETIC HORMONE-LIKE IMMUNOREACTIVITY

Relatively few cells (15 to 20 cells, total) in the entire mosquito nervous system are immunoreactive to the AKH antiserum (see Figure 2.1 for a summary) (13). One large positive cell is located dorsally in each side of the protocerebrum near the compound eye, and two other cells are located more posteriorly in each side. There are three to four immunoreactive cells in each side of the subesophageal ganglion, and their axons are interspersed in neuropil masses. No immunoreactive cells were found in the thoracic ganglia, whereas each abdominal ganglion contained a positive cell.

In the mosquito brain, all the cells that were immunoreactive to the AKH antiserum appeared to be conventional neurons. Therefore, the AKH-like peptides may function as neurotransmitters in the same way as other AKH-related peptides (93), rather than as neurohormones.

Co-localization of AKH-like and FMRFamide-like immunoreactivity is exhibited by some cells in the brain and by a cell in each abdominal ganglia (Figure 2.1). Preabsorption of the AKH or FMRFamide antisera with the reciprocal antigenic peptide does not affect binding of either antiserum to immunoreactive cells at either site. These results suggest that the cells may secrete both FMRFamide-like and AKH-like peptides, which are antigenically unrelated.

Other reports also indicate that neurons and neurosecretory cells in insects may secrete more than one type of chemical messenger (2, 107, 115). The ubiquity of this phenomenon is also supported by the demonstrated coexistence of different peptides, acetylcholine, amino acids, or biogenic amines in the same types of cells in molluscs and mammals (6, 50).

Isolation and Characterization of Neuropeptides Immunoreactive to FMRFamide Antiserum

As the first step toward defining the functional significance of FMRFamide-like peptides in mosquitoes, Matsumoto et al. (64) extracted and purified the immunoreactive peptides from more than 1 million mosquito heads. FMRFamide-like immunoreactivity in chromatography fractions was detected by RIA. After several steps of conventional liquid chromatography and HPLC, a sufficient quantity of the two immunoreactive peptides was obtained for amino acid composition and sequence analysis.

The two peptides are called *Aedes* Head Peptide I and II (Aea-HP-I and II) (Table 2.1). The chromatographic properties of the two native peptides are identical to those of synthetic peptides having the corresponding sequence, which thus confirms the structural analysis. As evident in the sequences, the C-terminal sequence of Arg-Phe-amide (RFamide) is the *only* structural element common to both the mosquito and molluscan peptides but is sufficient for recognition by the FMRFamide antiserum (77). A third immunoreactive peptide was also isolated, but structural analysis has not been completed.

TABLE 2.1. Amino acid sequences of insect RFamide peptides.

Peptide	Sequence
Aedes Head Peptide I (Aea-HP-I)	pGlu-Arg-Pro-Hyp-Ser-Leu-Lys-Thr-Arg-Phe-amide
Aedes Head Peptide II (Aea-HP-II)	Thr-Arg-Phe-amide
Leucomyosuppressin	pGlu-Asp-Val-Asp-His-Val-Phe-Leu-Arg-Phe-amide
Leucosulfakinin I	Glu-Gln-Phe-Glu-Asp-Tyr(SO_3H)-Gly-His-Met-Arg-Phe-amide
Leucosulfakinin II	pGlu-Ser-Asp-Asp-Tyr(SO_3H)-Gly-His-Met-Arg-Phe-amide

Characterization of the structure and function of three RFamide peptides from cockroaches (Table 2.1) supports, in part, our suggestion that Aea-HPs may have regulatory properties. Leucomyosuppressin isolated from the brain of *Leucophaea maderae* inhibits muscle contractions of isolated hindguts (49). Leucosulfakinin I and II stimulate hindgut muscle contractions and also are structurally similar to the gastrin/cholecystokinin family of vertebrate peptides (70, 71). The synthetic Aea-HPs are presently being screened for neurohormonal and neurotransmitter activity in the female mosquito.

Comparison of the Neuroendocrine and Midgut Endocrine Systems

Morphology and Anatomy

The neuroendocrine and midgut endocrine systems are made up of polarized cells that secrete chemical messengers. In neurosecretory cells, chemical messengers are synthesized in the cell body and transported via secretory granules to distant release sites along the axon. The messengers of midgut endocrine cells are also synthesized in the cell body and stored in secretory granules, but exocytosis occurs in the basolateral portion of the cell body. For both types of cells, the messengers are released either directly into the hemolymph for transport to distant target tissues or into the extracellular space for diffusion to nearby target cells.

The midgut endocrine system has a greater degree of cell disperal than the neuroendocrine system and endocrine glands of insects. Vigna (116) argues that the dispersal of gut endocrine cells is an important factor in the overall regulation of digestive processes in vertebrates. We believe that his idea is relevant to the midgut endocrine system of insects and have incorporated it in the following discussion. Food in the midgut lumen is "outside" the insect body. The midgut secretes enzymes that digest the food, it absorbs nutrients, and it moves food wastes through and out of the body. These potent enzymes could digest the midgut itself, if they were not released in specific quantities and at appropriate times. Therefore, detection of food quantity and quality within the midgut is required to regulate digestive and excretory processes to prevent self-digestion and to optimize nutrient absorption. Midgut endocrine cells may perform this function by acting as "sensory/effector" cells to assess food in the gut lumen and transmit this information to other cells via chemical messengers.

The diffuse distribution of endocrine cells in the midgut would facilitate an integrated sampling of the entire "external environment" in the lumen and thus titrate the signal to the insect. Hormone secretion could be related to nutrient load, and the digestive cells would respond to the graded hormone signal by releasing enzymes in the required amount. As nutrient-return from digestion diminished in the lumen, the dispersed endocrine cells could similarly signal the midgut musculature to move the food around, in, or out of the midgut.

The neuroendocrine system, which also has a significant degree of cell dispersal in the insect body, could affect regulation at two levels: (a) the medial and lateral neurosecretory cells could release chemical messengers directly into the general circulatory system, or (b) the perisympathetic organs, several ganglia (i.e., hypocerebral and ventricular ganglia), and many other neurosecretory cells distributed throughout the mosquito body could have axons extending to specific organs or tissues for local delivery of chemical messengers. All these neuroendocrine elements are innervated; therefore, secretion by these elements could be regulated by the nervous system in response to internal and external (sensory) cues.

Such endocrine glands as the corpora allata and prothoracic glands are organized into a cluster of nonpolarized cells. Regulation of secretion may be optimized by such clustering because all the cells are exposed to the same titers of circulating or locally released chemical messengers.

Chemical Messengers

Because of the shared morphological features of neurosecretory and midgut endocrine cells, it is not surprising that both the neuroendocrine and midgut endocrine systems utilize peptides as chemical messengers. Peptides have enormous structural diversity, which would have facilitated the evolution of a specific message from each peptide. The neuroendocrine system in insects is widely acknowledged to be an important source of neurohormones; a regulatory function for the midgut endocrine system, however, has yet to be demonstrated.

Isolating and characterizing such immunoreactive neuropeptides as the Aea-HPs may be easier than determining their function. For some immunoreactive factors, such as the insulin-like ones, insect and vertebrate peptides may regulate similar types of physiological processes in the organism. Immunoreactive factors from phylogenically distant species also may have very different functions; although they may share an antigenically related sequence of a few amino acids, they may also have different primary structures. Furthermore, identical or related peptides in an organism may be secreted by different regulatory systems and have different functions, as has been discovered for several brain-gut peptides, for example, cholecystokinin, in vertebrates (22, 120).

Immunocytochemical methods are used to identify cell and tissue sources of peptide messengers. In addition, RIAs can be used to quantify immunoreactive peptides in the hemolymph or tissues of insects during development or physiological or behavioral processes. This quantitative information, correlated with the occurrence of such events, may provide clues about the regulatory properties of such peptides.

Other methods, such as the identification of receptors for immunoreactive peptides (104, 105) in different tissues, may suggest other functions that could be subsequently explored. Muscle contraction assays may be useful for testing immunoreactive peptides for neurotransmitter activity, although this activity may be incidental to hormonal roles of peptides isolated from neuroendocrine or midgut endocrine systems.

The regulatory functions of factors secreted by the midgut endocrine system have yet to be defined. Because endocrine cells cannot be removed en masse, such manipulations as gland removal and implantation to define hormonal regulation of specific physiological processes cannot be performed. Therefore, assigning regulatory functions to chemical messengers from the midgut endocrine system will be difficult.

Another complication is that chemical messengers released from midgut endocrine cells may have paracrine effects on neighboring cells that differ from their endocrine effects on distant tissues. A good example of paracrine secretion in vertebrate gut endocrinology is the inhibitory effect of somatostatin on gastrin secretion (57). In the stomach, somatostatin is released from cells that are in contact with gastrin-secreting cells, a paracrine effect. Somatostatin also has a broad spectrum of endocrine effects, such as inhibition of growth hormone and pancreatic polypeptide secretion and reduction of gut motor activity (117).

We believe that the midgut endocrine system in the mosquito, in particular, and insects, in general, is an important source of chemical messengers, based upon the results of morphological and immunocytochemical studies. These messengers may regulate digestive and associated absorptive processes in epithelial cells, alert metabolic tissues (e.g., the fat body) to incoming nutrients, or modify feeding behavior in ways comparable to how gut hormones affect similar processes in vertebrates.

This review has presented several morphological studies of the midgut endocrine system in different insect species. Yet, as previously mentioned, no secretory products or regulatory functions for the midgut endocrine system have been identified. A similar situation existed earlier in this century when the evidence for neurosecretory hormones in all animals depended almost entirely on morphological studies. Since then, numerous neuroendocrine messengers have been identified, and functions have been ascribed to many of them. In time, the same will be true for the midgut endocrine system in insects.

Functional Relationships

Interactions between the neural, neuroendocrine, and midgut endocrine systems may coordinate digestive and excretory processes with feeding behavior. Chemical messengers released from neurosecretory and neuronal axons on the midgut exterior conceivably could affect secretion by endocrine cells; the opposite also could occur. An intriguing example of message partitioning was found in the mosquito, but its significance is not known. FMRFamide-like immunoreactivity is found only in neuroendocrine elements (the ventricular ganglia and axons) associated with the thoracic midgut, which is the primary tissue for nectar assimilation (96). In the abdominal midgut, the site of blood storage and protein digestion, such immunoreactivity was found only in the midgut endocrine system.

Host–Parasite Hormonal Interactions

Investigations of hormonal interactions that affect the development of parasites transmitted by mosquitoes and of the physiology of infected mosquitoes are warranted, since several studies have found that parasite invasion of the vertebrate gut alters the gut endocrine balance of the host (17). The midgut of the female mosquito is the first tissue barrier encountered by blood-borne parasites. Arboviruses and malaria cells enter and replicate in the midgut epithelium, and most nematodes penetrate the midgut, although they develop in other tissues. Factors from the neuroendocrine and midgut endocrine systems may directly or indirectly, through effects on midgut cells or on immune responses (i.e., encapsulation), affect parasite invasion, development, and replication. For example, a factor that stimulates malarial exflagellation in the lumen of the midgut is found in both midgut and head extracts (42, 74). The exflagellation factor may be released by endocrine cells into the lumen. In the vertebrate gut, several peptides, similarly released from endocrine cells, have been detected in luminal contents and may be functional (110).

Vertebrate-like peptides in mosquitoes reportedly affect parasite development. Terzian (109) has shown that the malarial infection level of *A. aegypti* was altered when mosquitoes were given insulin or adrenocorticotropic hormone (ACTH) in sugar solutions before and after feeding on an infected host. Those animals given insulin had increased numbers of oocysts on the midgut wall; those fed ACTH had decreased numbers of parasites. Both insulin- and ACTH-like immunoreactive factors have been identified in the midgut or the nervous system of several insects (56, 86). Attempts to identify such factors in mosquitoes have not been reported.

Future Directions

We expect that in the near future numerous peptides will be isolated for which structures but not functions are known. Many of these peptides may be immunologically related to peptide hormones found in other insects or in vertebrate and invertebrate species. Insect physiologists will need physiological and behavioral bioassays to test these "new" peptides. The basic information gained from such studies may lead to the use of insect model systems for vertebrate endocrinology (94). Advances in basic insect neurobiology may enhance the use of new biotechnological methods for insect pest and vector control as well (54).

Vertebrate endocrinology has been revolutionized as the functional significance of different peptides from the gut endocrine system has been demonstrated over the past three decades. Vertebrate neurobiology has been similarly affected, as the function of many neuropeptides structurally related to gut peptides was demonstrated. Related peptides in the gut endocrine system and nervous system may have the same or different functions, which enhances the complexity of interactions between these two regulatory systems. We believe that insect endocrinology and insect neurobiology await a comparable revolution. As more chemical messengers of the insect nervous system and midgut endocrine system are discovered, the variety of peptides known for vertebrates may, ultimately, seem insignificant compared to that of insects, the most numerous and diverse group of animals.

Acknowledgments. This work was supported by a grant (AI-17297) from the National Institutes of Health.

References

1. Andries, J.C., and Tramu, G., 1985, Ultrastructural and immunocytochemical study of endocrine cells in the midgut of the cockroach *Blaberus craniifer* (Insecta, Dictyoptera), *Cell Tissue Res.* **240**:323–332.
2. Andries, J.C., Belemtourgri, G., and Tramu, G., 1986, Coexistence of serotonin and cholecystokinin-like material in the nervous system of *Aeshna cyanea* (Insecta, Odonata), *Biogen. Amines* **3**:219–227.
3. Beyenbach, K.W., and Petzel, D.H., 1987, Diuresis in mosquitoes: Role of a natriuretic factor, *News Physiol. Sci.* **2**:171–175.
4. Billingsley, P.F., and Downe, A.E.R., 1986, Nondigestive cell types in the midgut epithelium of *Rhodnius prolixus* (Hemiptera: Reduvidae), *J. Med. Entomol.* **23**:212–216.

5. Boer, H.H., Schot, L.P.C., Veenstra, J.A., and Reichelt, D., 1980, Immunocytochemical identification of neural elements in the central nervous system of a snail, some insects, a fish, and a mammal with an antiserum to the molluscan cardio-excitatory tetrapeptide FMRFamide, *Cell Tissue Res.* **213:**21–27.
6. Boer, H.H., Schot, L.P.C., Steinbusch, H.W.M., Montagne, C., and Reichelt, D., 1984, Coexistence of immunoreactivity to anti-dopamine, anti-serotonin, and antivasotocin in the cerebral giant neuron of the pond snail *Lymnaea stagnalis, Cell Tissue Res.* **238:**411–412.
7. Borovsky, D., Thomas, B.R., Carlson, D.A., Whisenton, L.R., and Fuchs, M.S., 1985, Juvenile hormone and 20-hydroxyecdysone as primary and secondary stimuli of vitellogenesis in *Aedes aegypti, Arch. Insect Biochem. Physiol.* **2:**75–90.
8. Borovsky, D., and Thomas, B.R., 1985, Purification and partial characterization of mosquito egg development neurosecretory hormone: Evidence for gonadotropic and steroidogenic effects, *Arch. Insect. Biochem. Physiol.* **2:**265–281.
9. Bradley, T.J., 1987, Physiology of osmoregulation in mosquitoes, *Ann. Rev. Entomol.* **32:**439–462.
10. Briegel, H., and Lea, A.O., 1979, Influence of the endocrine system on tryptic activity in female *Aedes aegypti, J. Insect Physiol.* **25:**231–235.
11. Brown, M.R., Raikhel, A.S., and Lea, A.O., 1985, Ultrastructure of midgut endocrine cells in the mosquito, *Aedes aegypti, Tissue Cell* **17:**709–721.
12. Brown, M.R., Crim, J.W., and Lea, A.O., 1986, FMRFamide- and pancreatic polypeptide-like immunoreactivity in midgut endocrine cells of a mosquito, *Tissue Cell* **18:**419–428.
13. Brown, M.R., and Lea, A.O., 1988, FMRFamide- and adipokinetic hormone-like immunoreactivity in the nervous system of the mosquito, *Aedes aegypti, J. Comp. Neurol.* **270:**606–614.
14. Burgess, L., and Rempel, J.G., 1966, The stomodaeal nervous system, the neurosecretory system, and the gland complex in *Aedes aegypti* (L.) (Diptera: Culicidae), *Can. J. Zool.* **44:**731–765.
15. Carroll, L.S., Carrow, G.M., and Calabrese, R.L., 1986, Localization and release of FMRFamide-like immunoreactivity in the cerebral neuroendocrine system of *Manduca sexta, J. Exp. Biol.* **126:**1–14.
16. Cassier, P., and Fain-Maurel, M-A., 1977, Sur la présence d'un système endocrine diffus dans le mésenteron de quelques insectes, *Arch. Zool. Exp. Gen.* **118:**197–209.
17. Castro, G.A., 1981, Physiology of the gastrointestinal tract in the parasitized host, in Johnson, L.R. (ed): Physiology of the Gastrointestinal Tract, New York, Raven Press, 1492 p.
18. Childress S.A., and McIver, S.B., 1984, Morphology of the deutocerebrum of female *Aedes aegypti* (Diptera: Culicidae), *Can. J. Zool.* **62:**1320–1328.
19. Christophers, S.R., 1960, *Aedes aegypti:* Its Life History, Bionomics, and Structure. London, Cambridge University Press.
20. Clements, A.N., Potter, S.A., and Scales, M.D.C., 1985, The cardiacal neurosecretory system and associated organs of an adult mosquito, *Aedes aegypti, J. Insect Physiol.* **31:**821–830.
21. De Priester, W., 1971, Ultrastructure of the midgut epithelial cells in the fly, *Calliphora erythrocephala, J. Ultrastruct. Res.* **36:**783–805.

22. Dockray, G.J., 1982, The physiology of cholecystokinin in brain and gut, *Br. Med. Bull.* **38:**253–258.
23. Dockray, G.J., Vaillant, C., and Williams, R.G., 1981, New vertebrate brain–gut peptide related to a molluscan neuropeptide and an opioid peptide, *Nature* **293:**656–657.
24. Dockray, G.J., and Drimaline, R., 1985, FMRFamide- and gastrin/CCK-like peptides in birds, *Peptides* **6:**333–337.
25. Dow, J.A., 1986, Insect midgut function, *Adv. Insect Physiol.* **19:**187–328.
26. Duve, H., and Thorpe, A., 1980, Localization of pancreatic polypeptide (PP)-like immunoreactive material in neurones of the brain of the blowfly, *Calliphora erythrocephala* (Diptera), *Cell Tissue Res.* **210:**101–109.
27. Duve, H., and Thorpe, A., 1982, The distribution of pancreatic polypeptide in the nervous system and gut of the blowfly, *Calliphora vomitoria* (Diptera), *Cell Tissue Res.* **227:**67–77.
28. Duve, H., and Thorpe, A., 1984, Comparative aspects of insect-vertebrate neurohormones, in Borkovec, A.B., and Kelly, T.J. (eds): Insect Neurochemistry and Neurophysiology, New York, Plenum Press, pp. 171–195.
29. El-Salhy, M., Grimelius, L., Falkmer, S., Abou-Elella, R., and Wilander, E., 1980, Immunohistochemical evidence of gastro-entero-pancreatic neurohormonal peptides of vertebrate type in the nervous system of the larva of a diptera insect, the hover-fly, *Eristalis aeneus, Regul. Pept.* **1:**187–204.
30. El-Salhy, M., Falkmer, A., Kramer, K.J., and Speirs, R.D., 1983, Immunohistochemical investigations of neuropeptides in the brain, corpora cardiaca, and copora allata of an adult lepidopteran insect, *Manduca sexta* (L.), *Cell Tissue Res.* **232:**295–317.
31. Endo, Y., 1984, Ontogeny of endocrine cells in the gut of the insect *Periplaneta americana, Cell Tiss. Res.* **238:**421–423.
32. Endo, Y., and Nishiitsutsuji-Uwo, J., 1981, Gut endocrine cells in insects: The ultrastructure of the gut endocrine cells of the lepidopterous species, *Biomed. Res.* **2:**270–280.
33. Endo, Y., Nishiitsutsuji-Uwo, J., Iwanaga, T., and Fujita, T., 1982, Ultrastructural and immunohistochemical identification of pancreatic polypeptide-immunoreactive endocrine cells in the cockroach midgut, *Biomed. Res.* **3:**454–456.
34. Endo, Y., and Nishiitsutsuji-Uwo, J., 1982, Fine structure of developing endocrine cells and columnar cells in the cockroach midgut, *Biomed. Res.* **3:**637–644.
35. Endo, Y., and Nishiitsutsuji-Uwo, J., 1982, Exocytotic release of secretory granules from endocrine cells in the midgut of insects, *Cell Tissue Res.* **222:**515–522.
36. Endo, Y., Iwanaga, T., Fujita, T., and Nishiitsutsuji-Uwo, J., 1982, Localization of pancreatic polypeptide (PP)-like immunoreactivity in the central and visceral nervous systems of the cockroach *Periplaneta, Cell Tissue Res.* **227:**1–9.
37. Endo, Y., Sugihara, H., Fujita, S., and Nishiitsutsuji-Uwo, J., 1983, Kinetics of columnar and endocrine cells in the cockroach midgut, *Biomed. Res.* **4:**51–60.
38. Fuchs, M.S., and Kang, S-H., 1981, Ecdysone and mosquito vitellogenesis: A critical appraisal, *Insect Biochem.* **11:**627–633.

39. Glattli, E., Rudin, W., and Hecker, H., 1987, Immunoelectron microscopic demonstration of pancreatic polypeptide in midgut epithelium of hematophagus dipterans, *J. Histochem. Cytochem.* **35:**891–896.
40. Greenberg, M.J., and Price, D.A., 1983, Invertebrate neuropeptides: Native and naturalized, *Ann. Rev. Physiol.* **45:**271–288.
41. Grillot, J.P., 1977, Les organes périsympathiques des Diptères, *Int. J. Insect Morphol. Embryol.* **65:**303–343.
42. Gwadz, R.W., and Carter, R., 1981, Physiological and immunological factors controlling the infectivity of the sexual stages of malarial parasites to mosquitoes, in McKelvey, J.J. Jr., Eldridge, B.F., and Maramorosch, K. (eds): Vectors of Disease Agents, New York, Praeger, pp. 158–164.
43. Hagedorn, H.H., 1985, The role of ecdysteroids in reproduction, in Kerkut, G.A., and Gilbert, L.I. (eds): Comprehensive Insect Physiology, Biochemistry, and Pharmacology, Volume 8, New York, Pergamon Press, pp. 205–262.
44. Hanaoka, K., and Hagedorn, H.H., 1980, Brain hormone control of ecdysone secretion by the ovary in a mosquito, in Hoffman, J.A. (ed): Progress in Ecdysone Research, Amsterdam, Elsevier/North Holland, pp. 467–480.
45. Hansen, B.L., Hansen, G.N., and Scharrer, B., 1982, Immunoreactive material resembling vertebrate neuropeptides in the corpus cardiacum and corpus allatum of the insect *Leucophaea maderae, Cell Tissue Res.* **225:**319–329.
46a. Hayes, T.K., Pannabecker, T.L., Hinckley, D.J., Holman, G.M., Nachman, R.J., Petzel, D.H., and Beyenbach, K.W., 1989, Leucokinins, a new family of ion transport stimulators and inhibitors in insect Malpighian tubules, *Life Sci.* **44:**1259–1266.
46. Hecker, H., Freyvogel, T.A., Briegel, H., and Steiger, R., 1971, Ultrastructural differentiation of the midgut epithelium in female *Aedes aegypti* L. (Insecta, Diptera) imagines, *Acta Trop.* **28:**80–104.
47. Hecker, H., Freyvogel, T.A., Briegel, H., and Steiger, R., 1971, The ultrastructure of midgut epithelium in *Aedes aegypti* L. (Insecta, Diptera) males, *Acta Trop.* **28:**275–290.
48. Hecker, H., 1977, Structure and function of midgut epithelial cells in Culicidae mosquitoes (Insecta, Diptera), *Cell Tissue Res.* **184:**321–341.
49. Holman, G.M., Cook, B.J., and Nachman, R.J., 1986, Isolation, primary structure and synthesis of leucomyosuppressin, an insect neuropeptide that inhibits spontaneous contraction of the cockroach hindgut, *Comp. Biochem. Physiol.* **85C:**329–333.
50. Hokfelt, T., Everitt, B., Meister, B., Melander, T., Schalling, M., Johansson, O., Lundberg, J.M., Hulting, A-L., Werner, S., Cuello, C., Hemmings, H., Ouimet, C., Walaas, I., Greengard, P., and Goldstein, M., 1986, Neurons with multiple messengers with special reference to neuroendocrine systems, *Recent Prog. Horm. Res.* **42:**1–70.
51. Houk, E.J., and Hardy, J.L., 1982, Midgut cellular responses to bloodmeal digestion in the mosquito, *Culex tarsalis* Coquillet (Diptera: Culicidae), *Int. J. Insect Morphol. Embryol.* **11:**109–119.
52. Iwanaga, T., Fujita, T., Nishiitsutsuji-Uwo, J., and Endo, Y., 1981, Immunohistochemical demonstration of PP-, somatostatin-, enteroglucagon-, and VIP-like immunoreactivities in the cockroach midgut, *Biomed. Res.* **2:**202–207.

53. Jaffe, H., Raina, A.K., Riley, C.T., Fraser, B.A., Holman, G.M., Wagner, R.M., Ridgway, R.L., and Hayes, D.K., 1986, Isolation and primary structure of a peptide from the corpora cardiaca of *Heliothus zea* with adipokinetic activity, *Biochem. Biophys. Res. Comm.* **135:**622–628.
54. Keeley, L.L., and Hayes, T.K., 1987, Speculations on biotechnology applications for insect neuroendocrine research, *Insect Biochem.* **17:**639–651.
55. Kramer, K.J., 1985, Vertebrate hormones in insects, in Kerkut, G.A., and Gilbert, L.I. (eds): Comprehensive Insect Physiology, Biochemistry, and Pharmacology, New York, Pergamon Press, pp. 511–536.
56. Kramer, K.J., Childs, C.N., Spiers, R.D., and Jacobs, R.M., 1982, Purification of insulin-like peptides from insect hemolymph and royal jelly, *Insect Biochem.* **12:**91–98.
57. Larsson, L.I., Goltermann, N., De Magistris, L., Rehfeld, J.F., and Schwartz, T.W., 1979, Somatostatin cell processes as pathways for paracrine secretion, *Science* **205:**1393–1395.
58. Lea, A.O., 1967, The medial neurosecretory cells and egg maturation in mosquitoes, *J. Insect Physiol.* **13:**419–429.
59. Lea, A.O., 1972, Regulation of egg maturation in the mosquito by the neurosecretory system: The role of the corpus cardiacum, *Gen. Comp. Endocrinol.* (Suppl.) **3:**602–608.
60. Lea, A.O., and Van Handel, E., 1970, Suppression of glycogen synthesis in the mosquito by a hormone from the medial neurosecretory cells, *J. Insect Physiol.* **16:**319–321.
61. Lea, A.O., and Van Handel, E., 1982, A neurosecretory hormone-releasing factor from ovaries of mosquitoes fed blood, *J. Insect Physiol.* **28:**503–508.
62. Masler, E.P., Hagedorn, H.H., Petzel, D.H., and Borkovec, A.B., 1983, Partial purification of egg development neurosecretory hormone with reversed-phase liquid chromatographic techniques, *Life Sci.* **33:**1925–1931.
63. Matsumoto, S., Isogai, A., and Suzuki, A., 1985, N-terminal amino acid sequence of an insect neurohormone, melanization and reddish coloration hormone (MRCH): Heterogeneity and sequence homology with insulin-like growth factor II, *FEBS Lett.* **189:**115–118.
64. Matsumoto, S., Brown, M.R., Crim, J.W., Vigna, S.R., and Lea, A.O., 1989, Isolation and primary structure of neuropeptides from the mosquito, *Aedes aegypti,* immunoreactive to FMRFamide antiserum, *Insect Biochem.* **19:**277–283.
65. Matsumoto, S., Brown, M.R., Suzuki, A., and Lea, A.O., 1989, Isolation and characterization of ovarian ecdysteroidogenic hormones from the mosquito, *Aedes aegypti, Insect Biochem.* in press.
66. Meola, S.M., and Lea, A.O., 1972, The ultrastructure of the corpus cardiacum of *Aedes sollicitans* and the histology of the cerebral neurosecretory system of mosquitoes, *Gen. Comp. Endocrinol.* **18:**210–234.
67. Meola, R., and Readio, J., 1987, Juvenile hormone regulation of the second biting cycle in *Culex pipiens, J. Insect Physiol.* **33:**751–754.
68. Myers, C.M., and Evans, P.D., 1985, The distribution of bovine pancreatic polypeptide/FMRFamide-like immunoreactivity in the ventral nervous system of the locust, *J. Comp. Neurol.* **234:**1–16.
69. Myers, C.M., and Evans, P.D., 1985, An FMRFamide antiserum differenti-

ates between populations of antigens in the ventral nervous system of the locust, *Shistocerca gregaria, Cell Tissue Res.* **242:**109–114.

70. Nachman, R.J., Holman, G.M., Haddon, W.F., and Ling, N., 1986, Leucosulfakinin, a sulfated insect neuropeptide with homology to gastrin and cholecystokinin, *Science* **234:**71–73.
71. Nachman, R.J., Holman, G.M., Cook, B.J., Haddon, W.F., and Ling, N., 1986, Leucosulfakinin-II, a blocked sulfated insect neuropeptide with homology to gastrin and cholecystokinin, *Biochem. Biophys. Res. Comm.* **140:**357–364.
72. Nagasawa, H., Kataoka, H., Isogai, A., Tamura, S., Suzuki, A., Mizoguchi, A., Fujiwara, Y., Suzuki, A., Takahashi, Y., and Ishizaki, H., 1986, Amino amino acid sequence of a prothoracicotropic hormone of the silkworm *Bombyx mori, Proc. Natl. Acad. Sci. USA* **83:**5840–5843.
73. Nijhout, H.F., and Carrow, G.M., 1978, Diuresis after a bloodmeal in female *Anopheles freeborni, J. Insect Physiol.* **24:**293–298.
74. Nijhout, M.M., 1979, *Plasmodium gallinaceum:* Exflagellation stimulated by a mosquito factor. *Exp. Parasitol.* **48:**75–80.
75. Nishiitsutsuji-Uwo, J., and Endo, Y., 1981, Gut endocrine cells in insects: The ultrastructure of the endocrine cells in the cockroach midgut, *Biomed. Res.* **2:**30–44.
76. Normann, T.C., 1983, Cephalic neurohemal organs in adult Diptera, in Gupta, A.P. (ed): Neurohemal Organs of Arthropods, Springfield, IL, Charles C Thomas, pp. 454–480.

77a. Nyhof, J.M. and McIver, S.B., 1989, Structural organization of the brain and subesophageal ganglion of male *Aedes aegypti* (L.) (Diptera: Culicidae), *Int. J. Insect Morphol. & Embryol.* **18:**13–17.

77. O'Donohue, T.L., Bishop, J.F., Cronwall, B.M., Groome, J., and Watson, W.H., 1984, Characterization and distribution of FMRFamide immunoreactivity in the rat central nervous system, *Peptides* **5:**563–568.
78. O'Donohue, T.L., Chronwall, B.M., Pruss, R.M., Mezey, E., Kiss, J.Z., Eiden, L.E., Massari, V.J., Tessel, R.E., Pickel, V.M., DiMaggio, D.A., Hotchkiss, A.J., Crowley, W.R., and Zukowski-Grojec, Z., 1985, Neuropeptide Y and peptide YY neuronal and endocrine systems, *Peptides* **6:**755–768.
79. O'Shea, M., and Adams, M., 1987, Proctolin: From "gut factor" to model neuropeptide, *Adv. Insect Physiol.* **19:**1–28.
80. Penzlin, H., 1985, Stomatogastric nervous system, in Kerkut, G.A., and Gilbert, L.I. (eds): Comprehensive Insect Physiology, Biochemistry, and Pharmacology, Volume 5, New York, Pergamon Press, pp. 371–406.
81. Petzel, D.H., Hagedorn, H.H., and Beyenbach, K.W., 1986, Peptide nature of two mosquito natriuretic factors, *Am. J. Physiol.* **250:**R328–R332.
82. Pitman, R.M., 1985, Nervous system, in Kerkut, G.A., and Gilbert, L.I. (eds): Comprehensive Insect Physiology, Biochemistry, and Pharmacology, Volume 11, New York, Pergamon Press, pp. 5–54.
83. Pratt, S., and Pryor, S.C., 1986, Dopamine- and octapamine-sensitive adenylate cyclase in the brain of adult *Culex pipiens* mosquitoes, *Cell. Molec. Neurobiol.* **6:**325–329.
84. Price, D.A., 1986, Evolution of a molluscan cardioregulatory neuropeptide, *Am. Zool.* **26:**1007–1015.

85. Raabe, M., 1982, Insect Neurohormones, New York, Plenum Press, pp. 37–44.
86. Rafaeli, A., Moshitzky, P., and Applebaum, S.W., 1987, Diuretic action and immunological cross-reactivity of corticotropin and locust diuretic hormone, *Gen. Comp. Endocrinol.* **67:**1–6.
87. Raminani, L.N., and Cupp, E.W., 1978, Embryology of *Aedes aegypti* (L.) (Diptera: Culicidae): organogenesis, *Intl. J. Insect Morphol. Embryol.* **7:**273–296.
88. Rees, H.H., 1985, Biosynthesis of ecdysone, in Kerkut, G.A., and Gilbert, L.I. (eds): Comprehensive Insect Physiology, Biochemistry, and Pharmacology, Volume 7, New York, Pergamon Press, pp. 249–293.
89. Reinhardt, C.A., 1976, Ultrastructural comparison of the midgut epithelia of fleas with different feeding behavior patterns (*Xenopsylla cheopis, Echidnophaga gallinacea, Tunga penetrans,* Siphonaptera, Pulicidae), *Acta Trop.* **33:**105–132.
90. Roubos, E.W., 1984, Cytobiology of the ovulation-neurohormone producing neuroendocrine caudal-dorsal cells of *Lymnaea stagnalis, Int. Rev. Cytol.* **89:**295–346.
91. Rowell, H.F., 1976, The cells of the insect neurosecretory system: Constancy, variability, and the concept of the unique identifiable neuron, *Adv. Insect Physiol.* **12:**63–123.
92. Rudin, W., and Hecker, H., 1982, Functional morphology of the midgut of a sandfly as compared to other hematophagus Nematocera, *Tissue Cell* **14:**751–758.
93. Schaffer, M.H., 1986, Functional and evolutionary relationships among the RPCH-AKH family of peptides, *Am. Zool.* **26:**997–1005.
94. Scharrer, B., 1987, Insects as models in neuroendocrine research, *Ann. Rev. Entomol.* **32:**1–16.
95. Scharrer, B., 1987, Neurosecretion: Beginnings and new directions in neuropeptide research. *Ann. Rev. Neurosci.* **10:**1–17.
96. Schneider, M., Rudin, W., and Hecker, H., 1986, Absorption and transport of radioactive tracers in the midgut of the malaria mosquito, *Anopheles stephensi, J. Ultrastruct. Molec. Struct. Res.* **97:**50–63.
97. Shapiro, A.B., Wheelock, G.D., Hagedorn, H.H., Baker, F.C., Tsai, L.W., and Schooley, D.A., 1986, Juvenile hormone and juvenile hormone esterase in adult females of the mosquito *Aedes aegypti, J. Insect Physiol.* **32:**867–877.
98. Schooneveld, H., Romberg-Privee, H.M., Veenstra, J.A., 1985, Adipokinetic hormone-immunoreactive peptide in the endocrine and central nervous system of several insect species: A comparative immunocytochemical approach, *Gen. Comp. Endocrinol.* **57:**184–194.
99. Schooneveld, H., Romberg-Privee, H.M., and Veenstra, J.A., 1987, Phylogenetic differentiation of glandular cells in corpora cardiaca as studied immunocytochemically with region-specific antisera to adipokinetic hormone, *J. Insect Physiol.* **33:**167–176.
100. Schooneveld, H., van Herp, F., and van Minnen, J., 1987, Demonstration of substances immunologically related to the identified arthropod neuropeptides AKH/RPCH in the CNS of several invertebrate species, *Brain Res.* **406:**224–232.

101. Siegert, K.J., and Mordue, W., 1986, Elucidation of the primary structures of the cockroach hyperglycaemic hormones I and II using enzymatic techniques and gas-phase sequencing, *Physiol. Entomol.* **11:**205–211.
102. Solcia, E., Capella, C., Buffa, R., Usellini, L., Fiocca, R., and Sessa, F., 1981, Endocrine cells of the digestive system, in Johnson, L.R. (ed): Physiology of the Gastrointestinal Tract, Volume 1, New York, Raven Press, pp. 39–58.
103. Stanley, B.G., and Leibowitz, S.F., 1985, Neuropeptide Y injected in the paraventricular hypothalmus: A powerful stimulant of feeding behavior, *Proc. Natl. Acad. Sci. USA* **82:**3940–3943.
104. Stefano, G.B., and Scharrer, B., 1981, High affinity binding of an enkephalin analog in the cerebral ganglion of the insect *Leucophaea maderae* (Blattaria), *Brain Res.* **225:**107–114.
105. Stefano, G.B., Scharrer, B., and Assanah, P., 1982, Demonstration, characterization, and localization of opioid binding sites in the midgut of the insect *Leucophaea maderae* (Blattaria), *Brain Res.* **253:**205–212.
106. Tager, H.S., and Kramer, K.J., 1980, Insect glucagon-like peptides: Evidence for a high-molecular weight form in midgut from *Manduca sexta* (L.), *Insect Biochem.* **10:**617–619.
107. Takeda, S., Vieillemaringe, J., Geffard, M., and Remy, C., 1986, Immunohistochemical evidence of dopamine cells in the cephalic nervous system of the silkworm *Bombyx mori.* Coexistence of dopamine and α endorphin-like substance in neurosecretory cells of the subesophageal ganglion, *Cell Tissue Res.* **243:**125–128.
108. Teller, J.K., Rosinski, G., Pilc, L., Kasprzyk, A., and Lesicki, A., 1983, The presence of insulin-like hormone in head and midguts of *Tenebrio molitor* L. (Coleoptera) larvae, *Comp. Biochem. Physiol.* **74A:**463–465.
109. Terzian, L.A., Stahler, N., and Miller, H., 1953, A study of the relation of antibiotics, vitamins, and hormones to immunity to infection, *J. Immunol.* **70:**115–123.
110. Uvnas-Wallensten, K., Efendic, S., and Johansson, C., 1980, Intraluminal secretion of gastrointestinal hormones, *Front. Horm. Res.* **7:**65–71.
111a. Veenstra, J.A., 1988, Effects of 5-hydroxytryptamine on the Malpighian tubules of *Aedes aegypti, J. Insect Physiol.* **34:**299–304.
111. Veenstra, J.A., and Schooneveld, H., 1984, Immunocytochemical localization of neurons in the nervous system of the Colorado potato beetle with antisera against FMRFamide and bovine pancreatic polypeptide, *Cell Tissue Res.* **235:**303–308.
112. Veenstra, J.A., Romberg-Privee, H.M., Schooneveld, H., and Polak, J.M., 1985, Immunocytochemical localization of peptidergic neurons and neurosecretory cells in the neuroendocrine system of the Colorado potato beetle with antisera to vertebrate regulatory peptides, *Histochemistry* **82:**9–18.
113. Verhaert, P., Grimmelikhuijzen, C.J.P., and De Loof, A., 1985, Distinct localization of FMRFamide- and bovine pancreatic polypeptide-like material in the brain, retrocerebral complex and subesophageal ganglion of the cockroach *Periplaneta americana* L, *Brain Res.* **348:**331–338.
114. Verhaert, P., and De Loof, A., 1986, Substances resembling peptides of the vertebrate gonadotropin system occur in the central nervous system of *Periplaneta americana* L, *Insect Biochem.* **16:**191–197.

115. Vieillemaringe, J., Duris, P., Bensch, C., and Giraride, J., 1982, Colocalization of amines and peptides in same medial neurosecretory cells of locust, *Neurosci. Lett.* **31:**237–240.
116. Vigna, S.R., 1986, Gastrointestinal tract, in Pang, P.K.T., Schreibman, M.P., and Gorbman, A. (eds): Vertebrate Endocrinology: Fundamentals and Biomedical Implications, Volume 1, New York, Academic Press, pp. 261–278.
117. Walsh, J.H., 1981, Gastrointestinal hormones and peptides, in Johnson, L.R. (ed): Physiology of the Gastrointestinal Tract, Volume 1, New York, Raven Press, pp. 59–144.
118. Wheelock, G.D., and Hagedorn, H.H., 1985, Egg maturation and ecdysiotropic activity in extracts of mosquito (*Aedes aegypti*) heads, *Gen. Comp. Endocrinol.* **60:**196–203.
119a. Wheelock, G.D., Petzel, D.H., Gillet, J.D., Beyenbach, K.W., and Hagedorn, H.H., 1988, Evidence for hormonal control of diuresis after a blood meal in the mosquito *Aedes aegypti, Arch. Insect Biochem. Physiol.* **7:**75–89.
119. Whisenton, L.R., Kelly, T.J., and Bollenbacher, W.E., 1987, Multiple forms of cerebral peptides with steroidogenic functions in pupal and adult brains of the yellow fever mosquito, *Aedes aegypti, Mol. Cell. Endocrinol.* **50:**3–14.
120. Williams, J.A., 1982, Cholecystokinin: A hormone and a neurotransmitter, *Biomed. Res.* **3:**107–121.
121. White, K., Hurteau, T., and Punsal, P., 1986, Neuropeptide FMRFamide-like immunoreactivity in *Drosophila:* Development and distribution, *J. Comp. Neurol.* **247:**430–438.
122. Yui, R., Fujita, T., and Ito, S., 1980, Insulin-, gastrin-, pancreatic polypeptide-like immunoreactive neurons in the brain of the silkworm *Bombyx mori, Biomed. Res.* **1:**42–46.

3
Gametocytemia and Infectiousness in Falciparum Malaria: Observations and Models

Jerry Nedelman

Introduction

The interface between the mosquito vector of malaria and the human host is mediated in one direction by the sporozoites, whereby mosquitoes infect humans, and in the other by the gametocytes, whereby humans infect mosquitoes. This paper is about the gametocytes of *Plasmodium falciparum* and their environment in the human host: when and why gametocytes are produced; patterns of their appearance; the relationship between patent gametocytemia and infectiousness of the host to the vector; and the immune response of the host.

The biological literature has been reviewed by Carter and Gwadz (10). My review of the literature differs from their review in two ways: I present alternative interpretations of some results; and I include some new information that has appeared since 1980. Moreover, I also relate the review of the biology to mathematical and statistical models for malarial infection.

An important source of new information and ideas, which will be cited frequently, is the study of the epidemiology and control of malaria conducted by the World Health Organization (WHO) and the government of Nigeria in the Garki District of northern Nigeria (31). Lasting from 1969 to 1976, the Garki project included longitudinal parasitological follow-up of approximately 8000 individuals from whom thick blood smears were examined once every 10 weeks. Through the generosity of the WHO, the data from the project have been made available to researchers such as myself. Several new analyses of some of the data are reported below. The goal of the Garki study was to understand the factors determining the

---

Jerry Nedelman, Department of Mathematical Sciences, Clemson University, Clemson, South Carolina 29634-1907, USA. Present address: State University of New York at Albany, School of Public Health, 2523 Corning Tower, Empire State Plaza, Albany, New York 12237, USA.

transmission of malaria. Indeed, a primary task of the original research was to devise a mathematical model describing transmission (19). That model and its successors will be considered below with regard to gametocytemia and infectiousness in the human host.

Gametocytogenesis

Asexual merozoites in the peripheral blood of an infected human host give rise to male and female gametocytes; that process, still poorly understood, is called gametocytogenesis. These gametocytes are precursors of gametes. Gametogenesis, the process of gamete production, occurs in the mosquito's gut after the gametocytes have been ingested in a bloodmeal. Male and female gametes fuse in the mosquito's gut, producing a diploid zygote. After further development in the mosquito, haploid sporozoites appear in the mosquito's salivary glands, ready for inoculation into a human host upon the mosquito's next bloodmeal.

It is not known what stimulates the production of gametocytes during the course of asexual parasitemia. Indeed, whether any environmental stimulus exists is the subject of debate in the literature. In a recent report from a WHO Scientific Group (48), the WHO came down on the affirmative side. Reviewing the literature, they noted several reports of a correlation between gametocytogenesis and stress to the asexual parasite.

One such report was that of Miller (30), who noted a strong dependence of gametocytogenesis upon the occurrence of clinical symptoms. Also cited were reports of Smalley and Brown (43) and Smalley, Brown, and Bassett (44), who found that adding lymphocytes in low numbers from infected children to cultures of *Plasmodium falciparum* increased the rate of gametocyte production in vitro. Based on these reports, the WHO group identified the immune status of the host as an environmental determinant of gametocytogenesis. The group also identified "subcurative drug pressure . . . and metabolic stress." However, this opinion was not firm in the debate, since the group later summarized: "The evidence for the identification of specific inducers of gametocytogenesis is inconclusive."

Others have been even more skeptical about the evidence; thus, Sinden (41) conjectured that "a variety of conditions, all of which are adverse to the survival of the asexual parasite, are capable of triggering the parasite to a sexual path." Garnham (24) firmly denied the existence of any environmental stimulus, claiming that there is a regular time for the gametocytes' appearance. Bruce-Chwatt (8), contradicting Miller (30), reported no relationship between clinical episodes and gametocytemia.

Smalley et al. (44) claimed to have found evidence that the rate of production of gametocytes of *P. falciparum* increases with the duration of infection. But an observation regarding *P. falciparum* in Garki (31, pp. 109–172) suggests the opposite—that when transmission levels are high, lower rates of gametocytogeneisis might be associated with longer, more serious infections. In Garki, average parasite densities and recovery rates were compared among villages that differed with respect to vector density. As vector density increased, so did average density of asexual forms; but both the recovery rate from parasitemia and the average gametocyte density decreased. However, although the positive correlation between the average density of asexual forms and the vector density was statistically significant, the negative correlation involving gametocyte density was not. We will explore patterns involving prevalence of gametocytemia among villages in Garki further at the end of this chapter.

Patterns of Gametocytemia

The four species of *Plasmodium* that infect humans are divided into two subgenera that differ with regard to patterns of gametocyte development (42). The subgenus *Plasmodium* contains the species *malariae, vivax,* and *ovale,* in which gametocytes and asexual forms develop at equivalent rates. The subgenus *Laverania* contains the species *falciparum,* in which the sexual cycle takes up to six times longer than the asexual cycle. The gametocytes of *P. falciparum* also differ from those of other species in their crescentlike shape and in the wavelike pattern of their invasion of the bloodstream (see below) (45).

The time until patency of gametocytes in vivo (i.e., until gametocytes are observable in the blood) is variously measured from the onset of symptoms or from the onset of patent asexual parasitemia. Measuring from the onset of symptoms, Miller (30) reported an average of 11.7 days, with a range from 6 to 30 days, among the immune adults he studied in Liberia. Summarizing work done on nonimmune subjects with induced infections of many geographical strains, Shute and Maryon (40) reported that, when the attack was not treated, gametocytes first appear about the 10th day after the onset of fever.

Measured from the onset of asexual parasitemia, the first appearance of gametocytes in the blood has been reported to vary from 10 to 14 days in blood-induced infections of *P. falciparum* in nonimmune subjects (10).

Gametocytes of *P. falciparum* are not infective when the mature forms first appear in the blood. Only after a period of one to four days (29, 42) do they become capacitated.

The sojourn time of gametocytes in the peripheral blood was investigated by Smalley and Sinden (45). Their subjects, ranging in age from four months to seven years, were outpatients at their clinic in The Gambia who presented with *P. falciparum* infection. Such patients were given chloroquine, which eliminated all asexual forms of the parasites and immature gametocytes but not the older gametocytes. In 11 cases, gametocyte densities rose to a peak following drug administration and then declined.

(The increase in gametocyte density was attributed to the continued maturation of "those gametocytes that were beyond the chloroquine-susceptible stage.") These 11 were studied. Although peak gametocyte densities varied among the 11 cases, ranging from 622 to 6331/mm^3, the peak represented a definable point in time, not otherwise available among patients who had not been monitored from the onset of infection. When gametocyte densities were standardized as a percentage of the peak value, the decline in density was observed to occur in first-order (negative-exponential) fashion, with a half-life of 2.4 days. The sex ratio among the gametocytes was four female to one male throughout their sojourn of approximately 20 days; and, throughout that period, microgametocytes exflagellated.

For *P. falciparum,* the temporal pattern of trophozoite and gametocyte density in the peripheral blood has frequently been described as wavelike. Brumpt (9) wrote that after mature forms first appear in the blood, about 10 days after patency, they increase in density for a week, while asexual forms decline, sometimes almost to disappearance. When gametocyte numbers peak and start to decline, asexual forms again increase in number, followed by another wave of gametocytes. Brumpt claimed that falciparum infections "invariably exhibit" at least one of "these successive and consecutive waves of trophozoites and gametocytes."

Boyd (5), writing in the same collection as Brumpt, echoed Brumpt's description, and described the second and later waves of gametocytes as attaining maximum density after three to five days and then declining to disappearance in two to three weeks. That rate of decline is consistent with the rate of exponential decay of gametocytemia reported by Smalley and Sinden (45), but Boyd's accounting of sex ratios is different: He observed that at the tail end of a wave, all the gametocytes were macrogametocytes and the patient was noninfectious.

Eyles and Young (23) and Jeffery and Eyles (28) observed the wavelike gametocytemias among neurosyphilitic patients treated with two strains of *P. falciparum.* They reported that infections tended to divide temporally into two periods: an initial period of continuous parasitemia, ranging from 32 to 224 days, with an average of 121 days, followed by a terminal period of intermittent parasitemia, ranging from 0 to 283 days, with an average of 100 days. During the initial period, waves of gametocytes followed waves of asexual parasites in a cycle lasting approximately three weeks. Successive cycles produced lower densities, but neither asexual forms nor gametocytes disappeared altogether. Similar cycles occurred during the terminal period, but the densities were lower and the episodes of patency became briefer over time.

According to other observers, the wavelike nature of gametocytemia is not so pronounced. The differences may be due to the immune status of the subjects studied. Bruce-Chwatt (8), for example, describing the patterns observed among adult psychiatric patients in West Africa, identified two types of gametocytemia. In seven of 25 subjects exhibiting gametocytemia, "there were definite 'waves' of gametocyte production lasting for periods of three to 10 weeks in succession." But in the other 18 subjects, "gametocytes were seen only twice or at most three times, at irregular long intervals."

Related to the leapfrog relation between waves of asexual forms and gametocytes is the observation that bouts of gametocytemia tend to occur during periods of clinical remission following attacks (28, 40). This pattern may have serious epidemiological consequences, if patients who are released from hospitals or clinics without being cured later become infectious. Jeffery and Eyles, who worked with induced infections in neurosyphilitic patients in the United States, conjectured that, where malaria is endemic, asymptomatic carriers contribute significantly to transmission and that many of these asymptomatic carriers do not exhibit patent parasitemias.

Recognizing patterns of gametocytemia depends on observing and counting gametocytes, which, apparently, can be problematic with standard thick-film procedures when low-density infections are studied. Muirhead-Thomson cited Colbourne (14) as noting that the observed percentage of pregnant women in the Gold Coast harboring gametocytes increased from 6.9 to 20% when the number of thick fields examined increased from 100 to 1000. In the Garki study, thick blood smears were routinely examined for 200 fields, but a systematic sample of one-fifth of all blood smears were examined for 400 fields. Among smears examined for 200 fields, 12.7% were reported to contain gametocytes throughout the preintervention phase of the study; among those examined for 400 fields, 16.4% (31). These data are considered further in the section in the reservoir of infection.

Dowling and Shute (20) compared thick and thin blood films as applied to low-density infections. They worked in Nigeria with individuals "in different age groups . . . selected at random in clinics and in routine parasite surveys." They took paired films from infected individuals and examined thick films for 100 fields and thin films for 1000 fields. The total parasite densities per cubic millimeter were 2.5 times greater, on average, with the thin films. When they restricted their analysis to gametocytes, they found an average density of 71.3/mm^3 for thin films and 9.9 for thick films, an 86% loss for thick films (based on paired films from 46 individuals positive for gametocytes).

In a similar comparison of thick and thin films, Trape (47) did not find such discrepancies. Trape's study differed from that of Dowling and Shute in two and possibly three ways: Trape counted parasites for a fixed number of leukocytes, Dowling and Shute for a fixed number of fields. Trape stained the smears three to six days after collection; Dowling and Shute, within 24 hours of collection. Trape stated that he included only school children; Dowling and Shute do not mention whether the 46 individuals in their investigation of gametocyte densities provide a representative sample of the broader age distribution of the 233 individuals in their overall study. Trape cited the different times of staining as the most plausible explanation for the contradictory results.

Suppose the dramatic loss of gametocytes in the preparation of thick films, as observed by Dowling and Shute, commonly occurs. What are the implications of such an observation for data reported in the literature? Sometimes gametocytemia is assessed quantitatively in terms of parasite densities in the blood; other times only a qualitative judgment of the presence or absence of gametocytes is of interest. Consider quantitative measures. Computed densities depend on the volume of blood examined and the number of parasites observed. In the papers of Jeffery and Eyles (28) and Eyles and Young (23), for example, 0.1 mm^3 of blood was examined in thick films. Dowling and Shute calculated that 100 fields of average thickness of a thick film contain 0.16 mm^3, and 100 fields from the "best part" of a thick film contain 0.227 mm^3 of blood on average. For each of the three volumes, one observed gametocyte would correspond to 10, 6.25, or 4.4 mm^3, if there were no losses. Accounting for 86% loss, the corresponding figures would be 71.4, 44.6, and 31.4/mm^3. Variations in the degree of loss would cause a corresponding uncertainty about the true density.

Now consider the categorical determination of the presence or absence of gametocytes. Figure 3.1 shows the probabilities of finding a gametocyte as a function of the true density of gametocytes in the blood with and without a loss of 86% during slide preparation for 0.16 mm^3 of blood

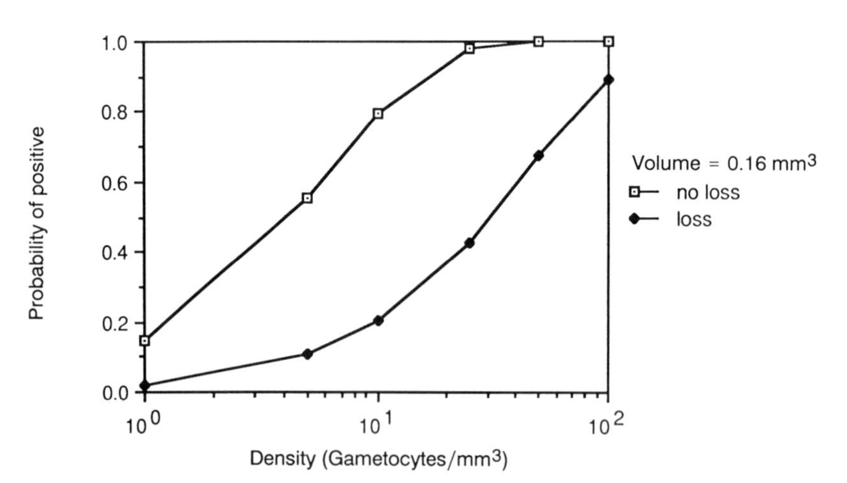

FIGURE 3.1. The probability of finding at least one gametocyte in 0.16 mm^3 of a thick smear vs. the true density of gametocytes in the blood, computed when there is no loss of gametocyes in slide preparation and when there is 86% loss of gametocytes in slide preparation.

(results are similar for 0.1 or 0.227 mm^3). A Poisson distribution is assumed for the number of gametocytes that would be present in the given volume of blood. With a low density of gametocytes, the chance of not finding one can be high. [The assumption of a Poisson distribution is justified at least through about 40 gametocytes per cubic millimeter (7).]

Keeping all these cautionary notes in mind, what kinds of gametocyte densities are reported? For nonimmune subjects, Shute and Maryon (40) reported initial peak gametocytemias ranging from 900 to 17,000/mm^3. Jeffery and Eyles (28) also observed cases in which gametocyte density exceeded 10,000/mm^3.

Studies with subjects living in areas of high malaria prevalence report similar densities for children but much lower densities for adults. Bruce-Chwatt (8), studying adults in Nigeria, reported the following distribution for 97 observations: eighty-eight slides, below 60/mm^3; six slides, 70–100/mm^3; and three slides, 120, 180, and 380/mm^3. The average was 39/mm^3. His summary of the literature indicated nearly unanimous agreement among surveys that falciparum gametocyte counts in the adult African average less than 100/mm^3. He cited references giving figures of 80/mm^3 in Lagos and 14/mm^3 in territories of the former French Equatorial Africa.

Circadian Rhythm

Conflicting reports have appeared in the literature regarding a circadian rhythm in gametocytemia that produces a peak in gametocyte density around midnight, just when mosquitoes would be most active. Originating with Hawking et al. (26, 27), and concerned primarily with simian malarias, the theory has been carefully reviewed by Carter and Gwadz (10). They found the evidence consistently favorable only for *P. knowlesi,* which generally has a highly synchronous, 24-hour asexual cycle. Bray et al. (6) were unable to demonstrate a circadian rhythm for *P. falciparum*.

Infectiousness Versus Gametocytemia

A necessary condition for the transmission of malarial infection from a human host to a mosquito is the presence of male and female gametocytes in the host's blood. However, that condition is not always sufficient, nor is detectability of gametocytes always necessary. Regarding insufficiency, we have already noted that gametocytes are not infectious for one to four days after they appear in the blood, and possibly not at the end of an infection if only macrogametocytes are present. Covell (15) and Carter and Gwadz (10) have provided excellent reviews of the relationship between gametocytemia and infectiousness. As Pampana (37, pp. 28ff) stated, some investigators find correlations between gametocyte density and infectiousness, whereas others do not; and "one cannot forget the old saying of S.P. James that for infectiousness to the mosquito the 'quality' of the gametocytes is more important than their quantity."

Since this subject has been well worked over in the above cited reviews, we will briefly consider only two important observations that appear in the literature: Individuals with very low-density gametocytemias are often infectious; and conversely, individuals with very high-density gametocytemias often are not infectious.

Regarding the latter observation, consider the work of Rutledge et al. (39). They studied naturally infected inhabitants of Thailand, aged 12 to 60 years with a median age of 20, to determine the effects of gametocyte density, asexual parasitemia, age, and season on infectiousness. They estimated densities of asexual parasites and gametocytes using thin films, and then allowed the mosquitoes to feed on the subjects. In every one of the 55 cases, gametocytes were found, but only 30 cases were infectious to mosquitoes. Of mosquitoes that fed on these 30 hosts, the average percentage becoming infected increased with the density of gametocytes to a maximum of approximately 60%, attained at approximately 3500 gametocytes/mm^3. Thus, only 30 out of 55 of the gametocyte carriers ever infected mosquitoes, and of those, even those with high-density gametocytemias could infect only 60% of the mosquitoes that bit them, on average. Subjects were least infective during the hot season, more infective during the rainy season, and most infective during the cool season. Asexual parasitemia was observed in 44 of the 55 cases, and those with "moderate numbers of asexual parasites tended to infect higher proportions of mosquitoes." There was no correlation between asexual parasitemia and gametocytemia, nor between age and infectivity.

Regarding the observation of infectiousness among infections with low-density gametocytemias, consider the work of Jeffery and Eyles (28). They induced infections of two strains of *P. falciparum* in neurosyphilitic patients in the United States. The overall pattern of their results was that infectivity increased with gametocyte density. However, they were surprised by the number of successes they had in attempts to infect mosquitoes by allowing them to feed on individuals whose gametocyte densities were 0 to 100/mm^3 during the later stages of infection.

These authors also had a statement to make about high-density gametocytemias: "Quite as notable was the number of failures encountered in lots fed when the gametocyte densities were between 100 and 1000 per cmm. Although in the higher ranges failures were comparatively few, it is obvious that presumably adequate gametocyte densities do not invariably guarantee good mosquito infections." The relative infrequency of noninfectiousness of high-density gametocytemias in this study involving neurosyphilitic patients in the United States was contrasted by Carter and Gwadz (10) with the relative frequency of such occurrences among the naturally infected, "semiimmune" residents of Thailand in the study of Rutledge et al. (39). Carter and Gwadz conjectured an immune mechanism for the difference.

Circumstantial evidence for the requirement of "quality" among gametocytes was found in the Garki project (31, pp. 109–172). The rate at which newborn infants contracted malaria was found to be much less than the rate at which sporozoite-positive bites were delivered by mosquitoes, and the difference between the two rates was greater in villages with higher mosquito densities. One possible explanation put forth for those observations was that infants may actually have received fewer bites than did the adult volunteers used to estimate the biting rates. Molineaux and Gramiccia (31) did not consider this explanation completely satisfactory, however, and preferred a theory that invokes higher levels of immunity among residents of villages where the mosquito densities were higher and transmission was thus more intense. This higher level of immunity decreased the infectiousness of mosquitoes, "presumably through a decrease in quantity or quality of gametocytes in the human population."

In the section on patterns of gametocytemia, we noted the wavelike appearance of bouts of gametocytemia. In this section, we have noted that the association between gametocytemia and infectiousness is not perfect. What, then, can be said about temporal patterns of infectiousness? That question was addressed by Eyles and Young (23), who presented results of serial feedings of mosquitoes on five of their neurosyphilitic patients experimentally infected with the South Carolina strain of *P. falciparum*. During the initial, continuous phase of parasitemia, infectivity tended to follow gametocyte density, although there were instances of failure to infect mosquitoes when the host's gametocyte density was high. "During the terminal intermittent parasitemias the infections in mosquitoes seemed to occur during and following the recurrent waves of gametocytes. . . . Infectivity during this period became progressively more sporadic as the infection became older."

Reservoir of Infection

The infectiousness of low-density gametocytemias generates a surprising distribution of contributors to the transmission of infection in areas where malaria is endemic. For such areas, it was long thought that the individuals most responsible for passing infections to mosquitoes were infants and young children. We shall refer to the relative frequency distribution of infectious individuals among age groups as the "reservoir of infection." Carter and Gwadz (10) found that, in fact, the reservoir of infection is a rather uniform distribution with approximately equal contributions from all age groups.

The primary evidence for these conclusions is seen in the work of Muirhead-Thomson (32, 33). In reference 32, he reported the results of a parasite survey, and subsequent mosquito-feeding experiments, conducted in Ghana between January and October of 1952. He examined 491 thick films from subjects ranging in age from 2 months to 75 years. The films were examined routinely for 200 fields, "but sometimes 400 or more were examined." In addition, 73 batches of Anopheles *gambiae* were fed on 42 gametocyte carriers of different ages. Data are summarized in Table 3.1. [A similar table appears in Carter and Gwadz (10). They, however, reported that infants (< 1 year) were not included in Muirhead-Thomson's 1957 study (33). In fact, there were 12 infants among the 28 in the youngest age group, but none was infectious.] Muirhead-Thomson was surprised by two observations: Young children with gametocytes were not less infectious to mosquitoes than infants with gametocytes (column 3); and some individuals with threshold-density gametocytemias were infectious to mosquitoes, whereas some with high-density gametocytemias were not (this latter observation, which echoes the previous section, is not evident from Table 3.1). He concluded that "one might go so far as to say that in trying to determine the proportion of malaria infectors in the African or native village community, a more accurate picture might be obtained if we forgot about blood films for a time and concentrated on test feeds carried out at random on a large scale."

In reference 33, Muirhead-Thomson followed his own advice. He fed mosquitoes on 347 subjects of all age groups in Liberia, where falciparum malaria was hyperendemic. Importantly, he did not look for gametocytes in his subjects; rather, he allowed 40 to 50 "hungry *gambiae*" mosquitoes to feed on each subject whose infection status was unknown. Infectiousness was then quantified by the number of mosquitoes infected and the number of oocysts per infected mosquito. Data are included in Table 3.1. Interestingly, the percentages of the various age groups infecting mosquitoes are very similar to the percentages of the age groups carrying gametocytes reported in the 1954 paper (32), with the possible exception of infants. Do the infectious low-density carriers and the noninfectious high-density carriers cancel each other out? Taking into account the age distribution of the Liberian population, as well as the age and infectiousness of his subjects, Muirhead-Thomson concluded that in a typical West African village population living under hyperendemic conditions, infectious infants and toddlers (< 5 years old) will comprise 4.2% of the entire population; infectious school children (5 to 14 years old) will comprise 3.0% of the entire population; and infectious adolescents and adults (> 15 years old) will comprise 3.3% of the entire population (column 6). Note that these are percentages of the total population. Thus, among those infectious to mosquitoes, the fractions in the three age groups are 0.40, 0.29, and 0.31. These latter fractions, comprising what we are calling the reservoir of infection, may be interpreted as probabilities that a mosquito

TABLE 3.1. Data on gametocytemia and infectiousness from Muirhead-Thompson (32, 33).

	1954		1957				
Age group	Gametocyte rate	Rate carriers infect mosquitoes	Rate individuals infect mosquitoes	Age group's proportion in population	Infectors' proportion in population	Proportion of infectors among infectors	Proportion mosquito batch with oocysts
0-1	21/52 = 0.40	2/13 = 0.15					
			8/28 = 0.28	0.15	0.042	0.40	30/157 = 0.19
2-4	16/62 = 0.26	3/8 = 0.37					
5-9	22/148 = 0.15	6/16 = 0.37					
			12/100 = 0.12	0.25	0.030	0.29	52/223 = 0.23
10-14	9/78 = 0.12	1/3 = 0.33					
≥15	9/151 = 0.06	0/2 = 0.00	12/219 = 0.06	0.60	0.033	0.31	48/239 = 0.20

who bites an infectious individual actually bites a member of the corresponding age group, assuming that all individuals are equally likely to be bitten. The oldest group is then seen to contribute about one-third to infectious bites. But, as Muirhead-Thomson pointed out (32), individuals are not equally likely to be bitten; mosquitoes bite the older age groups more frequently.

When infectiousness is assessed by the proportion of dissected, infected mosquitoes whose stomachs contained oocysts, the three age groups are not found to be significantly different (column 8). Importantly, no individual who was infectious infected all mosquitoes that bit him or her. Muirhead-Thomson (32) concluded that the "best infectors" will infect about one-third of the mosquitoes that feed on them; on the other hand, he wrote, "the carriers with the lowest threshold may still infect up to 10 per cent of the mosquitoes..."

Unfortunately, Muirhead-Thomson's study appears to be the only one of its kind. Considering the contradictions among other studies of other phenomena—e.g., triggers of gametocytemia, relationships between gametocytemia and infectiousness—caution may be warranted. The frequent observation of the infectiousness of low-density gametocytemias may be taken as indirect evidence supporting Muirhead-Thomson's results. Carter and Gwadz (10), however, pointed out that the infectiousness of an individual depends on the available mosquitoes as well as on the individual's infection status. Biological diversity demands that each host–parasite–vector system be studied to determine its own peculiarities.

In the Garki project, infectiousness was not tested by the experimental feeding of mosquitoes; but rather thick blood smears were examined routinely for either 200 or 400 fields, the latter figure corresponding to a systematic sample of 20% of the smears (31). If Muirhead-Thomson's results are representative, then we might not err too seriously if we take the observed prevalence of gametocytemia in Garki as a numerical estimate of the observed prevalence of infectiousness, even though there is not a one-to-one correspondence between operationally infectious individuals and observable gametocyte carriers. The larger size of the Garki study makes more detailed and reliable inferences possible.

The methods of contingency-table analysis may be applied to the Garki data. When a loglinear model (4) is used to assess the interaction between age group and number of fields examined among individuals positive for gametocytes, that interaction is judged to be weak. That is, although a larger fraction of individuals are found positive for gametocytes when 400 fields are examined than when 200 fields are examined, the distribution among age groups of those found positive does not depend on the number of fields examined. Table 3.2 shows fitted distributions from the model, one during a wet season and one during a dry season. The relative contribution from older individuals to the set of gametocyte carriers is

TABLE 3.2. Relative frequency distributions of gametocyte carriers among age groups, according to loglinear model fitted to Garki data.

Age group	Wet season	Dry season
<1	0.13	0.20
1–4	0.24	0.33
5–8	0.21	0.24
9–18	0.13	0.08
19–28	0.10	0.06
29–	0.19	0.10

seen to be higher during the wet season. If age groups are collapsed to permit a comparison with the results of Muirhead-Thomson, the age distribution of gametocyte carriers during the wet season is almost identical to the age distribution of infectious individuals reported by him.

Immune Response

The human host's immune response might be expected to affect gametocytes and infectiousness through any or all of the production, destruction, or infectiousness of gametocytes.

Opinions are divided regarding the effects of immunity on the production and destruction of gametocytes. The two processes combine, of course, to determine the density of gametocytes. Gametocyte density declines, on average, with age in endemic areas, but then so does the density of asexual parasitemia. Does the former decline depend only on the latter, or are there distinct immunological mechanisms affecting the gametocytes? Carter and Gwadz (10) opined that there are none, and cited Christophers (11) as evidence that gametocyte densities decline at a rate "similar to that of the asexual parasites." The data, however, may suggest otherwise (see below).

Miller (30), as we have seen, argued that clinical attacks are the stimulus to gametocytogenesis. It follows, he reasoned, that gametocyte densities decline with age because of the increasing ability of the adult to maintain parasite densities below the clinical threshold.

Bruce-Chwatt (8), as we have also seen, did not accept Miller's theory about the clinical trigger of gametocytogenesis. Unwilling to accept any definitive trigger, he reasoned "that as the mean number of crescents is roughly proportional to the mean number of parasites in the blood, the crescents will be detectable in approximately the same proportion of the population."

However, Bruce-Chwatt did cite Macdonald's (29) claim that immunity depresses gametocyte output. He admitted that "the restriction of gametocyte output may occur very early, before the curb of the production of asexual parasites becomes evident."

Indeed, the epidemiological evidence does suggest a brake on gametocyte production separate from any curbs on asexual parasites. Even though Christophers, cited by Carter and Gwadz, wrote among his conclusions that the "crescent output has a general correlation with asexual proliferation, the number and numerical value of crescent infections being greatest with the higher asexual infections and with the period of acute infestation," the observed gametocyte prevalence decreased among his subjects more rapidly with age than did the observed prevalence of asexual parasitemias. Such a pattern has been consistently observed in epidemiological studies, see Figure 3.2.

In Garki, Molineaux and Gramiccia (31, p.160) observed that, among infants, prevalence and density of gametocytes in first infections do not depend on the age at which that first infection occurs, unlike the density of asexual forms, which increases. They conjectured that asexual densities increase because of the gradual loss of passive immunity received from the mother, but that gametocyte prevalence and densities do not change becuase, as passive immunity is lost, immunity to gametocytes is quickly acquired. As further evidence, they noted that gametocyte densities are higher among infants than among children aged one to four years.

Attempts to prove directly that immune effectors either inhibit the production of gametocytes or facilitate their demise, however, have not been successful. Cohen et al. (13) transferred gamma globulin from immune adults to densely infected children in an experiment conducted in

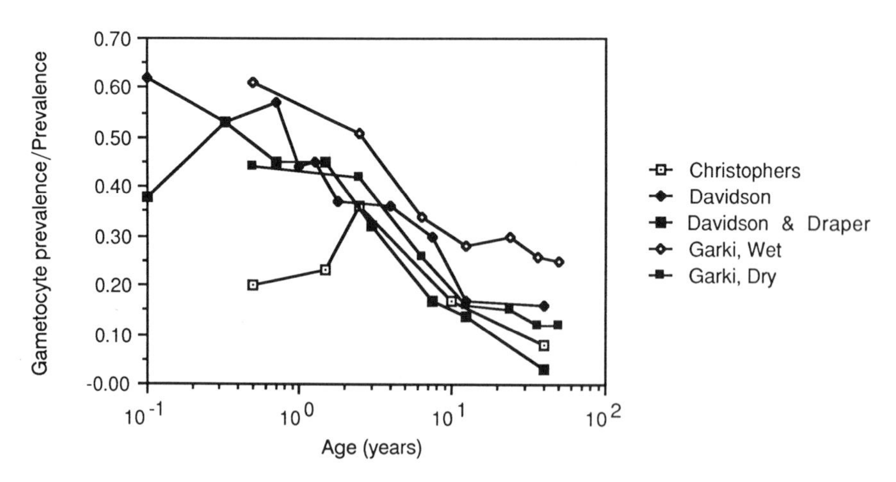

FIGURE 3.2. Reported ratios of age-specific gametocye prevalence to age-specific total parasite prevalence as a function of age from four sources: (11), (16), (17), (31, wet and dry seasons).

The Gambia. The gamma globulin successfully suppressed trophozoite densities. But the authors cite the record of one case in which gametocytes increased for six days during and following the decline of the asexual forms. Referring to this figure, the authors concluded that the gamma globulin probably does not affect gametocytes.

Smalley and Sinden (45), whose experiments were described in the section on patterns of gametocytemia, observed that four of their eleven patients developed antibody to gametocytes but that the presence of antibody had no effect on the rate of clearance of the gametocytes. The authors conjectured (a) that antibodies are produced by some but not all gametocyte carriers and (b) that antibodies, when present, have no role in removing gametocytes, but rather the gametocytes are removed in the spleen as part of the spleen's culling of erythrocytes.

Similarly, although epidemiological evidence may be interpreted as suggesting that immune responses reduce the "quality" of gametocytes (see the earlier discussion concerning evidence for "quality" in the Garki project), nonetheless, attempts to prove directly that immune effectors inhibit the infectiousness of *P. falciparum* gametocytes have failed. Carter and Gwadz (10) compared sera from a sample of Gambians of widely varying ages with sera from nonimmune Americans for their effects on the infectivity of *P. falciparum* gametocytes to *A. gambiae* mosquitoes feeding through membrances. Although all five Gambian sera suppressed oocyst densities relative to the nonimmune sera, the correlation between degree of suppression and antibody titers against the gametocytes was not significant. Carter and Gwadz postulated that the suppression observed was inadequate to account for the noninfectiousness of some gametocye carriers to mosquitoes, concluding that although most individuals in hyperendemic areas have high antibody titers, there is no evidence that these antibodies prevent infecton of mosquites. With regard to cellular effectors of the immune response, Carter and Gwadz noted evidence for the phagocytosis of extracelluar gametocytes; but they postulated that such processes, also, are probably not important in reducing infectivity to mosquitoes.

The WHO Scientific Group on malaria suggested (48, p.34) that natual antibodies have no effect on infectivity of gametocytes but that antibodies produced by immunization do.

Gametocytemia and Infectiousness in Mathematical Models of Malaria

On Modeling and Some Recent Models

Mathematical modeling of the transmission of malaria dates back to Ronald Ross himself (38). The purposes of such modeling have been many, but paramount have been the desire to understand the different epidemological patterns of transmission (e.g., stable endemicity versus isolated epidemics) and to eludicate quantitatively those factors in the transmission cycle of malaria most amenable to control. Generally, mathematical models are useful in science for representing hypotheses precisely and quantitatively, thereby allowing the evaluation of those hypotheses by comparing predictions with observations. This latter function of modeling will justify our attention here. We will focus on how various models deal with gametocytemia and infectiousness in the human host.

In the following discussion, the term *infected* will mean infected with falciparum parasites but not infectious to mosquitoes; for the latter condition, the term *infectious* will be used.

The complexity and sophistication of models for malaria transmisson have evolved over the past 80 years, and reviews of that evolution are many (see, for example, refs. 3, 18, and 35). For the first 60 years of work, the models did not distinguish between infectious and noninfectious individuals. Only within the past 15 years, has a distinction been made explicit. Three sets of models in the latter era are the model of Elderkin et al. (22); the model of Dutertre (21); and the family of models originating with that of Dietz, Molineaux, and Thomas (hereinafter referred to as DMT) (19).

The purpose of the model of Elderkin et al. (22) is to predict the change over time of the age distribution within the human host of three variables: the average density of asexual parasites; the death rate of asexual parasites; and the average density of gametocytes. The model consists of three, coupled, integrodifferential equations, the study of which requires the mathematical ingenuity demonstrated in the paper of Elderkin et al. Although the model is mathematically interesting and contains several important epidemiological ideas, such as an explicit representation of parasite density and of age dependence, it is, in its current form of little practical value.

Dutertre's model and DMT family of models, unlike that of Elderkin et al., are compartment models. They describe the change over time of the number of people, or fraction of the population, in each of several epidemiologically meaningful states, or compartments, corresponding to stages of malarial infection. These states represent various interpretations of susceptibility, infectiousness, infectedness, and immunity. Both models were designed to apply in hyperendemic regions of intense transmission of infection.

An understanding of these models requires familiarity with the notion of superinfection, which refers to an individual being reinfected by the same species of malaria parasite before he or she has recovered fully from a previous infection.

Figure 3.3a depicts Dutertre's model schematically. There are four compartments: uninfected and susceptible to infection (x); infectious

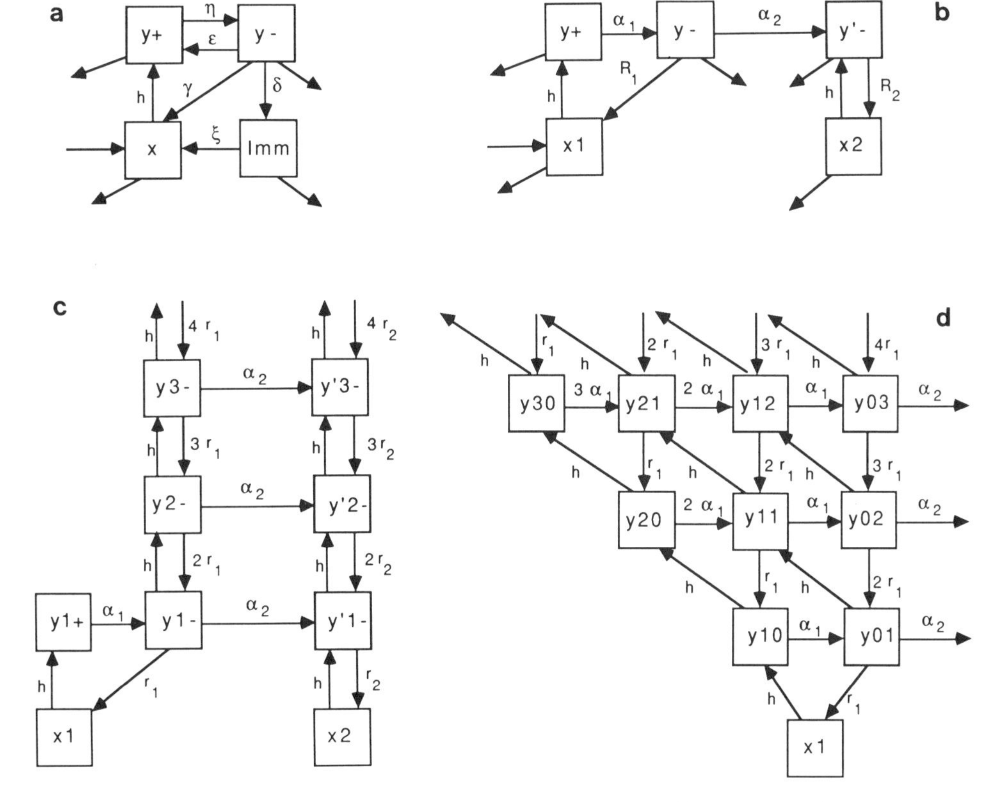

FIGURE 3.3. (a) Schematic diagram of Dutertre's model: (b) schematic diagram of DMT's model (adapted from 18); (c) the full representation of DMT's model (without birth and death transitions indicated), of which Figure 3.3b is an approximation (adapted from 18); (d) the representation, analogous to Figure 3.3c, of Nedelman's modification of DMT's model.

(y+); infected (y−); and uninfected but immune to infection (Imm). The arrow with no intitiating compartment and the arrows with no terminal compartments refer to births and deaths, respectively. Such a scheme does not uniquely define a model because it does not specify the mechanisms of transfer between compartments. Although parameters have been indicated adjacent to arrows, they do not all represent the same types of rate constants. In particular, the recovery from infectedness to susceptibility (y− to x) and the transition from immunity to susceptibility (Imm to x) depend on the rate at which superinfection occurs, so γ and ξ depend on h. The parameters h, ε, and δ, however, are simple, first-order rate constants.

Figure 3.3b (after Dietz, 18) schematically depicts DMT's model. There are five compartments: uninfected and susceptible to slowly recovering infection (x1); infectious (y+); infected and recovering slowly (y−); infected and recovering quickly (y′−); uninfected and susceptible to quickly recovering infection (x2). The distinction between slowly and quickly recovering infection is intended to model the observation that in Garki, directly estimated recovery rates increased with age. DMT interpreted this observation to mean that, in hyperendemic areas, repeated exposure to infection over many years elicits within an individual an immunity that facilitates clearing of parasites from the blood, even though it does not completely protect against infection. The transition from y+ to y− reflects the observation, discussed in the preceding section, that gametocytemia is suppressed before asexual parasitemia.

In the DMT model, R_1 and R_2 depend on h because of superinfection. Both R_1 and R_2 decrease montonically in h: As transmission intensity increases, increasing superinfection impedes recovery. The formula relating R_1 and R_2 to h is:

$$R_i = h/[\exp(h/r_i) - 1], \; i = 1,2,$$

where r_1 and r_2 are parameters estimated by fitting the model to data. This formula is intended to approximate the effects of superinfection according to Macdonald's hypothesis that reinfections develop independently and identically within the human host. DMT interpreted that hypothesis according to the schematic of Figure 3.3c. [This figure is after Dietz (18). To avoid clutter, arrows for births into x and deaths out of all compartments have not been included.] Reinfections occur at the same rate h as do first infections, but only the first infection renders the host infectious.

Nedelman (34) claimed that modeling reinfections as noninfectious violates Macdonald's hypothesis that reinfections are identical to the first infection. He modified DMT's model so that all slowly recovering infections independently and identically pass through infectious and noninfectious stages, corresponding to the schematic of Figure 3.3d. There, yij depicts the compartment of individuals having i infectious infections and j noninfectious infections.

In Nedelman's modification, transitions to the immune, quickly recovering, compartments take place only when all infections are noninfectious. Struchiner et al. (46; see also ref. 25) observed that this restriction means that when transmisson intensity is high, very few people will develop immunity! When h is large, people will be driven upward and leftward in the schematic of Figure 3.3d. The implication of this observation is that the following three premises, on which the original DMT model is based, are inconsistent: (1) immunity precludes infectiousness; (2) reinfections are independent and identical to the first infection; and (3) infectiousness is lost before immunity is acquired.

Struchiner et al. chose to drop premises 2 and 3. Their modification, in terms of Figure 3.3d, is to allow transitions to the quickly recovering compartments to take place out of any slowly recovering compartment, including those with infectious infections. Thus, infections are not independent, since all ongoing infectious infections can lose their infectiousness simultaneously by the transition to immunity; and infectiousness is lost simultaneously with, but not before, acquisition of immunity. Struchiner et al. (46) also assumed that the rate of transition to immunity increases with the total number, $i+j$, of infections.

Nedelman (34) compared his modification of DMT's model with DMT's original version on the basis of the models' ability to fit the data on prevalence from the base-line phase of the Garki study. In fact, he considered three modifications of DMT's model, which he called M1, M2, and M3. Common to all three were a change in the specification of how the inoculation rate is determined by the mosquito density and a change in the choice of free parameters to be used for fitting the models to the data; for further details, see Nedelman (34). Here, M1 is DMT's model, with only the changes of the previous sentence; M2 includes the modification described above about the dynamics of superinfection. In M3, superinfection is just ignored, rendering R_1 and R_2 simple first-order rate constants in Figure 3.3b. DMT's original version, M1, and M3 fit the data much better than did M2, but, although the three better ones captured the qualitative features of the data, neither one fits the data well by usual statistical measures of goodness-of-fit. It now appears that even the data do not fit the data—some estimates of prevalence are severely biased (36).

Let us compare the various models in light of the facts about gametocytemia and infectiousness surveyed in the first seven sections of this chapter. DMT's model and its modifications will be called "DMT's models" when the discussion involves features common to all of them. It should be stressed at the outset that mathematical models are developed for specific purposes and that modeling, by definition, ignores many details not needed for those specific purposes. Dutertre's model (21) and the DMT models were developed to predict changes in prevalence on time scales of a year or more.

Gametocytogenesis

None of the models specify a trigger for gametocytogenesis. Infectiousness is the first stage of infection for all first infections in Dutertre's model and for all first infections among nonimmune individuals in DMT's models.

Patterns of Gametocytemia

All models ignore the 10- to 18-day period of asexual parasitemia that precedes the appearance and capacitation of gametocytes. [All models also ignore the 14-day prepatent period of the asexual forms. The DMT model, however, did include that delay in its original presentation (19). Nedelman (34) argued that that delay could be ignored by interpreting inoculation to be coincident with patency.]

Dutertre's model permits transitions back and forth between infectiousness and noninfectious parasitemia, which is consistent with observations of wavelike patterns of gametocytemia and of infectiousness in individual infections. The probability of more than one bout of infectiousness, corresponding to a transition from y− to y+, depends on the transmission rate because superinfections will impede recovery from y− to x, thereby increasing the chance of a transition from y− to y+.

DMT's original model does not allow transitions back to infectiousness out of y−. Individuals may remain in y− a relatively long time, particularly when the transmission rate is high and superinfections are common, which would make R_1, through its dependence on h, small. However, the fact that such lengthy parasitemias do not revert to infectious in DMT's model is consistent with the observation made earlier about how gametocyte densities decreased in Garki when infections were longer lasting; indeed, that observation motivated this feature of the model (19).

In Nedelman's M2 and the modification of it by Struchiner et al. (46), reversions to infectiousness before recovery are possible in an individual because of superinfection. But superinfection is the only route to such a reversion.

Infectiousness versus Gametocytemia

Dutertre linked gametocytemia with infectiousness in describing his model. DMT explicitly eschewed that link, although, as we have seen, an observation from Garki about how gametocyte densities vary with vector densities motivated a key assumption about infectiousness in their model. As we have seen earlier, an individual who can infect some mosquitoes does not necessarily infect all mosquitoes. Transmission models sometimes include parameters that determine the susceptibilities of mosquitoes and humans to infectious bites on or from the other. The relevant parameter here is usually called c; it is the probability that an uninfected mosquito that bites an infectious human becomes infected. Dutertre explicitly assumed that $c = 1$; DMT, by ignoring c, implicitly did the same thing; Nedelman estimated c as 0.48. If immune responses affect infectiousness, then c may be a function of immunity.

Immune Response

In Dutertre's model, immunity prevents infection altogether, but it can be gradually lost in the absence of exposure. In DMT's original model and Nedelman's modifications, immunity precludes infectiousness and is not lost. This aspect of the DMT model is based on the observation, discussed earlier, that immunity affects gametocytemia more quickly than it does asexual parasitemia. After one acquires immunity, therefore, one no longer contributes to the reservoir of infection, in the DMT model. Another way this observation is reflected in the model is that infectiousness is lost before immunity is acquired—that is, the transition from y+ to y− precedes the transition from y− to y′−. Struchiner et al., who were studying the effects of vaccination programs, allowed immunity to be lost via a trasition from x2 to x1.

Theoretical and Empirical Investigations Regarding DMT's Models

This section considers DMT's models, as well as relevant data from the Garki study. Here, by "DMT's models," we will mean their original version and M3; M2 is ignored because of its poor performance (see ref. 34), although the modifications of Struchiner et al. may improve that performance. Because inoculation rates will not be functions of mosquito densities, DMT's original version and M1 will be essentially equivalent. Their parameter values differ, but the parameter values from reference 19 will be used.

Two sets of assumptions and predictions of these models will be considered: the effects of age and transmission intensity on infectiousness and immunity and the implications of the premise that infectiousness is lost before immunity is acquired. Comparisons with the Garki data will be hampered because gametocytemia, but not infectiousness, was observed in Garki. The results shown in the preceding section suggest that the age-specific prevalence of gametocytemia may be a good estimate of the age-specific prevalence of infectiousness, even though not all gametocyte carriers are infectious and vice versa. We will make a leap of faith, and also hope, that, for the analyses described below, the behavior of the set of individuals positive for gametocytes quantitatively mimics the behavior of the set of infectious individuals.

Age Distribution Associated with Infectiousness and Immunity

Dietz (18) simulated DMT's model with inoculation rates that he held constant within each simulation but he varied among simulation runs. Studying the predicted age-specific prevalence of infectiousness y+, he noticed that, in older age groups, the prevalence of infectiousness actually declines as h increases beyond a certain point. That is, the more intense the transmission of infection, the smaller the chance that an individual will be infectious. He noted that the observed effect is identical to one that had previously been studied by Aron and May (3) and Aron (1) for a similar model applied to a different system (resistance-conferring immunity boosted by reexposure).

There are two contributing causes to the observed effect. The first involves the behavior of the nonimmune compartments only. In DMT's model, the recovery rate R_1 from y− to x1 declines montonically with h. But α_1, the rate of transfer from y+ to y−, does not depend on h. As h increases, therefore, individuals flow out of x1 into y+ faster, and then out of y+ at an unchanging rate; but they tend to pile up in y−, prohibiting reentry into y+ via x1. The relative number of individuals in y+ thus will decrease as h increases.

The second contributing cause is that while individuals are in y−, they are capable of moving into y′−, thereby acquiring an immunity that precludes ever returning to y+. The larger h is, the longer individuals must wait in y− before they can return to x1, and so the greater chance they have of transferring to y′− and never returning to x1.

Which of the two contributing causes is the more important? The first cause is sufficient, because Aron and May's model exhibits the effect, and their model is formally equivalent to the nonimmune half of DMT's model. Is the second cause also sufficient? If so, than a similar effect might be observed in model M3. In that model, which ignores superinfection, R_1 does not depend on h. Although sojourn times in y− therefore do not increase with h, the number of visits to y− does, and so the chance of transferring to y′− increases with h.

Before answering the questions of the preceding paragraph, let us raise another set of questions. Age-specific prevalence of infectiousness and the reservoir of infection are like converses: The former expresses the probability that a person of a given age is infectious; the latter expresses the probability that an infectious person is of a certain age. What is the predicted behavior of the reservoir of infection as h varies?

The reservoir of infection depends not only on age-specific prevalences of infectiousness but also on the age distribution of the population, which may depend on h if mortality due to malarial infection significantly perturbs the population's vital statistics. DMT assumed that all individuals in their model have the same annual death rate of 36.5/1000. The work of Cohen (12), however, who based his analysis on the Garki data, suggests that, in fact, death rates vary importantly with age and with infection status. His designations of infection states do not correspond exactly to the compartments of the DMT models. However, by identifying his "never infected" and "previously infected" jointly with {x1, x2}, and his "fever cases" with y+, and his "currently infected" with {y−, y′−}, we can construct a schedule of age- and compartment-dependent death rates that at least will serve as a reasonable alternative to the constant value assumed by DMT. That schedule is displayed in Table 3.3.

To address all the above questions, DMT's model and the modification M3 of it that ignores superinfection were simulated with different constant values of the inoculation rate h, and with constant death rates or with

TABLE 3.3. Mortality rates by age and compartment (annual rate/1000).[a]

Compartment	Age group 0–4	5–18	≥19
x1	163	12	16.5
y+	253	19	24
y−	187	14	18
y′−	187	14	18
x2	163	12	16.5

[a] Adapted from Joel Cohen, "Estimating the effects of successful malaria control programmes on mortality," table 1, in the *Population Bulletin of the United Nations*, No. 25-1988 (United Nations publication, Sales No. E. 88, XIII. 6), pp. 6–26.

death rates as in Table 3.3. In Garki, inoculation rates vary seasonally; and if DMT's model and M3 incorporate that seasonal dependence of h, then they differ fundamentally because, in former, R_1 and R_2 will also vary seasonally, but not in the latter. For a given constant value of h, the two models are essentially equivalent except for parameter values, since the dependence of recovery rates on h is irrelevant if h does not vary. The parameter values, obtained by fitting the models to the Garki data, are (19,34)

$$\text{DMT: } \alpha_1 = 0.002/\text{day},\ \alpha_2 = 0.00019/\text{day},\ r_1 = 0.0023/\text{day},\ r_2 = 0.023/\text{day}$$

$$\text{M3: } \alpha_1 = 0.001/\text{day},\ \alpha_2 = 0.00014/\text{day},\ R_1 = 0.00024/\text{day},\ R_2 = 0.033/\text{day}$$

For simulations with different values of h, R_1 and R_2 will differ for the DMT model but not for model M3.

The results of the simulations are displayed in Table 3.4. We see that for all age groups except the first, age-specific prevalences peak and then decline as h increases. For the older age groups, the peak occurs for h smaller than 0.0007/day, so it is not evident. As an answer to the first set of questions above, the decline with h is also predicted by M3. The change with h for that model is much smaller than what is predicted by DMT's model, however; and it manifests itself at younger ages. As h increases, the suppression of infectiousness among the older ages is predicted to be much more pronounced with DMT's assumptions about superinfection.

As an answer to the second set of questions above, a similar decline with h is observed for the contribution to the reservoir of infection from the older age groups. Thus, older ages are more important contributors to the reservoir of infection (relatively, since the reservoir of infection is a relative frequency distribution the components of which sum to one) at lower inoculation rates. The two models differ from each other as they did

TABLE 3.4. Reservoir of infection and age-specific prevalence of infectiousness.

Reservoir of infection

Age group	h = 0.0007				h = 0.0041				h = 0.01			
	DMT		M3		DMT		M3		DMT		M3	
	μ con	μ var	μ con	μ var	μ con	μ var	μ con	μ var	μ con	μ var	μ con	μ var
<1	0.02	0.02	0.03	0.04	0.09	0.12	0.09	0.13	0.29	0.37	0.14	0.20
1–4	0.18	0.15	0.28	0.26	0.30	0.30	0.39	0.39	0.45	0.42	0.37	0.36
5–8	0.15	0.10	0.21	0.15	0.16	0.12	0.16	0.10	0.09	0.06	0.15	0.09
9–18	0.26	0.21	0.24	0.22	0.24	0.20	0.19	0.16	0.11	0.08	0.18	0.15
19–28	0.16	0.16	0.12	0.13	0.11	0.12	0.10	0.09	0.04	0.04	0.08	0.09
29–43	0.13	0.16	0.08	0.11	0.07	0.08	0.05	0.07	0.02	0.02	0.05	0.06
44–100	0.11	0.21	0.04	0.09	0.03	0.05	0.02	0.05	0.01	0.01	0.02	0.04

Age-specific prevalence of infectiousness

Age group	h = 0.0007				h = 0.0041				h = 0.01			
	DMT		M3		DMT		M3		DMT		M3	
	μ con	μ var	μ con	μ var	μ con	μ var	μ con	μ var	μ con	μ var	μ con	μ var
<1	0.09	0.08	0.10	0.09	0.35	0.34	0.39	0.38	0.53	0.52	0.60	0.59
1–4	0.20	0.19	0.28	0.26	0.34	0.35	0.46	0.47	0.23	0.25	0.45	0.46
5–8	0.20	0.19	0.23	0.23	0.22	0.21	0.21	0.20	0.06	0.05	0.20	0.19
9–18	0.18	0.18	0.14	0.14	0.16	0.16	0.13	0.13	0.03	0.03	0.13	0.13
19–28	0.16	0.15	0.14	0.14	0.11	0.11	0.09	0.09	0.02	0.02	0.09	0.08
29–43	0.13	0.13	0.07	0.06	0.07	0.06	0.06	0.05	0.01	0.01	0.06	0.05
44–100	0.09	0.09	0.03	0.03	0.02	0.02	0.02	0.02	0.00	0.00	0.02	0.02

Explanation of headings: h, inoculation rate (per day); DMT, DMT's model, recovery rates depend on h as in Figure 3; M3, modification of DMT's model with constant recovery rate; μ con, death rates constant, independent of age and infection status; and μ var, death rates varying, dependent on age and infection status.

for the prevalence of infectiousness. The effect of making death rates dependent on both age and infection status is to increase the contribution to the reservoir of infection from both extremes of the age distribution. There is a negligible effect on the age-specific prevalence of infectiousness, so the effect on the reservoir of infection is due to changes in the age structure of the population caused by differential mortality. The age distribution with constant death rates is a negative exponential distribution. Relative to it, the age distribution produced by death rates that depend on age and infection status drops off more sharply during infancy but then has a heavier tail through adulthood.

In Garki, entomological inoculation rates (EIR) were measured in eight of the twenty-two villages. To compare the above observations about the models with data, let us look at the gametocytemia age distribution and age-specific prevalence in these villages, hoping that those two observable features reflect the unobservable reservoir of infection and age-specific prevalence of infectiousness, respectively. The results are displayed in Table 3.5. The villages are ranked by cumulative EIR, that is, the observed total number of sporozoite-positive bites per person during the wet season of 1971, values of which range from 18 to 132. The data in the table represent summaries of all blood smears examined during the base-line phase of the study.

TABLE 3.5. Age distribution of gametocytemia and age-specific prevalence of gametocytemia, in villages of the Garki study, ranked by cumulative entomological inoculation rate.

a) Age-distribution of gametocytemia

Age group	Village #: / EIR[a]:	154 / 18	802 / 18	553 / 37	202 / 48	304 / 64	55 / 68	218 / 129	408 / 132
<1		0.13	0.18	0.20	0.15	0.16	0.14	0.16	0.19
1–4		0.28	0.34	0.32	0.31	0.20	0.20	0.33	0.31
5–8		0.21	0.15	0.25	0.25	0.10	0.22	0.16	0.24
9–18		0.18	0.10	0.09	0.07	0.20	0.15	0.12	0.09
19–28		0.06	0.05	0.03	0.06	0.18	0.12	0.05	0.06
29–43		0.10	0.14	0.06	0.10	0.11	0.11	0.11	0.06
44–		0.04	0.03	0.04	0.05	0.05	0.06	0.07	0.04

b) Age-specific prevalence of gametocytemia

Age group	Village #: / EIR[a]:	154 / 18	802 / 18	553 / 37	202 / 48	304 / 64	55 / 68	218 / 129	408 / 132
<1		0.36	0.34	0.45	0.30	0.36	0.36	0.40	0.34
1–4		0.47	0.33	0.31	0.36	0.30	0.33	0.37	0.37
5–8		0.26	0.22	0.21	0.19	0.15	0.26	0.23	0.24
9–18		0.21	0.16	0.09	0.09	0.12	0.18	0.12	0.12
19–28		0.07	0.05	0.03	0.06	0.09	0.08	0.07	0.05
29–43		0.07	0.08	0.03	0.04	0.04	0.05	0.05	0.04
44–		0.04	0.04	0.03	0.03	0.04	0.04	0.07	0.04

[a] EIR is the cumulative number of sporozoite bites received per person in the village during the wet season of 1971.

The patterns of Table 3.4 are not evident in Table 3.5; there is no systematic variation with cumulative EIR. The possible reasons are many. Gametocytemia is not equivalent to infectiousness; diagnosing gametocytemia is prone to error. The cumulative EIR is not necessarily an accurate measure of the inoculation rate. Indeed, just as not all uninfected mosquitoes biting infectious human become infectious, so, too, not all sporozoite-positive bites produce infections in humans. Nedelman (34) inferred from a model for the infection rate among infants that the probability of a sporozoite-positive bite producing an infection was higher in the village with the lowest cumulative EIR than in the village with the highest cumulative EIR, so that the two villages had cumulative infection-producing inoculation rates of 1.5 and 1.6, which are virtually identical (instead of the values of 18 and 132 for the cumulative EIRs reported in Table 3.5). However, Molineaux and Gramiccia (31) did report a significant positive correlation between mosquito densities and prevalences of asexual parasitemia among the villages. A contributing factor to the lack of any apparent pattern to the reservoirs of infection in Table 3.5 is that age distributions varied considerably among the eight villages. Since the data represent observations pooled from the entire base-line phase, these discrepancies may be due to different patterns of missing observations or other extraneous factors.

Aron (2) studied pattern of age-specific average density of parasites predicted by the model of Elderkin et al. (22). The model predicts that, for older age groups, gametocyte densities decline with h, but asexual parasite densities increase with h. The predicted behavior of gametocyte densities is due to a developmental period that the model posits for asexual parasites before they produce gametocytes. As transmission intensity increases, average levels of immunity increase, causing clearance rates of asexual parasites to increase. So, as transmission intensity increases, although there are more asexual parasites, fewer of them survive through the developmental period to produce gametocytes, according to the model.

Transitions Out of Infectiousness

The qualitative descriptions of patterns of gametocytemia presented earlier suggest that gametocytemia and infectiousness follow asexual parasitemia and may even signal the termination of an infection cycle. However, DMT's model posits that infectiousness occurs only at the beginning of an infection cycle and only for nonimmune individuals. Infectiousness in nonimmune individuals, according to DMT's model, is follwed by a period of noninfectious parasitemia that may last a rather long time, particularly when transmission is intense. Consider two individuals, one of whom is infectious and the other of whom is infected (but not infectious). DMT's model would ascribe to the first individual a higher probability of being infected at some fixed later point in time than it would to the second. From our earlier material (p. 63), we might expect the reverse ordering, although not necessarily—to be careful, we should construct a model. Let us focus on the predictions of DMT's model, and look at the Garki data for confirmation of refutation.

In the Garki study, individuals were followed up longitudinally, with blood smears taken at surveys once every 10 weeks. There were eight such surveys during the base-line (preintervention) phase of the study. Consider the ratio of the two probabilities discussed in the preceding paragraph, with the time interval being the time between two surveys. That is, consider a ratio the numerator of which is the probability that an individual positive for gametocytes at the current survey is infected or infectious at the next survey, and the denominator of which is the probability that an individual infected at the current survey is infected or infectious at the next survey. Such a ratio is called a relative risk. Call it ρ. Then, symbolically,

$$\rho = P(X(t_2) \in \{y+, y-, y'-\} | X(t_1) \in \{y+\}) / \\ P(X(t_2) \in \{y+, y-, '-\} | X(t_1) \in \{y-, y'-\})$$

where $X(t_1)$ and $X(t_2)$ are an individual's states at the current and next surveys, respectively. Both DMT's model and its M3 version without superinfection predict that $\rho > 1$ because of the logic described above. Moreover, both predict that the denominator of ρ will decrease with age, since as age increases, individuals are more likely to be in $y'-$ than $y-$ if they are in either one, and the rate of recovery out of $y'-$ is greater than out of $y-$. Hence, both models predict that ρ will increase with age.

The predictions of the two models are given in the final two columns of Table 3.6a. Those numbers come from simulations of the models with values of the parameters given above, and with h constant at 0.0041/day, typical of a yearly average inoculation rate in Garki. Recall that with a constant inoculation rate, the two models are equivalent except for their parameter values, since the inoculation-rate-dependent recovery rates of DMT's model do not change. The observed relative risks vary with the current survey, although not in any apparently systematic way. Values tend to be greater than one and are significantly so 13 times out of 49. Qualitatively then, the data seem consistent with the premise of the models. The variation of ρ with age appears to be quantitatively more consistent with the parameter values of M3.

For both DMT's model and M3, the rate of transition from $y+$ to $y-$ is so small that an individual who is infectious at one survey is very likely to be infectious at the next: The probability is 0.87 for DMT's parameters and 0.93 for M3's. Neither of these probabilities are well matched by the data—see Table 3.6b. The discrepancy between predictions and data may not be due to deficiencies of the models. The reason may be the nonidentity between infectiousness and gametocytemia; or the problem

TABLE 3.6. Investigation of transition rates out of y+.

(a) Relative risk of remaining infected at the next survey if in y+ at current survey, relative to y− or y′− at current survey.

	Observed current survey								Predicted	
Age	1	2	3	4	5	6	7	Av.	DMT	M3
<1	1.21[a]	1.11	1.00	1.21[a]	1.01	0.95	1.09	1.08	1.05	1.02
1–4	1.02	0.99	1.03	1.01	1.01	1.03	1.02	1.02	1.07	1.03
5–8	1.03	0.95	1.08[a]	1.02	1.05[a]	1.09[a]	1.07[a]	1.04	1.10	1.04
9–18	1.08	1.12	1.14	1.10[a]	1.10	1.18[a]	1.08	1.11	1.16	1.08
19–28	1.64[a]	1.31	1.16	0.99	0.99	1.35[a]	1.03	1.21	1.30	1.16
29–43	1.32	1.11	1.46[a]	1.09	1.10	1.10	1.19	1.20	1.50	1.30
44–	1.12	1.13	1.79[a]	1.82[a]	1.14	1.30	1.41	1.39	2.05	1.79

(b) Observed probabilities of being in y+ at the next survey if in y+ at current survey. Predicted = 0.870 (DMT) or 0.930 (M3).

	Current survey							
Age	1	2	3	4	5	6	7	Av.
<1	0.53	0.61	0.48	0.65	0.50	0.48	0.52	0.54
1–4	0.51	0.59	0.51	0.59	0.45	0.42	0.48	0.51
5–8	0.33	0.40	0.47	0.44	0.30	0.32	0.46	0.39
9–18	0.30	0.37	0.37	0.37	0.29	0.31	0.20	0.32
19–28	0.20	0.14	0.24	0.24	0.16	0.14	0.08	0.16
29–43	0.11	0.09	0.16	0.19	0.11	0.08	0.05	0.11
44–	0.03	0.06	0.00	0.21	0.13	0.12	0.05	0.09

[a] Significantly greater than 1 ($p < .025$).

may be failure to find gametocytes at the next survey in blood smears of individuals who harbor them in their blood. But the trend in the data with respect to age suggests that there may be an effect of age or immunity or both on the dynamics of infectiousness that the models have not covered.

Conclusions

Contradictions and uncertainties abound regarding the dynamics of gametocytemia and the relationship between gametocytemia and infectiousness. Debates continue about why production of sexual gametocytes is initiated during the course of infection with an asexually reproducing parasite. There is good evidence for the wavelike appearance of falciparum gametocytes, but perhaps not in immune individuals. The relationship between measurable densities of gametocytes in the blood and the ability of the host to infect mosquitoes is still not understood. Epidemiological evidence suggests that immune responses suppress gametocyte prevalence and density; but direct confirmation is lacking of immune effectors impeding gametocyte activity; and adults in endemic regions are important agents of transmission to the mosquito vector. At least there appears to be a consensus against the circadian-rhythm hypothesis for falciparum malaria.

The inability to identify individuals easily as infectious renders difficult the job of validating mathematical models that purport to describe the behavior of infectious individuals. Models have raised some interesting questions, answers to which would further our knowledge about gametocytemia and infectiousness: Just how do the reservoir of infection and age-specific prevalences of infectiousness vary with transmission intensity? How long do bouts of infectiousness last in individuals of different ages and infection histories? The kind of logical thought necessary for developing and analyzing mathematical models can be useful in organizing and verifying ideas about a complex disease, such as malaria.

Acknowledgments. I thank David Tonkyn for helpful comments.

References

1. Aron, J.L., 1983, Dynamics of acquired immunity boosted by exposure to infection, *Math. Biosci.* **64**:249–259.
2. Aron, J.L., 1988, Mathematical modeling of immunity to malaria, *Math. Biosci.* **90**:385–396.
3. Aron, J.L., and May, R.M., 1982, The population dynamics of malaria, in Anderson, R.M. (ed): Population Dynamics of Infectious Diseases, London, Chapman and Hall, pp. 129–179.
4. Bishop, Y.M.M., Fienberg, S.E., and Holland, P.W., 1975, Discrete Multivariate Analysis. Cambridge, MA, MIT Press, 557 p.
5. Boyd, M.F., 1941, Epidemiology of malaria: Factors related to the intermediate host, in Boyd, M.F. (ed): Malariology, Philadelphia, Saunders, pp. 551–607.
6. Bray, R.S., McCrae, A.W.R., and Smalley, M.E., (1976), Lack of a circadian rhythm in the ability of the gametocytes of *Plasmodium falciparum* to infect *Anopheles gambiae, Int. J. Parasitol.* **6**:399–401.
7. Broegger, S., Storey, J., and Bekessy, A., 1973, Some observations of malaria parasites in thick blood films, Technical Note No. 10, MPD/TN/73.1, WHO.
8. Bruce-Chwatt, L.J., 1963, A longitudinal survey of natural malaria infection in a group of West African adults. Part I, *W. Afr. Med. J.* **12**:141–173; Part II *W. Afr. Med. J.* **12**:199–217.
9. Brumpt, E., 1941, The human parasites of the genus *Plasmodium,* in Boyd, M.F. (ed): Malariology, Philadelphia, Saunders, pp. 65–121.
10. Carter, R., and Gwadz, R.W., 1980, Infectiousness and gamete immunization in malaria, in Kreier, J.P. (ed): Malaria, Volume 3, Immunology and Immunization, New York, Academic Press, pp. 263–298.
11. Christophers, S.R., 1924, The mechanism of immunity against malaria in communities living under hyperendemic conditions, *Indian J. Med. Res.* **12**:273–294.

12. Cohen, J.E., 1988, Estimating the effects of successful malaria control programs on mortality. *Popul. Bull. U. N.* **25:**6–26.
13. Cohen, S., McGregor, I.A., and Carrington, S., 1961, Gamma-globulin and acquired immunity to human malaria, *Nature* **192:**733–737.
14. Colbourne, M.J., 1956, The effect of prolonged examination of blood films on the parasite rate, *W. Afr. Med. J.* **1:**26–30.
15. Covell, G., 1960, Relationship between malarial parasitemia and symptoms of the disease, *Bull. WHO* **22:**605.
16. Davidson, G., 1955, Further studies of the basic factors concerned in the transmission of malaria, *Trans. R. Soc. Trop. Med. Hyg.* **49:**339–350.
17. Davidson, G., and Draper, C.C., 1953, Field studies of the basic factors concerned in the transmission of malaria, *Trans. R. Soc. Trop. Med. Hyg.* **47:**522–535.
18. Dietz, K., 1987, Mathematical models for transmission and control of malaria, in Wernsdorfer, W., and McGregor, I. (eds): Textbook of Malaria, Edinburgh, Churchill-Livingstone, Ch. 3.
19. Dietz, K., Molineaux, L., and Thomas, A., 1974, A malaria model tested in the African savannah, *Bull. WHO* **50:**347–357.
20. Dowling, M.A.C., and Shute, G.T.A., 1966, A comparative study of thick and thin blood films, *Bull. WHO* **34:**249–267.
21. Dutertre, J., 1976, Etude d'un modèle épidémiologique appliqué au paludisme. *Ann. Soc. Belge Med. Trop.* **56:**127–141.
22. Elderkin, R.H., Berkowitz, D.P., Gunn, C.F., Hickernell, F.J., Kass, S.N., Mansfield, F.I., and Taranto, R.G., 1977, On the steady state of an age dependent model for malaria, in Lakshikantham, V. (ed): Nonlinear Systems and Applications, New York, Academic Press, pp. 391–512.
23. Eyles, D.E., and Young, M.D., 1951, The duration of untreated or inadequately treated *Plasmodium falciparum* infections in the human host, *J. Natl. Malaria Soc.* **10:**327–336.
24. Garnham, P.C.C., 1966, Malaria Parasites and Other Haemosporidia, Oxford, Blackwell Scientific Publications. 1114 p.
25. Halloran, M.E., Struchiner, C.J., and Spielman, A., 1989, Modelling malaria vaccines II: Population effects of stage-specific malaria vaccines dependent on natural boosting, *Math. Biosci.* **94:**115–150.
26. Hawking, F., Worms, M.J., and Gammange, K., 1968, 24- and 48-Hour cycles of malaria parasites in the blood; their purpose, production, and control, *Trans. R. Soc. Trop. Med.Hyg.* **65:**549–559.
27. Hawking, F., Worms, M.J., Gammage, K., and Goddard, P.A., 1966, The biological purpose of the blood cycle of the malaria parasite *P. cynomolgi, Lancet* **ii:**422–424.
28. Jeffery, G.M., and Eyles, D.E., 1955, Infectivity to mosquitoes of *Plasmodium falciparum* as related to gametocyte density and duration of infection, *Am. J. Trop. Med. Hyg.* **4:**781–789.
29. Macdonald, G., 1957, The Epidemiology and Control of Malaria, London, Oxford University Press.
30. Miller, M.J., 1958, Malaria in the semi-resistant West African, *Trans. R. Soc. Trop. Med. Hyg.* **52:**152–168.
31. Molineaux, L., and Gramiccia, G., 1980, The Garki Project. Research on the

Epidemiology and Control of Malaria in the Sudan Savanna of West Africa, Geneva, WHO, 311 p.

32. Muirhead-Thomson, R.C., 1954, Factors determining the true reservoir of infection of *Plasmodium falciparum* and *Wuchereria bancrofti* in a West African village, *Trans. R. Soc. Trop. Med. Hyg.* **48:**208–225.
33. Muirhead-Thomson, R.C., 1957, The malarial infectivity of an African village population to mosquitoes (*Anopheles gambiae*), *Am. J. Trop. Med. Hyg.* **6:**971–979.
34. Nedelman, J., 1984, Inoculation and recovery rates in the malaria model of Dietz, Molineaux, and Thomas, *Math. Biosci.* **69:**209–233.
35. Nedelman, J., 1985, Some new thoughts about some old malaria models, *Math. Biosci.* **73:**159–182.
36. Nedelman, J., 1988, The prevalence of malaria in Garki, Nigeria: Double sampling with a fallible expert, *Biometrics* **44:**635–655.
37. Pampana, E., 1969, A Textbook of Malaria Eradication, London, Oxford University Press, 593 p.
38. Ross, R., 1911, The Prevention of Malaria, London, John Murray.
39. Rutledge, L.C., Gould, D.J., and Tantichareon, B., 1969, Factors affecting the infection of anophelines with human malaria in Thailand, *Trans. R. Soc. Trop. Med. Hyg.* **63:**613–619.
40. Shute, P.G., and Maryon, M., 1951, A study of gametocytes in a West African strain of *Plasmodium falciparum, Trans. R. Soc. Trop. Med. Hyg.* **44:**421–438.
41. Sinden, R.E., 1983a, Sexual development of malarial parasites, *Adv. Parasitol.* **22:**154–216.
42. Sinden, R.E., 1983b, The cell biology of sexual development in *Plasmodium.* in Whitfield, P.J. (ed): Symposia of the British Society for Parasitolgy, Volume 20, The Reproductive Biology of Parasites, published as *Parasitology* **86(4):**7–28.
43. Smalley, M.E., and Brown, J., 1981, *Plasmodium falciparum* gametocytogenesis stimulated by lymphocytes and serum from infected Gambial children, *Trans. R. Soc. Trop. Med. Hyg.* **75:**316–317.
44. Smalley, M.E., Brown, J., and Bassett, N.M., 1981, The rate of production of *Plasmodium falciparum* gametocytes during natural infections, *Trans. R. Soc. Trop. Med. Hyg.* **75:**318–319.
45. Smalley, M.E., and Sinden, R.E., 1977, *Plasmodium falciparum* gametocytes: Their longevity and infectivity, *Parasitology* **74:**1–8.
46. Struchiner, C.J., Halloran, M.E., and Spielman, A., 1989, Modelling malaria vaccines I: New uses for old ideas, *Math. Biosci.* **94:**87–114.
47. Trape, J.F., 1985, Rapid evaluation of malaria parasite density and standardization of thick smear examination for epidemiological investigations, *Trans. R. Soc. Trop. Med. Hyg.* **79:**181–184.
48. World Health Organization, 1987, The Biology of Malaria Parasites. Report of a WHO Scientific Group. World Health Organization Technical Report Series 743. Geneva, WHO.

4 Some Aspects of the Biology of Phlebotomine Sandfly Vectors

Richard D. Ward

Introduction

It is all too easy when embarking upon a description of recent advances in a topic to forget or ignore the foundations upon which it was built. Many who study phlebotomine sandflies throughout the world were inspired at some stage by the enthusiasm of the late D.J. Lewis. His work, which spanned some 60 years, resulted in outstanding contributions to the taxonomy of the group in virtually every zoogeographical zone. Crosskey (24) records that one of Lewis's fellow "Leishmaniacs" once wrote to him "the part you play in sandflies and leishmaniasis is so valuable, that I shudder to think what will happen when you finally hang up your microscope." Unfortunately, that time arrived at the end of 1986, and it remains for others to continue the work he so ably inspired.

Over the last 10 years, there have been substantial descriptions of the role of sandflies in the epidemiology of leishmaniasis in Lumsden and Evans (62), Molyneux and Ashford (72), Chang and Bray (17), and Peters and Killick-Kendrick (76). Less attention, however, has been specifically focused on current progress in the study of the subfamily, with one notable exception, when Killick-Kendrick (44) reviewed recent advances and outstanding problems in the study of phlebotomine sandflies. Ten years on, it is therefore again appropriate to examine the progress made and to look at those areas that have received insufficient attention.

In this chapter, I will focus on the application of new methods for the identification of phlebotomine sandflies and discuss how nearly every aspect of their study, has benefited from improved colonization techniques. New developments have been particularly rapid in our understanding of the developmental processes of *Leishmania* in the vector. Fundamental insight into the behavior of the flies has been made as a result of long-term studies of a limited number of species. There has also

---

Richard D. Ward, Department of Medical Entomology, Liverpool School of Tropical Medicine, Pembroke Place, Liverpool L35QA, United Kingdom.
 Advances in Disease Vector Research, Volume 6.

been increased attention paid to the development of hitherto unknown viruses in sandflies. Finally, the last decade has been notable for renewed attempts to test new control strategies and for the realization that sandflies are capable, like most other medically important Diptera, of developing resistance to insecticides.

Colonization

The ability to colonize sandflies has, without doubt, helped greatly in studying every aspect of their biology. Killick-Kendrick (44) noted that, by 1978, only two Old World and four New World species had been bred in good numbers as closed colonies over ten consecutive generations. In the ten years that have followed, at least eight more species in the New World fall into that category, including *Lutzomyia umbratilis, Lu. furcata,*[1] *Lu. intermedia, Lu. walkeri,* and *Lu. trapidoi* from Brazil, Colombia, and Panama (18, 74, 83, 93). Endris et al. (29) established colonies of *Lu. anthophora, Lu. cruciata,* and *Lu. shannoni* from flies collected in the southern United States. In the Old World, new colonies of *Phlebotomus ariasi, P. perfiliewi, P. perniciosus P. pedifer, P. langeroni, P. martini,* and *P. duboscqi* have been recorded, established with flies from France, Italy, Ethiopia, Egypt, and Kenya (6–8, 35, 46, 65). Beach et al. (6) were also the first to colonize flies of the mainly lizard-feeding genus *Sergentomyia* successfully. It is difficult to attribute the reasons for improved success in colonization to any single factor, although a number of new approaches have undoubtedly reduced problems of fungal contamination and made rearing less labor-intensive.

As a result, the reputation of sandflies as difficult insects to colonize is slowly losing credence, and the number of laboratories in which colonization is attempted has increased.

Climatic Control

The need for accurate climate control for sandfly colonization varies, depending upon the part of the world in which it is carried out and the type of experimentation envisaged. Temperatures at which tropical sandflies thrive range from 24 to 30 °C and will affect the speed of their development and influence the development of *Leishmania* within the fly (56). In southern Europe, however, to colonize species like *P. ariasi* successfully, it is necessary to maintain overwintering diapause larvae at 10 °C (46).

In many parts of the world, it is therefore possible to colonize sandflies in a normal laboratory with no climate control. Alternatively, a whole

---

[1] The abbreviation *Lu.* is used for the genus *Lutzomyia* to avoid confusion with the genus *Leishmania*

room may be adapted as an insectary with controlled temperature and humidity, which may then serve for sandflies and any other insects being studied. Despite the economic advantages of such systems, they are less adaptable than controlled climate cabinets that allow some small-scale degree of experimentation (Figure 4.1a). Apart from the expense, the disadvantage of cabinets in the tropics is often the lack of maintenance facilities, spare parts, and service engineers familiar with the equipment. In addition, when climate cabinets are used, problems may be experienced with condensation on the walls of the glass or plastic rearing equipment. To avoid the drowning of early stages under such conditions, it is advisable to line the inner surfaces with some absorbent material.

Maintenance of a high humidity (70% ± 5% RH) is essential for the well-being of adult sandflies and such flies as *Lutzomyia flaviscutellata* from forests that are less resistant to desiccation than are such species as *Phlebotomus papatasi* from arid semidesert regions of the world. High humidity for adults can be maintained by the enclosure of Barraud cages draped with a damp cloth or paper toweling within closed plastic bags. Even though we routinely rear our colonies of sandflies in a humidified insectary, this toweling dries out unless moistened in this way.

Most sandflies can be colonized under an alternating 12-hour light and dark cycle, with either tungsten or fluorescent illumination. There appears to be no need to provide a twilight period either to stimulate feeding or reproductive behavior, although the effect of the provision of such conditions at dusk and dawn on sandfly behavior has never been properly investigated. Longer daylight periods of 17 hours alternated with 7 hours of darkness have been used in colonizing Palaearctic sandflies, which, when used in combination with high temperatures (28 °C), simulates the conditions of a Mediterranean summer and prevents the onset of larval diapause (84).

The Immature Stages

Larval-rearing containers have ranged from unused water tanks, Petri dishes lined with filter paper, earthenware bean pots, and plastic screw-capped pots of various shapes and sizes. The most widely used system consists of a plastic screw-capped pot, lined in the base with 0.5 to 1 in. of plaster of Paris (Figure 4.1b). The bottom of the pot is perforated in several places and the plaster is kept damp by standing it on a layer of damp paper or sponge in an airtight container, like a plastic sandwich box. Blood-fed females are introduced into the pots through a screw-cap lid, lined with a layer of fine netting or bolting silk. Oviposition then takes place on the surface of the plaster and larval food can be added following hatching. The early instars are very susceptible to fungal contamination, although the use of autoclaved polymethyl pentene pots reduces this type of problem (71). Vinyl plastics containing N-butyl phthalate are toxic to insects and should therefore be avoided (29).

a

b

FIGURE 4.1. (a) Climate cabinet containing plastic sandwich boxes for holding larval rearing pots and cages for adult sandflies; (b) larval rearing pots for sandflies.

Larval foods vary from laboratory to laboratory, with different recipes enthusiastically promoted by their inventors. Some favor the use of a single ingredient like dried *Daphnia* or prawns (4), while others have adopted mixes of animal feces, laboratory chow, and dried liver powder. These mixtures are often moistened and left to mature over several months until all fungal growth has ceased and only a friable loam remains. The disadvantage of such matured foods is the offensive odors produced during their prolonged production. We have found the most useful food is a modification of that described by Ready and Croset (84), which consists, by weight, of 25% pet rabbit feces, 25% plant potting compost, 25% dried *Daphnia,* and 25% sand. This mixture is not only cheap and quick to prepare but also has the advantage that it can be used immediately following preparation. When it is used sparingly in the early stages of larval development, no problems of fungal growth are experienced, and if mistakes are made with overfeeding then the addition of further sand to the pot helps break up any hyphal mats. Using different combinations of these techniques, several laboratories are producing several thousand adults per week and have considerably reduced the amount of daily maintenance required (6, 29, 71).

Adults

Adults are most easily released into cages if the sleeves are made sufficiently large to place the entire pot inside, and then remove the lid. The alternative is to aspirate emerging flies, either mechanically or manually, from the pots and then blow them into the cage. The latter technique risks physical damage to these delicate insects and manual suction inevitably results in the inhalation of sandfly body hairs, which can prove irritating to the throat. Barraud cages draped with damp toweling and enclosed in a plastic bag are most widely favored. Concentrated sucrose solution absorbed on a pledge of cotton wool or slices of apple are provided daily for both males and females. There is evidence that, in the wild, sandflies feed on natural sugars of plants, or honeydew from aphids or coccids, although the adoption of what has been learned from these studies has yet to be incorporated into routine colonization. Killick-Kendrick (76) has, however, reported that *P. papatasi* survive less well when provided with the caper plant *Capparis spinosa* on which it feeds in the wild (97) than when given concentrated sucrose.

Anesthetized hamsters have been used for bloodmeals for our laboratory colonies of female *Lu. longipalpis,* although for *P. ariasi,* rabbits have been used (46); for *P. pedifer,* Gemetchu (35) used mice and guinea pigs. Although membrane feeders have been employed using baudruche, chick, and rabbit skin to infect flies with *Leishmania,* they have not been used routinely to maintain colonies, as is the case with tsetse flies. Oviposition is the stage of the life cycle when the highest mortalities usually occur, although some species like *Lu. flaviscutellata* and *P. duboscqi* are unusual in this respect, and survival rates as high as 75% during the primary oviposition have been observed (7). It might be assumed that mortalities as high as 70% seen in laboratory populations of *Lu. longipalpis* may be a reflection of natural mortality. Some simple manipulation of oviposition surfaces has, however, been shown to increase the number of females laying eggs, to increase the number surviving oviposition, and to increase the average number of eggs laid per female. It was found that a higher proportion of females laid their eggs when confined in vials with a folded filter paper substrate; also more eggs per female were laid than when a flat paper surface was used. An average increase of 15.5 eggs per female was achieved, although, despite this improvement, the length of postoviposition survival was not prolonged. Some of the improvements in oviposition performance were attributable to an increase in surface area for egg laying (113). Due to the pressure for results from grant-aided research, most workers are content to accommodate to the inadequacies in their colonies, provided sufficient flies can be produced to allow for experimentation. Despite the progress of the last decade, there is nonetheless a need for more quality control and experimentation in the various aspects of sandfly colonization. Sandflies may then be reared with the same efficiency and ease with which genetecists breed such insects as *Drosophila*.

Sandflies as Reservoirs and Vectors of Viruses

Sandflies have long been associated with the sandfly fever virus (27), and in 1938, Petrischeva and Alymov showed that transovarial transmission occurred in *Phlebotomus papatasi*. It was not, however, until the early 1980s that the studies of Jennings and Boorman and Tesh et al. began into the behavior of phleboviruses and other genera in sandflies. The principal reason that experimental work of this type had not been pursued widely before was the lack of productive sandfly colonies (discussed in the preceding section). In addition, tissue culture, simple assay techniques, and the laboratory culture of phleboviruses were not available (106). Within the family Bunyaviridae, 37 different *Phlebovirus* serotypes are now recognized, seven of which have been associated with human illness (104). In addition, other viruses isolated from phlebotomine sandflies include members of the vesiculoviruses and orbiviruses.

Experimental Infection

In a few cases in which sandfly–host associations are suspected, laboratory-reared insects have been fed on experimentally infected animals. Thus, Tesh et al. (104) fed four- to seven-day-old *Lutzomyia gomezi* on the opossum *Didelphys*, which had been experimentally infected with Arboledas virus from Colombia. More frequently, oral infection with viruses has been investigated using the membrane-feeding technique, which allows for an exact dose of virus to be delivered and does not depend upon an adequate viremia in an experimental animal. Oral infections are, however, often unsuccessful and the use of intrathoracic inoculation has been widely employed. Jennings and Boorman (40) found that 95% of female *Lu. longipalpis* and 75% of the males survive intrathoracic inoculation. The use of moist carbon dioxide for anesthetizing the insects was shown to improve the survival rate. Techniques for parenteral infection have been described by Rosen and Gubler (92) and Boorman (10). Lastly, a sandfly cell line, designated LL5, has been established from *Lu. longipalpis* eggs and used to assay the growth of 29 viruses (105). Of the viruses tested, only 13 multiplied in the cells and, unexpectedly, most of the phleboviruses failed to replicate. Six out of seven of the rhabdoviruses developed in the sandfly cells, but only Changuinola, an orbivirus, had any cytopathic effect.

Sandfly Susceptibility

A number of studies have been carried out in which laboratory-reared sandflies have been infected with strains of the virus they transmit in nature. The purpose of such investigations is to monitor the replication of the virus, to attempt transmission by bite, and to determine if transovarial transmission is a feature of the life cycle. For example, in Florida, Endris et al. (30) investigated the role of *Lutzomyia anthophora* in the transmission of Rio Grande virus to wood rats. This sandfly lives in large numbers in these rodent burrows in South Texas and is assumed to be the vector. Following intrathoracic inoculation, the virus increased more than 10,000-fold in one week, after which it appeared to stabilize. Over 54% of the F_1 progeny of some of these flies that took a bloodmeal were shown to contain Rio Grande virus (males and females included). Oral infection was not attempted due to the difficulties in obtaining a sustained viremia in laboratory animals. Even in experimentally infected wood rats, a very short, low-titer viremia is produced, which, in view of the small bloodmeal taken by *Lu. anthophora*, led Endris et al. (30) to conclude that Rio Grande virus may be maintained partially, or even exclusively, by transovarial transmission. Similarly, Vesiculovirus-Indiana has been isolated from naturally infected *Lu. trapidoi* in Panama and, after oral infection in the laboratory, this fly was shown to transmit the virus to its offspring (102). In contrast, *Lu. gomezi* and *Lu. sanguinaria* failed to transmit the virus transovarially, which indicates the presence of some virus–vector specificity. In the Old World, Sicilian and Naples sandfly fever viruses have been used to infect their natural vector *P. papatasi* experimentally. A low level of replication for Sicilian virus (20-fold) was observed eight days after inoculation, and inconsistent results were obtained with Naples (41). It was suggested that these unexpected results were due to reduced viral infectivity, induced by long-term passage through the brains of mice. The two phleboviruses Arbia and Toscana have also been used to infect their presumed vector, *P. perniciosus,* in Italy. Low levels of viral multiplication were observed after intrathoracic inoculation and only some insects became infected after oral ingestion. However, the concentrations of virus observed were similar to those seen in natural infections and the presence of antibodies in human sera suggest that these levels are sufficient to allow transmission by bite (21). Tesh and Modi (106) also studied the growth of Toscana in its natural vector *P. perniciosus,* as well as *P. papatasi,* which is not considered a vector of this virus. They showed that although Toscana multiplies in both species of sandfly, higher levels of replication were seen in the natural vector.

Apart from attempts to match appropriate vectors and viruses, many experimental infections have been carried out in which a sandfly that exists in a colony that has been infected with viruses have an overlapping geographical distribution. Thus, Jennings and Boorman (40) infected *Lu. longipalpis* with the Blue Tongue virus, which is normally transmitted by *Culicoides*. Although *Lu. longipalpis* does not occur in places where Blue Tongue is endemic, it was chosen because a colony existed and because other species of sandflies do occur in zones where Blue Tongue exists. The virus multiplied in the flies after intrathoracic inoculation but not after ingestion. Parenterally infected flies failed to transmit the virus through a membrane six to nine days after infection. Similarly, Hoch et al. (38) infected *Lu. longipalpis* with the African phlebovirus Rift Valley fever, which is normally associated with mosquitoes, although some virus has been isolated from blackflies and biting midges. Sandflies fed on Syrian hamsters inoculated intraperitoneally with the virus were unable to transmit it, in contrast to those inoculated intrathoracically, of which 6/326 infected hamsters by bite after a five-day incubation. From this preliminary evidence, the authors concluded that the vector potential of sandflies from enzootic areas of Sub-Saharan Africa should also be evaluated. Tesh and Modi (106) also compared the growth of Pacui virus, normally transmitted by *Lu. flaviscutellata* (37) (Figure 4.2a, b) in the Old World sandfly *P. papatasi* and the New World sandfly *Lu. longipalpis*. Nearly 33% of the F_1 offspring of *Lu. longipalpis* were infected transovarially, in contrast with only 2% of *P. papatasi*.

Parenteral infection of sandflies with viruses usually results in some degree of replication, although there is considerable variation depending upon the vector and virus combination used. Oral infections are less successful, and it has been calculated by Tesh and Modi (106) that the viremia of a vertebrate host would need to be about $10^{4.0}$ infective units to infect feeding sandflies. The sandflies' small intake of blood (0.0003 to 0.0005 ml) and limited flight range would be factors limiting the occasions

a

b

FIGURE 4.2. (a) *Lutzomyia flaviscutellata*, the vector of Pacui virus; (b) Igapo (wet swamp forest) the breeding site of *Lutzomyia flaviscutellata* and habitat of the reservoir spiny rat *Proechimys*—Catu, near Bélem, Pará, Brazil.

when flies would encounter hosts with sufficiently high viremias to infect them. This evidence, and the isolation of viruses from wild-caught male and female flies, has led to much interest in transovarial transmission, which, it is believed, may be the main mechanism of maintenance for many viruses found in sandflies.

Transovarial Transmission

In the laboratory, there is considerable variability in the ability of sandflies to infect their offspring with viruses and a range of possible explanations for this are cited in the literature. For example, it is suggested that viral changes may take place due to frequency of passage in laboratory hosts (41). In addition, genetic bottlenecks in laboratory-reared sandfly populations may lower the susceptibility to viral infection (21). There is evidence from work on the La Crosse and San Angelo virus in mosquitoes that, following oral infection, the virus is not transmitted to the larvae produced in the first gonotrophic cycle. Subsequent cycles do, however, produce transovarial transmission (104). In view of the high mortality rates during the first ovarian cycle in many sandfly colonies, it is reasonable to suspect that this might often be the cause of our inability to demonstrate vertical transmission. In such cases, it would be of considerable interest to use flies like *Lu. flaviscutellata* and *P. duboscqi,* which are capable of repeated gonotrophic cycles in the laboratory. When inappropriate sandfly–virus combinations are used, then a gut barrier, inappropriate pH, or lack of receptor sites may interfere with viral development (16). Tesh (103) suggested that the reason for less successful experimental transovarial transmissions may be that the viruses fail to infect the oogonia and establish a stabilized infection. In laboratory infections, nonstabilized infections may result from the virus infecting only some oocytes. Evidence for this type of mechanism came initially from experimentation on the mode of vertical transmission of Sigma virus in *Drosophila*. In addition, there are now many natural examples of transovarial transmission of viruses and other microorganisms through cytoplasmic inheritance. These germ-line infections in the wild result in continued high levels of infection in the progeny of each subsequent generation. Sandflies with stabilized viral infections may represent a subpopulation that fluctuates in relative size, but that represents the true reservoir of infection. It is possible that, on occasion, new females might enter that subpopulation by feeding on viremic hosts, or by venereal infection. Circumstantial evidence that transovarial transmission is regularly augmented by oral infection following viral amplification in vertebrates was recently presented by Tesh and Modi (107). They found that a Toscana virus, parenterally infected line of *P. perniciosus* in the laboratory was able to transmit the virus transovarially over 13 consecutive generations. The virus remained unchanged and flies from the F_7 generation were able to transmit the virus by bite to newborn mice. There was, however, a steady decline in the percentage infection rate from 86% in the F_5 generation to 5.8% in the F_{12} generation. In mosquitoes, venereal transmission is known to occur, although low experimental transovarial transmission rates were observed from females infected in this way (26). Tesh and Modi (107) were unable to show any evidence of venereal transmission of Toscana virus in sandflies.

Isolation of New Viruses

In addition to studies on the behavior of known viruses in their sandfly vectors, new isolations have been made in the Neotropics. Travassos da Rossa et al. (108) described eight new phleboviruses, two from sandflies including *Lu. umbratilis,* and six others from rodents, marsupials, and people in the Amazon area of Brazil. The stimulus for this work came from the extensive development of a new road network in the region, followed by government-sponsored settlement of new inhabitants, mainly from other states of Brazil. Monitoring of potential health hazards to these nonimmune settlers has been carried out over a wide area of development, including large-scale mining developments in the south of Pará State. Of the eight viruses, only Alenquer was found in humans; it caused an infection similar to sandfly fever. Nonetheless, the antibody rate in the people examined proved to be low, and it seems likely that this virus is not a public health problem. Continued monitoring of this area will, however, undoubtedly reveal further new viruses in view of the rich sandfly and vertebrate fauna present in the forests. Two new vesiculoviruses, Carajas and Maraba, have also been isolated from sandflies by Travassos da Rossa et al. (109). The new viruses replicated in experimentally infected *Lu. longipalpis* and were transmitted transovarially following intrathoracic infection. In pigs and cattle, some vesiculoviruses cause illnesses that are difficult to distinguish from foot-and-mouth disease, while in humans, the effects may range from flulike symptoms to encephalitis. Maraba virus is antigenically close to three other vesiculoviruses that can produce vesicular disease in domestic animals. The veterinary or medical importance of these two new viruses is as yet unknown, since they were isolated from a new iron mining area in which there is little farming and a relatively small human population. Recently, one further phlebovirus (Arboledas), while apparently does not pose a public health problem, was described from the coffee-growing area of Norte de Santander, Colombia (104). Antibodies were found in only 5.2% of people examined, although 29% of opposums sampled were antibody-positive. Arboledas virus was not found in sandflies from other areas of Colombia, suggesting that its geographical distribution is limited.

Age Determination

The longevity of sandflies affects their ability to transmit pathogens, and because of this, many attempts have been made to determine the approximate age of individual females, by reference to changes observed during their gonotrophic cycle.

Accessory Glands

Adler and Theodor (2) were the first to draw attention to accessory gland production of granules in fed *P. perniciosus,* and they noted that some of the granular material was retained following oviposition. The presence of this material was used as a criterion to distinguish between nulliparous and parous flies. Using this method, they commented that less than 5% of wild caught unfed females captured in Malta had laid eggs. This method was also used by Lewis and Minter (60) in Kenya, Lewis (61) in Belize, and Chaniotis and Anderson (19) in California. Johnson and Hertig (42) and Ward (112), however, drew attention to the fact that accessory gland secretion may occur before a bloodmeal, making it impossible to recognize nulliparous flies by this criterion alone.

Follicular Relics

Lewis (61), in an attempt to find an alternative to accessory gland granules for determining parity in sandflies, examined the ovariolar stalks of *Lutzomyia cruciata* and found dilatations and follicular relics, a technique that had been used successfully to age-grade *Anopheles* mosquitoes. Because dissection of each fly took five minutes, he concluded that "the method is not quick enough for routine use." It was not until 1980 that renewed attempts were made to apply Polovodova's follicular relic technique to determine the age of phlebotomines. Wilkes and Rioux (121) captured *P. ariasi* from a farmhouse wall and light traps in the Cévennes of southern France. Ovarian dissection showed a higher proportion of parous flies in the light trap and one individual with three dilatations. In the same area during 1979, Guilvard et al. (34) captured sandflies in light traps and dissected 1570 *P. ariasi* between the middle of June and the middle of September. The results showed that, in early summer, the proportion of parous flies is low and gradually increases toward the end of August to mid-September. They concluded that the maximum risk period for leishmaniasis infection is in late summer. In an attempt to estimate the life expectancy of *P. ariasi* in seven sites in the Cévennes and Garrigues, Dye et al. (28) then dissected over 10,000 sandflies during the summers of

1985–1986. In the first year, flies were classified by ovarian dissection into nulliparous, one, two, and three parous, and during 1986 into nulliparous, parous, and gravid. They showed that although it is easy to distinguish nulliparous from parous flies, the presence of ovarioles with sacs in individuals that have recently oviposited makes the distinction between one, two, and three parous flies inaccurate. Estimates of *P. ariasi* life expectancy were 1.54 (SE 0.04) ovarian cycles; these estimates showed little variation between subsequent years, or between the Cévennes and the Garrigues. The authors concluded, therefore, that regional differences in the prevalence of *Leishmania* infections in dogs and *P. ariasi* population size are not due to differences in survival rates. In the second year, some local parous rate variation was noted between light traps, which appeared to be related to the proportion of males in traps. It is suggested that the proximity of the traps to breeding sites affects observed parous rates. If trap location and selectivity for older flies, as suggested by the results of Wilkes and Rioux (121), affect the observed parous rate, then the simultaneous use of a number of catching techniques may be necessary to compensate for such variables. The biting periodicity of parous sandflies has also been shown to differ from that of nulliparous flies in the forests of northern Brazil. Wilkes et al. (122) captured *Psychodopygus wellcomei* in the Serra dos Carajas, Pará State and found that flies caught in the morning on human bait consisted of 35 : 37 nulliparous : parous individuals. In contrast, when a Shannon light trap was used in the evening, 68 : 23 nulliparous : parous flies were caught. Differences in the catching techniques employed does, however, raise the interesting possibility that the differences are due to trapping selectivity and not to the time of capture, although it should be added that crepuscular peaks of nulliparous biting activity have also been recorded for *Anopheles darlingi* in Brazil (20).

Due to high mortality at egg laying in most laboratory colonies of sandflies, the accuracy of Polovodova's technique for sandflies could not, until recently, be checked beyond a single oviposition. Thus, Magnarelli et al. (63) were only able to demonstrate single relics and sac stages in colonized *P. papatasi*. However, some Neotropical species (*Lu. flaviscutellata and Lu. furcata*), now colonized, often survive multiple ovipositions. Ready et al. (87) carried out "blind" trials in Brazil, using nulliparous–three parous laboratory flies to study the accuracy of the technique. A count of follicular relics in both species resulted in the misclassification of over 40% of the individuals, but if the flies were classified as nulliparous or parous, then only a 5% error was recorded. Dissection times were long at four to six minutes and a higher proportion of laboratory flies than field flies were in the sac stage. This was because the flies had to be dissected within 24 hours of oviposition, since they died beyond this time if not offered another bloodmeal.

Daily Growth Layers

The limitations of the various age-grading techniques for sandflies, have led to further attempts to find alternative methods. Some Diptera have been correctly classified for age by counting the daily growth layers that form on the chitinous inner muscle attachments, or apodemes. Using one- to ten-day-old laboratory-bred *P.papatasi,* Yuval and Schlein (123) found that they could determine the age of individual flies with an 80% accuracy rate. In field-caught flies, the growth lines were found to be more distinct than in laboratory flies and they speculate that this is due to marked daily fluctuations in temperatures in the Jordan Valley. The disadvantages of the technique appear to be that it takes several hours to process the flies and that the dissection and examination require considerable practice to achieve proficiency. It is not yet known if the technique is applicable to flies from tropical climates in which daily temperature variations are more subtle.

Sugar Feeding

Less than 10 years ago, although it was known that sandflies took sugar meals, the origin of the sugar was unknown. Subsequently, studies in France and Israel have significantly advanced our knowledge of this aspect of fly behavior. The considerable interest in sugar feeding by sandflies first arose as the result of experimental transmission studies carried out in India. Attempts to transmit *Leishmania donovani* by *Phlebotomus argentipes* were unsuccessful until infected flies were supplied with raisins as a sugar source (100,101). Several workers, using different methods, then recorded the natural presence of sugars in the crops of flies from Sudan, Belize, and France. In Panama, flies were offered a choice of different sugars in the laboratory, and in Kenya, Minter and Wijers (70) observed sandflies apparently probing the leaves of plants. Nonetheless, Killick-Kendrick (45) predicted that honeydew excreted by coccids and aphids might be a more probable sugar source, although he thought that the mouth parts of male sandflies might be inadequate to pierce plants.

Honeydew

The prediction of Killick-Kendrick, referred to above, was confirmed recently for *P. ariasi* in the Cévennes, southern France, when it was shown that the honeydew of the oak-feeding aphid *Lachnus roboris* is avidly consumed by sandflies (47). The honeydew of other aphids, on other plants, was either less attractive or was not consumed. Flies feeding on droplets of honeydew on the leaves remained for a prolonged period of time with the tip of the proboscis firmly applied to the leaf surface, so that it looked as if the proboscis were piercing the leaf. Analysis of the honeydew in the crops of *P. ariasi* showed that it consisted mainly of the trisaccharide melezitose and its hydrolysis products, turanose, glucose, and fructose (73). Laboratory-reared *P. papatasi* fed on pure melezitose were shown to rapidly hydrolyze it to turanose, glucose, and fructose.

Plant Feeding

Plant feeding by sandflies has also been enthusiastically pursued, and in laboratory experiments, Schlein and Warburg (97) exposed *P. papatasi* to 19 different plants of the species found in the habitats of this Phlebotomine in the Jordan Valley. Two of the plants offered were contaminated with the excretions of the coccid *Icerya purchasei* and the aphid *Aphis nerii*. Some of the plants were avoided by the sandflies, others elicited indifferent behavior, while a few produced excitement, probing, and feeding. Most of the plant-feeding behavior took place at night or under subdued lighting. *Phlebotomus papatasi* of both sexes were observed to pierce leaves or stems and take sap from the plants. Particular preference was shown for the caper plant *Capparis spinosa; Solanum luteum,* a member of the family Solanaceae; and *Malva nicaeensis,* of the family Malvaceae. The plants contaminated with honeydew were only fed on when contaminated. During surface feeding, sugars were directed to the crop; in contrast, when plant tissues were pierced, sap was subsequently found in the midgut. The authors conclude their paper with the interesting comment that, in view of their observations, sandflies should now be considered as potential vectors of plant pathogens.

Field studies in the Jordan Valley on the behavior of *P. papatasi* leaving and entering the burrows of the fat sand rat *Psammomys obesus* has also contributed to our knowledge of sugar feeding. Yuval and Schlein (124) placed, oil-paper–lined, exit and entrance traps at the mouths of rodent burrows and observed the different physiological states of flies, before and after midnight. The majority of active flies before midnight were leaving the rodent burrows; in contrast, the most active flies after midnight were entering the burrows. Of the flies leaving the burrows before midnight, 39 to 59% were positive for fructose by the anthrone test; after midnight, 79 to 81% entering the burrows were shown to contain sugars. It is suggested that, apart from dispersal and the search for bloodmeals, much of the nocturnal flight activity is related to seeking carbohydrate sources. Trapping experiments were then carried out to demonstrate if the plants indicated as a sugar source were attractive under natural conditions. Schlein and Yuval (98) baited CDC light traps from which the bulbs had been removed with branches of 12 different plants, belonging to a range of seven families. The branches were supported in sealed flasks containing water, with their leaves and stems projecting beneath the dorsal trap covers and adjacent to the trap entrance. The trapping was carried out 30 km north of Jericho, at the center of a bare ploughed field. An unbaited control and a trap with an aritificial damp filter paper "plant," were also set for comparison. Of the twelve plants tested, nine attracted more *P. papatasi* than the unbaited control and the most attractive plant was *Capparis spinosa var. aravensis*. An attraction index was calculated by dividing the mean number of sandflies captured in the baited trap by the mean number in the control. The index for *Capparis spinosa var. aravensis* was 3.9. Although the attraction index for *Solanum nigrum, Prosopsis farcta,* and *Atriplex halimus* were also high at 2.1, the wet filter paper-baited trap was marginally more attractive, with an index of 2.2. Laboratory tests with the plants tested in the field showed that over 50% (52.1 to 86) of *P. papatasi* fed on four of the plants *(Capparis spinosa, Solanum nigrum, Ricinus communis,* and *Prosopis farcta).* None of the flies, however, fed on some of the plants *(Atriplex halimus,* which had proven attractive in the field). Attraction to the plants in these field trials may therefore be mediated by the search for sugars, humidified shelter, carbon dioxide, or even potential breeding sites. For example, *Atriplex halimus,* a common plant that grows on or near the burrows of the fat sand rat, also serves as a shelter for sandflies.

Potential Applications

To determine if the sugar-feeding behavior of sandflies could be used to mark sandfly populations, Schlein (95) sprayed dyed sugar bait on vegetation above fat sand rat burrows. Over a period of six nights, after the application of colored sugar solution to the plants, he used CDC light traps without a light source to capture 3,881 *P.papatasi*. On the first night, 25.9% of the adults captured were marked quite clearly with the dye and this proportion rose to 50% on the third night. On the sixth night, only 32.9% were marked and the intensity of the blue food coloring used had begun to diminish. Apart from its potential application in studying dispersal of marked flies, Schlein (95) also suggested that future addition of insecticides or bacterial contaminants to sugars might be developed as a novel means of sandfly and *Leishmania* control.

Role of Sugars in the Vector

In addition to the role of sugars in sandflies as a source of energy, it has been suggested that the presence of sugar in plants may inhibit bacterial growth. *Leishmania* do not develop in the presence of bacteria or in alimentary canals of contaminated sandflies. In a series of experiments, the pooled contents of five crops from *P. papatasi* absorbed onto sterilized 2.3 mm diameter, filter paper discs were placed in Petri dishes with cultures of *Staphylococcus aureus, Streptococcus* group A, *Shigella sonnei, Pseudomonas aeruginosa*, and *Escherichia coli* (96). For up to eight hours, the discs were shown to inhibit bacterial growth by the presence of a six- to ten-millimeter diameter clear zone around the discs. Control discs with water, saline, or other sandfly tissues did not have the same inhibitory effect on bacterial growth. It was suggested that protein (cercropin) bacterial growth inhibitors may be produced in the saliva of the flies, possibly from the maxillary gland. Warburg and Schlein (111) investigated the effect of post-bloodmeal nutrition on the development of *Leishmania major* in *P. papatasi*. Flies were fed on 10% sucrose, trehalose, albumin, and sucrose plus albumin. Ten to twelve days later, the infected flies were force fed by placing microcapillaries over their mouth parts and folding back the labium. Transmission of parasites into the microcapillaries was evaluated by counting the number of flies that transmitted parasites and the number of parasites egested. Of those flies that had been maintained on sugars and albumin alone, between 11.4 to 20.0% (4/35 to 11/55) transmitted parasites. However, in the group maintained on albumin and sucrose, 34.8% (23/66) egested parasites into the microcapillaries. About three-quarters of all transmitting flies egested from seven to one hundred parasites, though nearly 5% expelled over 1000 *Leismania* into the microcapillaries. Despite the differences in transmission rates observed, no differences were seen between the four groups of flies in the numbers of parasites that grew, or in their development. It was suggested that the higher transmission rate by the albumin plus sugar-fed group of flies was due to a greater restriction of the passage through the stomodeal valve by attachment of parasites. This partial blockage might then lead to a backwash of blood plus parasites following a buildup of pressure in the esophagus, as the fly attempted to feed. This result is clearly of considerable interest in understanding transmission mechanisms, and it remains to be seen if a similar outcome might be obtained using an animal model, in place of the artificial microcapillary system.

Species Groups and Complexes

During this century, the taxonomy of sandflies has evolved through three phases. Flies were distinguished initially by external characters only. For example, Newstead (75) originally described *P. perniciosus* from Malta as follows: "thorax and coxae pale, translucent ochreous; abdomen similar, but sometimes pale smoky grey. Hairs pallid. Wings faintly irridescent in strong light." Adler and Theodor (1) introduced the use of internal structures for the first time when they indicated the "diagnostic importance of the pharynx, buccal cavity and spermathecae." Stimulated by the need to differentiate between individuals of species groups and complexes, genetic techniques, biochemistry, and molecular biology have recently led, for example, to *P. papatasi* being characterized by Ready et al. (86) as follows: "on agarose gels, Southern blotted and then hybridized to 28S-18S rDNA of *Drosophila melanogaster* (completed gene repeat, $_{p}$Dm 238). Restriction site polymorphisms were found that characterized three different laboratory populations." The systematics of sandflies are, therefore, currently witnessing the enthusiastic application of new technology to long-standing problems. Nonetheless, it is well to remember that the present state of sandfly taxonomy has been compared to that of mosquitoes in the 1940s and 1950s (49), so that much remains to be resolved if parity with other vector groups is to be achieved.

New Approaches: Crossing Characteristics

Some of the earliest crossing experiments to define mosquito species complexes were carried out on members of the *Anopheles maculipennis* complex in the early 1930s. The difficulties of sandfly colonization, however, delayed this type of experimentation with sandflies until 1980-1981, when Ward et al. (119) crossed populations of *Lu. longipalpis*. This species had been considered to be a possible species complex by the Brazilian entomologist Mangabeira (64), who had noticed morphological variation in the males. The abdominal tergites of males of some populations bear a pair of pale patches dorsolaterally on the fourth segment. Others have an additional pair of spots on the third segment (Figure 4.3). Females do not have these abdominal spots. The third-segment spots are reduced in size in some populations by as much as one-quarter to one-half. The initial crosses were carried out to determine if allopatric populations of different morphology were conspecific. A population of flies from Lapinha Cave near Belo Horizonte, Minas Gerais, southern Brazil, with a single pair of pale spots was crossed with one from Morada Nova, Ceará, northeast Brazil, which has two pairs of spots. When males of the Lapinha Cave population were crossed with the females from the northeast, hatch rates of eggs were as low as 0.3 to 5.3% (3/933 to 38/714). In the cross between Morada Nova males and Lapinha Cave females, hatch rates were higher, at 10.2 to 20.6% (33/323 to 76/369), which was less than one-half the rate observed in control crosses. The spermathecae were dissected from females in some further experimental crosses, and it was found that although courtship and copulation had taken place, in nearly all cases, no sperm was transferred. During copulation, females frequently attempted to repulse males by lowering the abdomen (decamping) and by moving away from the male, often dragging him along the wall of the cage. Comparisons of copulation durations in experimental and control crosses showed significant differences.

Later crosses between sympatric populations of one-spot and two-spot flies from Sobral, Ceará, northeast Brazil, also showed reproductive incompatibility, with hatch rates as low as 0 to 1.9% (0/99 to 3/152). From

FIGURE 4.3. A male *Lutzomyia longipalpis* showing the third and fourth pale tergal spots, the sites of pheromone release.

these initial results it began to appear that male morphology was a reliable guide to the recognition of two reproductively isolated siblings. Some further crosses, however, contradicted this view. It was found, for example, that there was also reproductive incompatibility between the one-spot population from the Lapinha Cave and a population of similar morpholgy from Marajo island, Pará, northern Brazil. When male Marajo flies were crossed with females from the Lapinha Cave, none of the observed 215 eggs hatched; in the reciprocal cross, 10.2% (26/254) were viable. It was also later shown that flies of dissimilar morphology (Marajo and Morada Nova) might also be reproductively compatible. Eight crosses between different Brazilian populations of *Lu. longipalpis* have now been carried out and have been summarized by Ward et al. (117).

Semiochemical Identification

The pale tergal spots then became the focus for further investigation, when Lane and Ward (55) examined the spots by scanning electron microscopy. It was shown that only the cuticle of the spots bore numerous small convex papules 3 to 3.5 μ in diameter. There was a pore of about 0.25 μ in diameter at the center of each papule. The presence of the papules, and the agitated wing fluttering of flies observed during courtship and mating, was indicative of a source of pheromones, which it was thought might play a role in sexual communication.

Bernades and Lane (unpublished observations) found a gland beneath the papules, which they sectioned and examined by transmission electron microscopy. They described large columnar cells, the lower halves of which were highly vacuolated, and beneath the cuticle they found an end apparatus consisting of a spherical array of radiating microvilli. The cells appeared to release the contents of their vacuoles from the end apparatus onto the surface of the papules through a duct. A pheromone gland of similar structure has been described in the males of the Australian scorpion fly *Harpobittacus australis* (25). The gland of the sandfly was subsequently shown to undergo a developmental cycle. In males, up to six hours after emerging from the pupa, the gland is composed of undifferentiated cells. By 24 hours, columnar cells are formed and, within a week, become enlarged with numerous vacuoles, giving a spongy appearance, which has been likened to fat body (12).

Further proof that the gland is a site of pheromone production was obtained by Lane et al. (53) when they extracted gland segments from the abdomen with hexane and analyzed the extract by gas chromatography and mass spectroscopy. Lapinha Cave flies with one pair of spots were found to contain a pheromone-like chemical with a molecular formula $C_{16}H_{26}$. The molecular weight and apparent structure most resembled a substance called farnasene ($C_{15}H_{24}$), which is produced by some ants and termites as a trail pheromone. In contrast, the Morada Nova flies with two pairs of spots were found to contain a chemical with a longer carbon chain ($C_{20}H_{32}$), which is thought to be a diterpenoid. It was concluded, therefore, that the mating incompatibilities observed between populations of flies with dissimilar morphology were most likely due to these differences in their means of chemical communication. The results did not, however, explain why flies of similar morphology might be incompatible or why flies with different abdominal spots were sometimes compatible. Four more populations of *Lu. longipalpis* from Brazil were examined by Phillips et al. (80) to determine their pheromonal identity. They found that one-spot fly populatons from Santarem, Pará and Marajo, Pará also produced diterpenoids identical to the two-spot population from Morada Nova. It thus became clear that the number of tergal spots did not correlate with the semiochemical identity of the secretions of the glands beneath them. During an examination of another four populations (117), including flies from Colombia, Bolivia, and Brazil, a population from Jacobina, Bahia, eastern Brazil was found with two pairs of abdominal spots and a gland producing a farnasene-like pheromone.

In summary, all possible combinations of the different morphologies and the two pheromone types have been found after examining 10 *Lu. longipalpis* populations. No apparent geographical or ecological separation between the two pheromone types is as yet clear, but it may become evident when a larger number of populations has been sampled.

Specific Song Patterns

The circadian rhythm of hatching, locomotor activity, and the oscillation in the interval between the bursts of sound that male fruit flies (*Drosophila*) make during courtship are all controlled by the period (*per*) gene. The temporal characteristics of the songs produced by different species of fruit fly are highly species-specific (5). Both sexes of the sandfly *Lu. longipalpis* flutter their wings in what appears to be a series of pulses. It was decided, therefore, to determine if the two sympatric populations from Sobral, of different pheromone type, also differed in their song characteristics. Individuals were recorded either singly or in pairs in an anechoic chamber, as described previously by Cowling and Burnet (22). When couples were employed, the wings of one individual were removed so that the sexual identity of the song could be determined. It was found that flies produce bursts of sound as polycyclic wave trains of differing lengths. The two populations examined differ in burst repetition rate, and only in males are there differences in intra-burst frequencies (117). It has been established, therefore, that some sandfly populations produce species-specific songs, which may have a role in courtship and possibly in territorial rivalry for space or partners.

Cuticular Hydrocarbons

The cuticular hydrocarbons in the waxy cuticle of insects help prevent desiccation and are also involved in chemical communication. These hydrocarbons act as the female sex pheromone in the stable fly *Stomoxys calcitrans* (36) and are known to vary in quantity and/or structure between several members of medically important vector species (79). The technique was first successfully applied to the phlebotomines by Ryan et al. (94), who used it to identify members of the *squamiventris* series from Brazil. The two species concerned, *Psychodopygus wellcomei* and *Psychodopygus complexus,* occur sympatrically in the Serra dos Carajas in north Brazil, where *Leishmania braziliensis braziliensis* is endemic.

The males of the two species have very characteristic genitalia, which facilitates their identification. In contrast, the females cannot be separated on morphological criteria, even by multivariate analysis (54). Using gas chromatography and mass spectroscopy, Ryan et al. (94) found that the peaks representing C_{17}, C_{24}, C_{27}, and C_{35} compounds differ quantitatively between the two species. The *squamiventris* series of species are thought to be vectors in many parts of South America, and this new technique may find wider application in the future. An alternative approach has been to examine populations of the same species captured from different habitats.

In the Cévennes area of southern France, Kamhawi et al. (43) have compared *Phlebotomus ariasi* from oak woodlands with flies from inside houses situated only 900 m away. Cluster and discriminant analysis of gas liquid chromatography peaks showed that the two populations were significantly different and that specimens identified "blind" could be correctly typed with a 92% success rate. The possibility that differences in age between the two populations caused the observed results was eliminated by examining laboratory-reared individuals of known age. The marked separation of the silvatic and domestic populations prompted the authors to suggest that there may not be free gene flow between them. It is evident that further studies on crossing relationships and relative susceptibility to *Leishmania infantum* infection are required before the full significance of these results can be appreciated. Phillips et al. (78) have also examined allopatric populations of *P. perfiliewi* from either side of the Appenine Mountains in Central Italy. Populations from the Adriatic and Tyrrhenian sides had previously been compared by isoenzyme methodology and marked allele frequency differences had been recorded (114). The gas chromatography study distinguished between populations of this species from three areas of Italy.

Chromosomes for Species Differentiation

Suspecting that *Lu. longipalpis* might be a species complex, Carvalho et al. (14) made larval salivary gland preparations of flies from Minas Gerais and Ceará states in Brazil. Their polytene preparations, however, had poor banding patterns and they concentrated on the metaphase chromosomes of the cerebral ganglia, ileum, and egg. They noted slight size differences between the two populations in some pairs of the third and fourth metaphase chromosomes. White and Killick-Kendrick (120) similarly compared *Lu. longipalpis* populations and examined the premeiotic metaphase chromosomes of the pupal gonads. They did not, however, confirm the size differences reported by Carvalho et al. (14). White and Killick-Kendrick (120) also prepared larval salivary gland polytene chromosomes, although there was much fragmentation and ectopic pairing of the individual arms. Nonetheless, they obtained a provisional map of one chromosome, which indicated 15 marked segments they thought might form the basis of a formal notation.

No further results of polytene chromosome studies have been published since, although other studies of sandfly karyotypes have been carried out on *P. papatasi* (9) and *P. perniciosus* (110). In addition, Kreutzer et al. (48) have described the brain cell karyotypes of *P. colabaensis, P. argentipes,* and *Lu. spinicrassa*. They concluded from this limited survey that the genus *Phlebotomus* is chromosomally more polymorphic than *Lutzomyia* and predicted that sandfly karyotypes might be useful for the elucidation of relationships among phlebotomine species.

Application of Electrophoresis

The use of isoenzyme electrophoresis to distinguish between members of species complexes and study genetic variation between populations has been widely employed in the study of mosquitoes and blackflies (66, 69). Since 1978, when Miles and Ward (68) first studied the isoenzymes of *Lu. flaviscutellata* in Brazil, several attempts have been made to use this technique to investigate sandfly species groups or possible geographical variations. Some investigations have shown that certain diagnostic alleles are suitable for distinguishing between different species. For example, Ward et al. (116) could distinguish *P. perfiliewi* from *P. perniciosus* in Tunisia by variations in the phosphoglucomutase locus. Peterson (77) was similarly able to distinguish *Lu. ylephiletor* from other Panamanian species by a unique hexokinase allele. These results were, however, of little epidemiological benefit, since these species could already be identified on morphological criteria. The results were, nonetheless, seen as encouraging indications that isoenzyme methodology may solve species-complex problems.

In other studies, allele frequency variations were seen between allopatric populations of the same species (67, 77, 114). This type of variation is well known in other organisms, such as fish and butterflies, and is thought to be adaptive to climatic and other physical factors. Alternatively, quite small allele frequency differences in populations of *Drosophila* have been associated with various degrees of speciation (91). To date, however, there have been no studies in which sibling phlebotomine species have been detected by the use of electrophoresis. Attempts to differentiate species within groups have also been made in cases in which one sex is easily identifiable by morphological differences but the other sex is not. For example, Ready et al. (85) used 16 enzymes but were unsuccessful in separating the females of *Psychodopygus wellcomei* from those of *Psychodopygus complexus*. Recently, Caillard et al. (13) were able to discriminate between the females of *Psychodopygus carrerai* and *Psychodopygus yucumensis* in Bolivia. Eleven enzymes were used, of which peptidase and xanthine dehydrogenase were both diagnostic.

These two species are sympatric in an area of *Leishmania braziliensis* transmission and have both been found to be naturally infected with promastigotes. However, the identity of *L.braziliensis* was only confirmed in *P. yucumensis* (59). It is evident, therefore, that isoenzyme studies are potentially useful in the taxonomy of sandflies, although their practical relevance has yet to be fully realized.

DNA Probes

Of the new methods available, the use of DNA probes is probably the most powerful because such probes can be used to study the fundamental basis of genetic variation. This can be expressed as differences in the number of copies of a particular sequence that is present or as differences in its position in the chromosome. Differences may also be found in the order of bases within a sequence. Sequences that are shown to contain such differences, or that vary in this way between populations, can be used as diagnostic probes to characterize unidentified specimens. The advantages of DNA hybridization methods are (a) field specimens may be preserved in alcohol; (b) DNA does not change throughout the life cycle of the insect; and (c) DNA sequences remain constant in vivo regardless of environmental conditions. DNA probes have already been developed and used successfully to study the *Simulium damnosum* complex of blackflies (81, 82) and the *A. gambiae* complex of malaria vectors from Africa (32, 33).

The application of DNA probes to sandfly populations is still at an early stage; it was begun by Ready et al. (86), who compared *P. papatasi* from India, Iraq, and Saudi Arabia. Their probe was a ribosomal DNA complete gene repeat from *Drosophila melanogaster*. It was shown that restriction-site polymorphisms characterized the three populations. In these studies and in more recent crossing experiments, it was found, however, that there were no reproductive barriers between any of six populations, including *P. papatasi* from Cyprus, India, and Italy. Furthermore, Ready and Smith (88) subsequently cloned a segment of the rDNA gene repeat of *P. papatasi* that distinguishes it from *P. duboscqi,* a closely related species. In addition, the probe does not hybridize with eight other species of sandfly that live in a Tunisian focus of *Leishmania major*. Ready et al. (89) then developed a multicopy genomic sequence probe for the kDNA of *L. major* to identify the presence of promastigotes in squash blots of infected *P. papatasi*. The potential advantages of such a system applied to epidemiological investigations are considerable. For example, the DNA is stable on the dry nylon filters used for the blots, which will allow specimens to be stored for subsequent investigation in laboratories with the technical facilities for such work.

Ready and Smith (88) used the intergenic "nontranscribed" spacer rDNA as their vector probe because the multiple RNA polymerase I recognition sites that usually occur in this region had been shown to be species-specific in *Drosphila*.

An alternative approach is being used by Crampton et al. (23), who are differentially screening sandfly genomic libraries to find diagnostic probes that will identify females of *Lu. longipalpis* complex. This method was favored in preference to that of Ready and collegues because the use of variant spacer sequences of ribosomal genes did not distinguish between closely related siblings of blackflies and mosquitoes. The advantage of genomic library screening is that the repetitive sequences it detects are generally present in very high copy number, thus allowing very small aliquots of the insect to be used. Consequently, parts of the remaining specimen can be used for other types of analysis. In addition, repetitive sequences tend to be short, which makes the sequencing and development of synthetic probes a practical possibility. In their preliminary studies, Crampton et al. (23) examined six *Lu. longipalpis* populations of known pheromone type. Using dot blots, they were able to demonstrate a probe (MN7B) that is specific for a Bolivian population from Chijchipa. This probe did not, however, hybridize with individuals for Lapinha, Morada Nova, and Sobral in Brazil. If a combination of such probes is used, it may soon become possible to identify unknown females from areas where the two main forms are sympatric. It may also be possible to develop nonradioactive probes that can be used in field laboratories without sophisticated facilities.

Morphological Investigation

Despite the innovative application of new technology, there is still an interest in searching for new morphological features that may be critical in separating members of species groups and complexes. The subgenus Larroussius includes 26 known species, which are frequently separated on the basis of number of spermathecal segments. This feature and the proportion of the pharynx that is armed with fine teeth are used as important diagnostic characteristics of the group. However, Leger et al. (58) have shown by careful dissection of the spermathecal individual and common ducts that the bases of these ducts differ structually between species. For example *P. longicuspis, P. perfiliewi, and P. perniciosus* all have very characteristic lateral pockets of different shapes, which lie adjacent to the genital atrium where the ducts arise. Similarily, a problem existed, until recently, in distinguishing females of *P. papatasi* from females of *P. bergeroti* that occur sympatrically in Saudi Arabia. Previous methods included the examination of spermathecae, which often proved unreliable, and the examination of pharyngeal armature, which is frequently indistinct and difficult to use. Lane and Fritz (52) emphasized the importance of looking at allopatric and sympatric populations of these two species to eliminate the possibility of clinal variation in morphology.

They examined in detail the antenna, labrum, spermatheca, wing, and femur of both species from 12 countries and found that the antennal ascoid on segment four differs significantly between the two species. Although intraspecific variation was found for many characters between samples; there were no clinal trends in the ascoids. Renewed study of the morphology of different geographical forms, or "subspecies," of *P. major* has also shown that there is clinal variation in the features used to define them. Hence, subspecies such as *P.m. krimensis, P.m. neglectus,*and *P.m. syriacus* are now regarded as unacceptable taxa (50, 57).

As an aid in detecting micromorphological features previously overlooked, a number of workers have begun to use scanning electron microscopy (SEM) on adult and immature stages of sandflies. In the Soviet Union, Chaika (15) examined the antennae of *P. papatasi* using SEM and showed the presence of trichoid and basiconic sensillae. Similarly, Boufana and Ward (11) studied different populations of *Lu. longipalpis* to characterize the antennal sensilla present and to determine if there were differences between the two siblings. Although some new sensilla were found, no apparent differences were detected. The surface structures of eggs and larval stages have also been observed by SEM, but have, as yet, not revealed any features that could be utilized to define members of species complexes (31, 39, 51, 118, 125).

Chemically Mediated Behavior

Aggregation Pheromones

Recent interest in cuticular hydrocarbons and pheromones for the purposes of identification has also led to investigations into the role of these chemicals in the behavior of sandflies. Aggregation pheromones have been found in a number of species of ticks and mosquitoes; pheromones are released by feeding individuals and appear to function to attract others to the same host. For example, Rechav et al. (90) described a pheromone produced by male *Amblyomma,* which has the dual role of host and mate finding. Similarly, Ahmadi and McClelland (3) were able to demonstrate that feeding female *Aedes sierrensis* flies produce a pheromone that attracts more females to the food source.

When female sandflies feed, they frequently appear to cluster together, and it was this behavior in *P. papatasi* that prompted Schlein et al. (99) to look for evidence of chemically mediated aggregation. Using the membrane-feeding technique with saline-filled feeders maintained at 28 °C, they offered female flies a choice between membranes that had been fed upon 10 minutes earlier by other flies and unused membranes. Over three times as many flies were attracted to used membranes, as unused. The membranes were placed on the outer side of a mesh cage so that the feeding females had to feed through the mesh. In a second experiment, an unused membrane was placed at the same site where feeding had occurred 10 minutes earlier and was compared with an unused membrane, on a previously unused site. Again there were three times as many feeding flies recorded at the used site, compared with the unused. In a repeat of this experiment in which unused membranes were placed 0.5 cm from the mesh surface, a significant response to the used surface was also observed. The attractivity of used membranes and sites was transient, and was not evident when membranes were reused 30 minutes later. To identify the source of chemical attraction, Schlein et al. (99) obtained hexane extracts of different body parts of males and females, which they used to treat unused membranes. They compared extracts of (a) female palps plus mouth parts, (b) female headless bodies, and (c) male palps plus mouth parts, in turn, with leg extracts on membranes, at 28 °C. Of these, the membrane treated with the extract of female palps plus mouth parts was shown to be three times more attractive than the membrane treated with leg extract. The headless female bodies and male palps plus mouth parts had no significant effect. At the higher temperature of 36 °C, membranes treated with female palps plus mouth parts extract did not attract significantly more *P. papatasi.* The authors were of the opinion that the aggregation pheromone was being produced from large vacuolated cells, which they observed in the third and fourth palpal segments of females. Unfortunately, no attempts were made to identify the chemical component of palpal extracts by gas chromatography and mass spectroscopy.

Sex Pheromones

The discovery of two compounds in the tergal glands of different siblings of male *Lu. longipalpis,* which appear to control sexual compatibility, has also led to an interest in their role in behavior. Just as female sandflies when feeding have been observed to cluster together, it has also been observed that *Lu. longipalpis* males congregate on hosts, often before the females arrive. These gatherings of males are typified by groups of agitated individuals, jostling other males and courting arriving females. This type of behavior is seen in the laboratory when hosts are placed in cages and has also been observed in South American sites endemic for kala-azar. Based on these observations, Ward et al. (115) decided to investigate if males and host odor, in combination, were more attractive to hungry females than a host on its own. For this purpose, a three-cage olfactometer was constructed, by connecting a response-fly cage to experimental and control cages, via two 23-cm lengths of 4.5-cm diameter tubing. Sandflies from the Morada Nova colony and a diterpenoid pheromone were used in the experiments. At the start of the experiments, responder females were placed in the response cage and could then choose to fly toward an anesthetized hamster in the presence of males or an isolated hamster. A small fan at the response cage end of the apparatus was used to draw a current of air (34 to 37 cm/sec) toward the females from the experimental and control cages. Of 384 responder females, 221 (57.6%) were attracted to the host in the presence of males, in contrast to only 40 (10.4%) that chose the host without male flies. Using the same apparatus, live males were replaced by an extract of tergal pheromone gland absorbed on to a 2-cm diameter filter paper disc. This disc was placed on the abdomen of the anesthetized host in the experimental cage. In this experiment, eight third and fourth male abdominal tergites were dissected and extracted in 200 μl of hexane for 10 minutes. Once again, very similar results were obtained, when 226/417 (54.2%) of responder flies chose the hamster plus pheromone disc, in contrast with 28/417 (6.7%) attracted to a hamster in the presence of a solvent disc only.

Females responding to the pheromone extract discs were observed to hop excitedly on and off the discs and often to remain on them for some minutes, vibrating their wings intermittently. To simplify the bioassay for further investigations, we decided to monitor the number of contacts females made with a disc on the abdomen of a host. In these disc-contact experiments, responder flies were placed in a single cage containing two anesthetized hosts, one with a pheromone disc and the other with the control solvent disc. Contacts were monitored for periods of 30 minutes. The responses of females to discs prepared with eight males per extract were compared with those responding to the extract of one male only. Of the 219 contacts recorded when the eight male extract was used against a control disc, 196 (89%) were on the pheromone disc and 23 (11%) were on the control. When the single male extract was tested, 88/115 (77%) of the contacts were to the pheromone and 27/115 (23%) were to the control. Further experiments of this type were also carried out to determine the effects of using pheromone discs that had been stored for up to six days. To our surprise, the total response to the discs increased with storage time, and it is thought that this may be due to some oxidative process enhancing the attractivity of the pheromone. Thus, although freshly extracted discs elicited 112 contacts, those stored for six days resulted in over twice as many, with 236 contacts.

The response of females in the olfactometer and disc contact experiments were over relatively short distances (up to 60 cm). Morton et al. (unpublished observations) therefore decided to determine if the male pheromone acts to attract females over distances of up to 2.4 m. For this purpose, a 2.4 × 0.4 × 0.4-m cage was constructed with a dexion frame support. Female flies were released at one end of the cage and the time taken to reach an observation area around a host, with and without a pheromone disc, was monitored. The presence of a pheromone disc was shown to decrease the t_{50} (time when 50% of flies reached the observation area) by 12.6 minutes, indicating that the pheromone is active over greater distances than previously shown.

Acknowledgments. I am indebted to Mr. M. Whiting for help in preparing the figures and, in particular, to Mr. P. Smith for his assistance with literature sources and the final revision of the manuscript.

References

1. Adler, S., and Theodor, O., 1926, On the *minutus* group of the genus *Phlebotomus* in Palestine, *Bull. Ent. Res.* **16:**399–405.
2. Adler, S., and Theodor, O., 1935, Investigations on mediterranean kala-azar. III. Further observations on mediterranean sandflies, *Proc. R. Soc.* (London) **116:**505–515.
3. Ahmadi, A., and McClelland, G.A.H., 1985, Mosquito-mediated attraction of female mosquitoes to a host, *Physiol. Entomol.* **10:**251–255.
4. Anez, N., Oviedo, M., 1985, Two new larval diets for rearing sandflies in the laboratory, *Trans. R. Soc. Trop. Med. Hyg.* **79:**739–740.
5. Ashburner, M., 1987, *Drosophila* genetics, love-song and circadian rhythm, *Nature* **326:**741.
6. Beach, R., Young, D.G., and Mutinga, M.J., 1983, New Phlebotomine sandfly colonies: Rearing *Phlebotomus martini, Sergentomyia schwetzi,* and *Sergentomyia africana* (Diptera: Psychodidae), *J. Med. Entomol.* **20:**579–584.
7. Beach, R., Young, D.G., and Killu, G., 1986, New Phlebotomine sandfly colonies. II. Laboratory colonization of *Phlebotomus dubosqui* (Diptera: Psychodidae), *J. Med. Entomol.* **23:**114–115.
8. Beier, J.C., El Sawaf, B.M., Morsy, T.A., Merdan, A.I., Rifaat, M.M., and El Said, S., 1986, Sandflies (Diptera: Psychodidae) associated with visceral leishmaniasis in El Agamy, Alexandria Governorate, Egypt. III. Experimental *Leishmania* infections and vector status, *J. Med. Entomol.* **23:**616–621.
9. Bhat, U.K.M., and Modi, G.B., 1976, Karyotype of the sandfly *Phlebotomus papatasi, Curr. Sci.* **45:**265–266.
10. Boorman, J., 1975, Semi-automatic device for inoculation of small insects with viruses, *Lab. Practice* **24:**90.
11. Boufana, B., and Ward, R.D., 1987, Preliminary studies on the antennal sensilla of *Lutzomyia longipalpis* (Diptera: Psychodidae), *Trans. R. Soc. Trop. Med. Hyg.* **81:**509.
12. Boufana, B., Ward, R.D., and Phillips, A., 1986, Development of the tergal "pheromone" gland in male *Lutzomyia longipalpis* (Diptera: Psychodidae), *Trans. R. Soc. Trop. Med. Hyg.* **80:**333–334.
13. Caillard, T., Tibayrenc, M., Le Pont, F., Dujardin, J.P., Desjeux, P., and Ayala, F.J., 1986, Diagnosis by isozyme methods of two cryptic species *Psychodopygus carrerai* and *P. yucumensis* (Diptera: Psychodidae), *J. Med. Entomol.* **23:**489–492.
14. Carvalho de, H.C.A., Falcao, A., and Schreiber, G., 1962, Cariotipo dos *Phlebotomos, Cienc. Cult. (São Paulo),* **14:**38.
15. Chaika, S.Y., 1975, Electron microscopic investigation of the olfactory sensilla of the sandfly (Diptera: Phlebotomidae) (in Russian), Proc. 2nd All-Union Symp. on Insect Chemoreception, Vilnius, pp. 69–75.
16. Chamberlain, R.W., and Sudia, W.D., 1961, Mechanism of transmission of viruses by mosquitoes, *Ann. Rev. Entomol.,* **6:**371–389.

17. Chang, K.P., and Bray, R.S. (eds), 1985, *Human Parasitic Diseases: Leishmaniasis,* Volume 1, Amsterdam, New York, Oxford, Elsevier, 490 p.
18. Chaniotis, B.N., 1986, Successful colonization of the sandfly *Lutzomyia trapidoi* (Diptera: Psychodidae), with enhancement of its gonotrophic activity, *J. Med. Entomol.* **23:**163–166.
19. Chaniotis, B.N., and Anderson, J.R., 1967, Age structure, population dynamics and vector potential of *Phlebotomus* in northern California, *J. Med. Entomol.* **4:**251–254.
20. Charlwood, J.D., and Wilkes, T.J., 1979, Studies on the age-composition of samples of *Anopheles darlingi* Root (Diptera: Culicidae) in Brazil, *Bull. Entomol. Res.* **69:**337–342.
21. Ciufolini, M.G., Maroli, M., and Verani, P., 1985, Growth of two Phleboviruses after experimental infection of their suspected sandfly vector, *Phlebotomus perniciosus* (Diptera: Psychodidae), *Am. J. Trop. Med.* **34:**174–179.
22. Cowling, D.E., and Burnet, B., 1981, Courtship songs and genetic control of the acoustic characteristics in sibling species of the *Drosophila melanogaster* subgroup, *Animal Behav.* **29:**924–935.
23. Crampton, J.M., Knapp, T., and Ward, R.D., 1989, DNA probes for vector taxonomy, NATO ASI Series, Zakinthos, Greece. Plenum Press, New York.
24. Crosskey, R., 1987, David James Lewis (1909–1986)–A Remembrance, *Antenna* **11:**54–56.
25. Crossley, A.C., and Waterhouse, D.F., 1969, The ultrastructure of a pheromone secreting gland in the male scorpion fly *Harpobittacus australis* (Bittacidae: Mecoptera), *Tiss. Cell* **1:**273–294.
26. DeFoliart, G.R., Grimstead, P.R., and Watts, D.M., 1987, Advances in mosquito-borne arbovirus/vector research, *Ann. Rev. Entomol.* **32:**479–505.
27. Doerr, R., Franz, K., and Taussig, S., 1909, *Das Pappatacifieber.* Leipzig, Vienna, Franz Deuticke, 166 p.
28. Dye, C., Guy, M.W., Elkins, D.B., Wilkes, T.J., and Killick-Kendrick, R., 1987, The life expectancy of phlebotomine sandflies: First field estimates from southern France, *Med. Vet. Entomol.* **1:**417–425.
29. Endris, R.G., Perkins, P.V., Young, D.G., and Johnson, R.N., 1982, Techniques for laboratory rearing of sandflies (Diptera: Psychodidae), *Mosquito News* **42:**400–407.
30. Endris, R.G., Tesh, R.B., and Young, D.G., 1983, Transovarial transmission of Rio Grande virus (Bunyaviridae: Phlebovirus) by the sandfly, *Lutzomyia anthophora, Am. J. Trop. Med.* **32:**862–864.
31. Endris, R.G., Young, D.G., and Perkins, P.V., 1987, Ultrastructural comparison of egg surface morphology of five *Lutzomyia* species (Diptera: Psychodidae), *J. Med. Entomol.* **24:**412–415.
32. Gale, K.R., and Crampton, J.M., 1987, DNA probes for species identification of mosquitoes in the *Anopheles gambiae* complex, *Med. Vet. Entomol.* **1:**127–236.
33. Gale, K.R., and Crampton, J.M., 1987, A DNA probe to distinguish the species *Anopheles quadriannulatus* from other species of the *Anopheles gambiae* complex, *Trans. R. Soc. Trop. Med. Hyg.* **81:**842–846.
34. Guilvard, E., Wilkes, T.J., Killick-Kendrick, R., and Rioux, J.A., 1980, Ecologie des leishmanioses dans le sud de la France. 15. Déroulement des

cycles gonotrophiques chez *Phlebotomus ariasi* Tonnoir, 1921 et *Phlebotomus mascitti* Grassi, 1908 en Cévennes. Corollaire épidémiologique, *Ann. Parasitol.* **55**:659–664.
35. Gemetchu, T., 1976, Laboratory culture and biology of *Phlebotomus pedifer* (Diptera: Psychodidae), *Ethiop. Med. J.* **15**:1–4.
36. Harris, R.L., Oehler, D.D., and Berry, I.L., 1976, Sex pheromone of the stable fly: Effect on cuticular hydrocarbons of age, sex, species, and mating, *Environ. Entomol.* **5**:973–977.
37. Hervé, J.P., Travassos da Rossa, A.P.A., Sá Filho, G.C., Travassos da Rossa, J.F., and Pinheiro, F.P., 1984, Mise en évidence de la transmission transovarienne du virus Pacui chez *Lutzomyia flaviscutellata* (Phlebotominae). Conséquences épidémiologiques, *Cah. O.R.S.T.O.M. Sér. Entomol. Méd. Parasitol.* **22**:297–212.
38. Hoch, A.L., Turell, M.J., and Bailey, C.L., 1984, Replication of Rift Valley fever in the sandfly *Lutzomyia longipalpis, Am. J. Trop. Med.* **33**:295–299.
39. Irungu, L.W., Mutinga, M.J., and Kokwaro, E.D., 1986, Chorionic sculpturing of eggs of some Kenyan phlebotomine sandflies, *Insect Sci. Applic.* **1**:45–48.
40. Jennings, M., and Boorman, J., 1980, The susceptibility of the sandfly *Lutzomyia longipalpis* (Lutz & Neiva), Diptera, Phlebotomidae, to laboratory infection with Bluetongue virus, *Arch. Virol.* **64**:127–131.
41. Jennings, M., and Boorman, J., 1983, Laboratory infection of the sandfly *Phlebotomus papatasi* Scopoli (Diptera, Psychodidae) with three Phleboviruses, *Trans. R. Soc. Trop. Med. Hyg.* **77**:62–64.
42. Johnson, P.T., and Hertig, M., 1961, The rearing of *Phlebotomus* sandflies. II. Development and behavior of panamanian sandflies in laboratory culture, *Ann. Entomol. Soc. Am.* **54**:764–776.
43. Kamhawi, S., Molyneux, D.H., Killick-Kendrick, R., Milligan, P.J.M., Phillips, A., Wilkes, T.J., and Killick-Kendrick, M., 1987, Two populations of *Phlebotomus ariasi* in the Cévennes focus of leishmaniasis in the south of France revealed by analysis of cuticular hydrocarbons, *Med. Vet. Entomol.* **1**:97–102.
44. Killick-Kendrick, R., 1978, Recent advances and outstanding problems in the biology of phlebotomine sandflies. *Acta Tropica* **35**:297–313.
45. Killick-Kendrick, R., 1979, The biology of *Leishmania* in phlebotomine sandflies, in Lumsden, W.H.R., and Evans, D.A. (eds): Biology of the Kinetoplastida, Volume 2, London, New York, San Fransisco, Academic press, pp. 395–460.
46. Killick-Kendrick, R., and Killick-Kendrick, M., 1987, The laboratory colonization of *Phlebotomus ariasi* (Diptera: Psychodidae), *Ann. Parasitol. Hum. Comp.* **62**:354–356.
47. Killick-Kendrick, R., and Killick-Kendrick, M., 1987, Honeydew of aphids as a source of sugar for *Phlebotomus ariasi, Med. Vet. Entomol.* **1**:297–302.
48. Kreutzer, R.D., Modi, G.B., Tesh, R.B., and Young, D.G., 1987, Brain cell karyotypes of six species of new and old world sand flies (Diptera: Psychodidae), *J. Med. Entomol.* **24**:609–612.
49. Lane, R.P., 1986, Recent advances in the systematics of phlebotomine sandflies, *Insect Sci. Applic.* **7**:225–230.
50. Lane, R.P., 1988, Geographic variation in Old World phlebotomine sandflies,

in Service, M.W. (ed): Biosystematics of Haematophagous Insects, Oxford, Oxford University Press.

51. Lane, R.P., and El Sawaf, B., 1986, The immature stages of *Phlebotomus langeroni* (Diptera: Psychodidae), *J. Med. Entomol.* **23:**263–268.
52. Lane, R.P., Fritz, G.N., 1986, The differentiation of the leishmaniasis vector *P. papatasi* from the suspected vector *P. bergeroti* (Diptera: Phlebotominae), *System. Entomol.* **11:**439–445.
53. Lane, R.P., Phillips, A., Molyneux, D.H., Procter, G., and Ward, R.D., 1985, Chemical analysis of the abdominal glands of two forms of *Lutzomyia longipalpis:* Site of a possible sex pheromone? *Ann. Trop. Med. Parasitol.* **79:**225–229.
54. Lane, R.P., and Ready, P.D., 1985, Multivariate discrimination between *Lutzomyia wellcomei,* a vector of mucocutaneous leishmaniasis, and *Lu. complexus* (Diptera: Phlebotominae), *Ann. Trop. Med. Parasitol.* **79:**469–472.
55. Lane, R.P., and Ward, R.D., 1984, The morphology and possible function of abdominal patches in males of the two forms of the leishmaniasis vector *Lutzomyia longipalpis, Cah. O.R.S.T.O.M. Sér. Ent. Méd. Parasitol.* **22:**245–249.
56. Leaney, A.J., 1977, The effect of temperature on *Leishmania* in sandflies, *Parasitology,* **75:**28–29.
57. Legér, N., Pesson, B., and Madulo-Leblond, G., 1985, A propos des problèmes posés par la taxonomie des vecteurs de leishmanioses; le cas de *Phlebotomus major* Annandale (Diptera, Psychodidae), *Bull. Soc. Fr. Parasitol.* **2:**187–197.
58. Legér, N., Pesson, B., Madulo-Leblond, G., and Abonnenc, E., 1983, Sur la Différenciation des femmelles du sous-genre *Larrousius* Nitzulescu, 1931 (Diptera-Phlebotomidae) de la région méditerranéenne, *Ann. Parasitol. Hum. Comp.* **58:**611–623.
59. Le Pont, F., and Desjeux, P., 1986, Leishmaniasis in Bolivia. II. The involvement of *Psychodopygus yucumensis* in the silvatic transmission cycle of *Leishmania braziliensis braziliensis* in a lowland subandean region, *Mems. Inst. Oswaldo Cruz* **81:**311–318.
60. Lewis, D.J., and Minter, D.M., 1960, Internal structural changes in some African phlebotomines, *Ann. Trop. Med. Parasitol.* **54:**351–365.
61. Lewis, D.J., 1965, Internal structural features of some Central American phlebotomine sandflies, *Ann. Trop. Med. Parasitol.* **59:**375–385.
62. Lumsden, W.H.R., and Evans, D.A. (eds), 1979, Biology of the Kinetoplastida, Volume 2, London, New York, San Francisco, Academic Press, 738 p.
63. Magnarelli, L.A., Modi, G.B., and Tesh, R.B., 1984, Follicular development and parity in phlebotomine sandflies (Diptera: Psychodidae), *J. Med. Entomol.* **21:**681–689.
64. Mangabeira, O., 1969, Sobre a sistemática e biologia dos *Phlebotomos* do Ceará, *Rev. Bras. Malar. Doenç. Trop.* **21:**3–26.
65. Maroli, M., Fiorentino, S., and Guandalini, E., 1987, Biology of a laboratory colony of *Phlebotomus perniciosus* (Diptera: Psychodidae), *J. Med. Entomol.* **24:**547–551.
66. Meredith, S.E.O., and Townson, H., 1981, Enzymes for species identification in the *Simulium damnosum* complex from West Africa, *Tropenmed. Parasit.* **32:**123–129.

67. Mery, A., Pasteur, N., Trouillet, J., and Vattier-Bernard, G., 1983, Polymorphisme des phosphoglucoisomerases chez quelques Phlébotomes Africains, *Biochem. Syst. Ecol.* **11**:63–66.
68. Miles, M., and Ward, R.D., 1978, Preliminary isoenzyme studies on phlebotomine sandflies (Diptera: Psychodidae), *Ann. Trop. Med. Parasitol.* **72**: 398–399.
69. Miles, S.J., 1979, A biochemical key to adult members of the *Anopheles gambiae* group of species (Diptera: Culicidae), *J. Med. Entomol.* **15**:297–299.
70. Minter, D.M., and Wijers, D.J.B., 1963, Studies on the vector of kala-azar in Kenya. IV. Experimental evidence. *Ann. Trop. Med. Parasitol.* **57**:24–31.
71. Modi, G.B., and Tesh, R.B., 1983, A simple technique for mass rearing *Lutzomyia longipalpis* and *Phlebotomus papatasi* (Diptera: Psychodidae) in the laboratory, *J. Med. Entomol.* **20**:568–569.
72. Molyneux, D.H., and Ashford, R.W., 1983, The Biology of *Trypanosoma* and *Leishmania,* Parasites of Man and Domestic Animals, London, Taylor & Francis, 294 p.
73. Moore, J.S., Kelly, T.B., Killick-Kendrick, R., Killick-Kendrick, M., Wallbanks, K.R., and Molyneux, D.H., 1987, Honeydew sugars in wild caught *Phlebotomus ariasi* detected by high performance liquid chromatography (HPLC) and gas chromatography (GC), *Med. Vet. Entomol.* **1**:427–434.
74. Morales, A., Carrasquilla, C.F., and Rodriguez, C.I., 1984, Establecimiento de una colonia de *Lutzomyia walkeri* (Newstead, 1914) (Diptera: Phlebotiminae), *Biomedica* **4**:37–41.
75. Newstead, R., 1911, The papataci flies *Phlebotomus* of the Maltese islands, *Bull. Entomol. Res.* **2**:47–78.
76. Peters, W., and Killick-Kendrick, R., 1987, The Leishmaniases in Biology and Medicine, Volumes 1 & 2, London, Orlando, Academic Press, 430 p., 300 p.
77. Peterson, J., 1982, Preliminary survey of isoenzyme variation in anthropophilic Panamanian *Lutzomyia* species, in Chance, M.L., and Walton, B.C. (eds): Biochemical Characterisation of Leishmania, UNDP/World Bank/WHO, Geneva, pp. 104–114.
78. Phillips, A., Kamhawi, S., Milligan, P.J.M., and Molyneux, D.H., 1989, Cuticular hydrocarbon analysis as a tool in sandfly identification, NATO ASI Series, Zakinthos, Greece. New York, Plenum Press.
79. Phillips, A., and Milligan, P., 1986, Cuticular hydrocarbons distinguishing sibling species of vectors, *Parasitol. Today* **2**:180–181.
80. Phillips, A., Ward, R.D., Ryan, L., Molyneux, D.H., Lainson, R., and Shaw, J.J., 1986, Chemical analysis of compounds extracted from the tergal "spots" of *Lutzomyia longipalpis* from Brazil, *Acta Tropica* **43**:271–276.
81. Post, R., 1985, DNA probes for vector identification. *Parasitol. Today* **1**:89–90.
82. Post, R., and Crampton, J.M., 1988, The taxonomic use of variation in repetitive DNA sequences in the *Simulium damnosum* complex, in Service, M.W. (ed): Biosystematics of Haematophagous Insects, Oxford, Oxford University Press, pp. 245–256.
83. Rangel, E.F., De Souza, N.A., Wermelinger, E.U., and Barbosa, A.F.,

1985, Establecimento de colônia, em laboratôrio, de *Lutzomyia intermedia* Lutz and Neiva, 1912 (Diptera, Psychodidae, Phlebotominae), *Mem. Inst. Oswaldo Cruz. Rio de Janerio,* **80:**219–226.

84. Ready, P.D., and Croset, H., 1980, Diapause and laboratory breeding of *Phlebotomus perniciosus* Newstead and *Phlebotomus ariasi* Tonnoir (Diptera: Psychodidae) from southern France, *Bull. Entomol. Res.* **70:**511–523.
85. Ready, P.D., and Da Silva, R.M.R., 1984, An alloenzymic comparison of *Psychodopygus wellcomei*—an incriminated vector of *Leishmania braziliensis* in Pará State, Brazil—and the sympatric morphospecies *Ps. complexus* (Diptera, Psychodidae), *Cah. O.R.S.T.O.M. sér. Entomol. Méd. Parasitol.* **22:**3–8.
86. Ready, P.D., Killick-Kendrick, R., Smith, D.F., and Bailly, M., 1986, rDNA structural polymorphisms as diagnostic markers for populations of *Phlebotomus papatasi* from areas endemic and non-endemic for zoonotic cutaneous leishmaniasis, *Trans. R. Soc. Trop. Med. Hyg.* **80:**341–342.
87. Ready, P.D., Lainson, R., Wilkes, T.J., and Killick-Kendrick, R., 1984, On the accuracy of age-grading neotropical phlebotomines by counting follicular dilatations: First laboratory experiments, using colonies of *Lutzomyia flaviscutellata* (Mangabeira) and *L. furcata* (Mangabeira) (Diptera: Psychodidae), *Bull. Entomol. Res.* **74:**641–646.
88. Ready, P.D., and Smith, D.F., 1989, DNA sequence polymorphisms as genotypic markers for phlebotomine vectors of *Leishmania,* NATO ASI Series Zakinthos, Greece. New York, Plenum Press.
89. Ready, P.D., Smith, D.F., and Killick-Kendrick, R., 1988, DNA hybridizations on squash-blotted sandflies to identify both insect vector and infecting *Leishmania, Med. Vet. Entomol.* **2:**109–116.
90. Rechav, Y., Parolis, H., Whitehead, G.B., and Knight, M.M., 1977, Evidence for an assembly pheromone(s) produced by males of the bont tick, *Amblyomma hebraeum* (Acarina: Ixodidae), *J. Med. Entomol.* **14:**71–78.
91. Richmond, R.C., 1972, Enzyme variability in the *Drosophila willistoni* group. III. Amounts of variability in the subspecies *D. paulistorum, Genetics* **71:**87–112.
92. Rosen, L., and Gubler, D., 1974, The use of mosquitoes to detect and propagate dengue viruses, *Am. J. Trop. Med. Hyg.* **23:**1153–1160.
93. Ryan, L., Lainson, R., Shaw, J.J., and Fraiha, H., 1986, Ecologia de Flebotomíneos (Diptera: Psychodidae: Phlebotominae) Na Região Amazonica, in *Instituto Evandro Chagas. 50 Anos De Contribuição Às Ciências Biológicas E À Medicina Tropical,* Volume 1, pp. 307–320. Fundação Serviços de Saúde Pública, Brazil, 529 p.
94. Ryan, L., Phillips, A., Milligan, P., Lainson, R., Molyneux, D.H., and Shaw, J.J., 1986, Separation of female *Psychodopygus wellcomei* and *P. complexus* (Diptera: Psychodidae) by cuticular hydrocarbon analysis, *Acta Tropica* **43:**85–89.
95. Schlein, J., 1987, Marking of *Phlebotomus papatasi* (Diptera: Psychodidae) by feeding on sprayed, coloured sugar bait: a possible means for behavioral and control studies, *Trans. R. Soc. Trop. Med. Hyg.* **81:**599.
96. Schlein, J., Polacheck, I., and Yuval, B., 1985, Mycoses, bacterial infections and antibacterial activity in sandflies (Psychodidae) and their possible role in the transmission of leishmaniasis, *Parasitology* **90:**57–66.

97. Schlein, J., and Warburg, A., 1986, Phytophagy and the feeding cycle of *Phlebotomus papatasi* (Diptera: Psychodidae) under experimental conditions, *J. Med. Entomol.* **23:**11–15.
98. Schlein, J., and Yuval, B., 1987, Leishmaniasis in the Jordan Valley. IV. Attraction of *Phlebotomus papatasi* (Diptera: Psychodidae) to plants in the field, *J. Med. Entomol.* **24:**87–90.
99. Schlein, J., Yuval, B., and Warburg, A., 1984, Aggregation pheromone released from the palps of feeding female *Phelbotomus papatasi* (Psychodidae), *J. Insect Physiol.* **30:**153–161.
100. Shortt, H.E., 1945, Recent research on kala-azar in India. *Trans. R. Soc. Trop. Med. Hyg.* **39:**13–41.
101. Smith, R.O.A., Halder, K.C., and Ahmed, I., 1941, Further investigations on the transmission of kala-azar. Part VI. A second series of transmissions of *L. donovani* by *P. argentipes, Indian J. Med. Res.* **29:**799–802.
102. Tesh, R.B., 1975, Multiplication of Phlebotomus fever group arboviruses in mosquitoes after intrathoracic inoculation, *J. Med. Entomol.* **12:**1–4.
103. Tesh, R.B., 1984, Transovarial transmission of arboviruses in their invertebrate vectors, in Harris, K.F. (ed): Current Topics in Vector Research, New York, Praeger, pp. 57–76.
104. Tesh, R.B., Boshells, J., Young, D.G., Morales, A., Corredor, A., Modi, G.B., Carrasquilla, C.F., Rodriguez, C., and Gaitan, M.O., 1986, Biology of arboledas virus, a new Phlebotomus fever serogroup (Bunyavirdae: Phlebovirus) isolated from sandflies in Colombia, *Am. J. Trop. Med. Hyg.* **35:**1310–1316.
105. Tesh, R.B., and Modi, G.B., 1983, Development of a continuous cell line from the sandfly *Lutzomyia longipalpis* (Diptera: Psychodidae), and its susceptibility to infection with arboviruses, *J. Med. Entomol.* **20:**199–202.
106. Tesh, R.B., and Modi, G.B., 1984, Studies on the biology of Phlebovirues in sandflies (Diptera: Psychodidae). I. Experimental infection of the vector, *Am. J. Trop. Med. Hyg.* **33:**1007–1016.
107. Tesh, R.B., and Modi, G.B., 1987, Maintenance of Toscana virus in *Phlebotomus perniciosus* by vertical transmission, *Am. J. Trop. Med. Hyg.* **36:**189–193.
108. Travassos da Rossa, A.P.A., Tesh, R.B., Pinheiro, F.P., Travassos da Rossa, J.F.S., and Peterson, N.E., 1983, Characterization of eight new Phlebotomus fever serogroup arboviruses (Bunyaviridae: Phlebovirus) from the Amazon region of Brazil, *Am. J. Trop Med. Hyg.* **32:**1164–1171.
109. Travassos da Rossa, A.P.A., Tesh, R.B., Travassos da Rossa, J.F., Hervé, J.P., and Main, A.J., 1984, Carajas and Maraba viruses, two new vesiculoviruses isolated from Phlebotomine sandflies in Brazil, *Am. J. Trop. Med. Hyg.* **33:**999–1006.
110. Troiano, G., 1982, The karyotype of *Phlebotomus perniciosus* with some observations on cytogenetics of Phlebotominae in relation to other Psychodidae, *Parassitologia* (Rome) **24:**231–236.
111. Warburg, A., and Schlein, J., 1986, The effect of post-bloodmeal nutrition of *Phlebotomus papatasi* on the transmission of *Leishmania major, Am. J. Trop. Med. Hyg.* **35:**926–930.
112. Ward, R.D., 1974, Granule formation in the accessory glands of a laboratory

strain of *Lu. longipalpis* (Diptera: Phlebotomidae) from Ceará State, Brazil, *Trans. R. Soc. Trop. Med. Hyg.* **68:**171.

113. Ward, R.D., 1974, *Studies on the Adult and Immature Stages of some Phlebotomid Sandflies (Diptera: Psychodidae) in Northern Brazil.* Ph.D. Thesis, University of London.
114. Ward, R.D., Bettini, S., Maroli, M., McGarry, J.W., and Draper, A., 1981a, Phosphoglucomutase polymorphism in *Phlebotomus perfiliewi perfiliewi* Parrot (Diptera: Psychodidae) from central and northern Italy, *Ann. Trop. Med. Parasitol.* **75:**653–661.
115. Ward, R.D., Morton, I., Lancaster, V., Smith, P., and Swift, A., 1989, Bioassays as an indicator of pheromone communication in *Lutzomyia longipalpis* (Diptera: Psychodidae), NATO ASI Series, Zakinthos, Greece. New York, Plenum Press.
116. Ward, R.D., Pasteur, N., and Rioux, J.A., 1981, Electrophoretic studies on genetic polymorphism and differentiation of phlebotomine sandflies (Diptera: Psychodidae) from France and Tunisia, *Ann. Trop. Med. Parasitol.* **75:** 235–245.
117. Ward, R.D., Phillips, A., Burnet, B., and Marcondes, C.B., 1988, The *Lutzomyia longipalpis* complex: reproduction and distribution, in Service, M.W. (ed): Biosystematics of Haematophagous Insects, Oxford, Oxford University Press.
118. Ward, R.D., and Ready, P.A., 1975, Chorionic sculpturing in some sandfly eggs (Diptera: Psychodidae), *J. Entomol. (A)* **50:**127–134.
119. Ward, R.D., Ribeiro, A.L., Ready, P.D., and Murtagh, A., 1983, Reproductive isolation between different forms of *Lutzomyia longipalpis* (Lutz and Neiva), (Diptera: Psychodidae), the vector of *Leishmania donovani chagasi* Cunha and Chagas, and its significance to kala-azar distribution in South America, *Mem. Inst. Oswaldo Cruz* **78:**269–280.
120. White, G.B., and Killick-Kendrick, R., 1975, Polytene chromosomes of the sand fly *Lutzomyia longipalpis* and the cytogenetics of Psychodidae in relation to other Diptera, *J. Entomol.* **50:**187–196.
121. Wilkes, T.J., and Rioux, J.A., 1980, The application of Polovodova's technique for the age determination of *Phlebotomus (Larroussius) ariasi, Trans. R. Soc. Trop. Med.Hyg.* **74:**119.
122. Wilkes, T.J., Ready, P.D., Lainson, R., and Killick-Kendrick, R., 1984, Biting periodicities of nulliparous and parous females of *Psychodopygus wellcomei, Trans. R. Soc. Trop. Med. Hyg.* **78:**846–847.
123. Yuval, B., and Schlein, Y., 1987, Age determination of *Phlebotomus papatasi* by detection of cuticular growth lines, *Trans. R. Soc. Trop. Med. Hyg.* **81:**166–167.
124. Yuval, B., and Schlein, Y., 1986, Leishmaniasis in the Jordan Valley. III. Nocturnal activity of *Phlebotomus papatasi* (Diptera: Psychodidae) in relation to nutrition and ovarian development, *J. Med. Entomol.* **23:**411–415.
125. Zimmerman, J.H., Newson, H.D., Hooper, G.R., and Christensen, H.A., 1977, A comparison of the egg surface structure of six anthropophilic phlebotomine sandflies *(Lutzomyia)* with the scanning electron microscope (Diptera: Psychodidae), *J. Med. Entomol.* **13:**574–579.

5
Vector–Spirochete Relationships in Louse-Borne and Tick-Borne Borrelioses with Emphasis on Lyme Disease

Willy Burgdorfer and Stanley F. Hayes

Introduction

The discovery in 1981 of the spirochete, now known as *Borrelia burgdorferi,* as the etiological agent of Lyme disease in the United States and of associated clinical manifestations in Europe (11, 12) has rekindled interest in arthropod-borne spirochetoses. Since then, hundreds of publications have appeared that deal not only with the complex clinical aspects of Lyme disease but also with the natural history of this agent, and particularly its relationship to its arthropod vectors—ticks of the genus *Ixodes*. In the United States, where Lyme disease is now considered the most prevalent tick-borne disease (16), *B. burgdorferi* is associated with at least three *Ixodes* species, namely *Ixodes dammini* in the northeastern and midwestern regions of the country, *Ixodes pacificus* in the West, and possibly *Ixodes scapularis* in the Southeast. In Europe (55), the sheep or castor bean tick, *Ixodes ricinus,* is the vector, whereas in Asia *Ixodes persulcatus* is said (17) to be involved in the maintenance and distribution of the Lyme disease spirochete.

Borrelia burgdorferi may have eluded investigators since the beginning of the century because of its unique parasite–vector relationship, which, as we shall see, differs from that of most other arthropod-borne spirochetes, and because ticks of the genus *Ixodes* had never been suspected to serve as vectors of spirochetes.

Before the 1981 discovery of *B. burgdorferi* in *Ixodes* ticks (11), most arthropod-borne spirochetes, as summarized in Table 5.1, were known to

Willy Burgdorfer, Department of Health and Human Services, Public Health Service, National Institutes of Health, National Institute of Allergy and Infectious Diseases, Laboratory of Pathobiology, Rocky Mountain Laboratories, Hamilton, Montana 59840, USA.
Stanley F. Hayes, Department of Health and Human Services, Public Health Service, National Institutes of Health, National Institute of Allergy and Infectious Diseases, Laboratory of Pathobiology, Rocky Mountain Laboratories, Hamilton, Montana 59840, USA.

TABLE 5.1. Characteristics and distribution of arthropod-borne borreliae.[a]

Borrelia sp.	Arthropod vector	Animal reservoir	Distribution	Disease
B. recurrentis (syn. *B. obermeyeri, B. novyi*)	*P. humanus humanus*	Humans	Worldwide	Louse-borne, epidemic relapsing fever
B. duttonii	*O. moubata*	Humans	Central, Eastern, and Southern Africa	East African tick-borne, endemic relapsing fever
B. hispanica	*O. erraticus* (large variety)	Rodents	Spain, Portugal, Morocco. Algeria, Tunisia	Hispano-African, tick-borne relapsing fever
B. crocidurae, B. merionesi B. microti, B. dipodilli	*O. erraticus* (small variety)	Rodents	Morocco, Libya, Egypt, Iran, Turkey, Senegal, Kenya	North African, tick-borne relapsing fever
B. persica	*O. tholozani* (syn, *O. papillipes, O. crossi*?)	Rodents	From West China and Kashmir to Iraq and Egypt, USSR, India	Asian-African, tick-borne relapsing fever
B. caucasica	*O. verrucosus*	Rodents	Caucasus to Iraq	Caucasian, tick-borne relapsing fever
B. latyschewii	*O. tartakowskyi*	Rodents	Iran, Central Asia	Caucasian, tick-borne relapsing fever
B. hermsii	*O. hermsi*	Rodents, chipmunks, tree squirrels	Western United States	American, tick-borne relapsing fever
B. turicatae	*O. turicata*	Rodents	Southwestern United States	American, tick-borne relapsing fever

B. parkeri	*O. parkeri*	Rodents	Western United States	American, tick-borne relapsing fever
B. mazzottii	*O. talaje* (*O. dugesi*?)	Rodents	Southern United States, Mexico, Central and South America	American, tick-borne relapsing fever
B. venezuelensis	*O. rudis* (syn. *O. venezuelensis*)	Rodents	Central and South America	American, tick-borne relapsing fever
B. burgdorferi	*I. dammini*	Rodents	Eastern United States	Lyme disease and related disorders
	I. pacificus	Rodents	Western United States	Lyme disease and related disorders
	I. ricinus	Rodents	Europe	Lyme disease and related disorders
	I. persulcatus	Rodents	Asia, China, Japan	Lyme disease and related disorders
	Possibly other hematophagous arthropods	Possibly other reservoirs (deer, birds)	Worldwide(?)	
B. coriaceae	*O. coriaceus*	Rodents (deer?)	Western United States	Epizootic bovine abortion (?)
B. theileri	*Rhipicephalus* spp. *Boophilus* spp.	Cattle, horses, sheep (?)	Worldwide	Bovine borreliosis
B. anserina	*Argas* spp. (mites?)	Fowl	Worldwide	Avian borreliosis

[a]Spirochete-tick associations of unknown or little health significance not included.

be associated with soft-shelled ticks of the genus *Ornithodoros* or *Argas*. Exceptions were *Borrelia recurrentis,* the louse-borne, relapsing fever spirochete, and *Borrelia theileri,* a spirochete maintained and distributed among cattle, horses, and sheep by ticks of the genus *Rhipicephalus* and *Boophilus*.

After the human body louse *Pediculus humanus* (53) and the African soft-shelled tick *Ornithodoros moubata* (20) were found to be the vectors of the relapsing fever spirochetes (known today as *Borrelia recurrentis* and *Borrelia duttonii,* respectively) the development of these microorganisms in their vectors and the modes of their transmission to humans, was intensively studied. The purpose of this chapter is to review the most salient findings of these studies and to compare them with observations made so far on *B. burgdorferi,* in its tick vectors.

The Behavior of Louse-Borne and Tick-Borne Spirochetes in Their Vectors

Shortly after Sergent and Foley (53) confirmed that the human body louse *P. humanus humanus* was the vector of the European relapsing fever spirochete, in 1910, Nicolle et al. (45) studied the behavior of a North American strain of *B. recurrentis* in lice and noted that the spirochetes had disappeared from the midgut 24 hours after they had been ingested; they were not detectable again until days six through eight, when they suddenly began to reappear in the hemolymph. A similar "negative phase" for *B. duttonii* in *O. moubata* had been observed previously by several investigators, including Dutton and Todd (21), Leishman (38), Fantham (22), and Hindle (33) and later also by Hatt (30) and Nicolle et al. (44).

These authors found that ingested spirochetes invade the gut epithelium, where they lose their motility and, after three to four days, develop into spherules or gemmae that contain varying numbers of granules or chromatin bodies. According to Dutton and Todd (21), these spherules are formed by a protuberance (aneurysm) of the periplasmic membrane; they may form at any point along the spirochete. Some time after the spherules form, they are said to burst and to release their contents. The tenth day after an infectious feeding, morphologically typical spirochetes were no longer found (21); Dutton and Todd found large numbers of granules from which new spirochetes eventually developed, provided the ticks were maintained at temperatures above 25 °C.

Hindle (33) reported similar findings. In infected ticks held at 21 °C, the spirochetes disappeared from the midgut by the tenth day after infectious feeding. They could no longer be detected either in the gut or in the tissues of various organs. Triturates of such ticks injected into mice, however, regularly proved infectious, and an increase in temperature to 35 °C led to the reappearance of morphologically typical spirochetes.

This "granulation theory" received a significant boost in 1950, when Hampp (29), of the National Institute of Dental Research, showed by stained smears and darkfield and electron microscopy that oral treponemes and *Borrelia vincenti* in cultures produced blebs and granules, which he considered to be possible germinative units. His hypothesis was supported by the observation that 31-month-old cultures containing only granules invariably produced typical spirochetes upon transfer to fresh medium.

Similar observations were also reported by DeLamater et al. (18). Their data supported the occurrence of a complex life cycle in the pathogenic and nonpathogenic strains of *Treponema pallidum*. This spirochete was said to multiply (a) by transverse or binary fission and (b) by the production of gemmae in which either a single granule or masses of granules appeared to be the primordia of daughter spirochetes.

Many investigators, including Wittrock (59), Kleine and Eckard (35), Kleine and Krause (34), Feng and Chung (23), and Burgdorfer (9), who conducted studies on the developmental dynamics of various species of borreliae in lice or ticks, however, found no evidence of a "negative" phase. Although most of these authors verified the existence of blebs or gemmae on spirochetes, they considered them to be products of degeneration—a conclusion also reached by Pillot et al. (48), who published an extensive electron microscopic study of these degenerating forms on cultured *T. pallidum, B. duttonii, B. hispanica,* and *Leptospira icterohaemorrhagiae*.

In 1947, Garnham et al. (26) suggested that louse-borne spirochetes could be divided into those that have a negative phase (Europe, North Africa, and Kenya) and those that do not have a negative phase (Abyssinia and China). In 1962, however, Heisch—who had worked with Garnham in 1947—and Harvey (31) called the negative phase an artifact. Thus, the question of a complex developmental cycle for borreliae appeared settled, and it was generally accepted that *B. recurrentis* spirochetes in the body louse *P. humanus humanus,* after ingestion of a patient's blood, arrive in the midgut where most are destined to die. Those that survive pass through the gut wall into the hemolymph where they multiply by binary fission. They then invade the neural ganglia and the muscles of the head and thorax. They were never observed in the salivary gland tissues or in other organs. Thus, the transmission of *B. recurrentis* to humans does not occur by bite, that is, via saliva, but rather by contamination of the bite wound with infectious hemolymph of lice crushed or wounded by a person's scratching.

Similarly, *Ornithodoros moubata* and other relapsing fever ticks, during their mostly short feeding (10 to 30 minutes), ingest spirochetes via the pharynx and esophagus into the midgut where they can be found in gradually decreasing numbers for about 14 days. Within hours after a blood meal, spirochetes accumulate in the intercellular spaces of the tick's gut epithelium. As early as 24 hours after ingestion, they penetrate the basement membrane to enter the body cavity where they undergo intensive multiplication by binary fission. From here, the spirochetes invade the various tissues, particularly those of the central ganglion, the coxal organs, and the genital system. They also thrive in the connective tissue surrounding the Malpighian tubules. Only in nymphal ticks do salivary gland tissues also become heavily infected. Once an infected *O. moubata* reaches the adult stage, the salivary glands are spirochete-free, or only mildly infected. Thus, the mechanisms for transmission of *B. duttonii* by *O. moubata* vary, and depend on the developmental stage of the tick. Nymphs transmit by spirochetes (a) bite, that is, via spirochete-containing saliva, and by (b) spirochete-containing coxal fluid, excreted shortly before feeding ends, which washes the spirochetes directly into the bite wound. Adult ticks, however, transmit spirochetes primarily by infected coxal fluid, and only rarely by infectious saliva (9).

The invasion of spirochetes into germinal cells, that is, oogonia and early oocytes, invariably leads to transovarial infection, with filial infection rates as high as 90%. It is said (1) that spirochetes remain relatively passive during the early phases of oogenesis but leave the ooplasm to invade the developing embryonic nervous tissue system. Only from this tissue do spirochetes invade other tissues, such as the salivary glands, during the tick's postembryonic stage.

The attraction of spirochetes to nervous system tissue of their arthropod vectors has been observed for *B. recurrentis* in the body louse and also for many Ornithodoros–spirochete associations. In fact, microscopic examination of central ganglion tissues is by far the most dependable way to determine the infection status of field-collected, relapsing fever ticks.

The dynamic development of spirochetes, which is similar to that outlined above for *B. recurrentis* in *P. humanus humanus* and for *B. duttonii* in *O. moubata,* has been also recorded for other tick–spirochete associations, including *Borrelia crocidurae* in *Ornithodoros erraticus* (25) and *Borrelia anserina* in *Argas persica,* and three other *Argas* spp. (19, 60). For most relapsing fever tick–spirochete associations, numerous references describe the vector's efficiency in transmitting spirochetes and in passing borreliae transovarially, that is, via eggs to the progeny (10). But vertical transmission does not occur in every tick–spirochete association. It has been reported for the genus *Ornithodoros,* in *O. moubata, O. erraticus* (both varieties), *O. tholozani, O. tartakovskyi, O. verrucosus, O. turicata,* and *O. hermsi,* but not in *O. parkeri, O. talaje,* and *O. rudis.* The percentage of infected female ticks passing spirochetes via eggs (transovarial transmission rate) varies greatly, as does the percentage of filial ticks (filial infection rate) that become infected. The efficiency of these phenomena appears to depend on the degree of spirochetal infections in ovarial tissue and germinal cells.

Although transovarial transmission is an effective way to infect ticks, it may render strains of borreliae nonpathogenic if they are maintained continuously without occasional passage in a susceptible host (28). This was considered responsible for a dramatic drop in the incidence of relapsing fever in Tanzania, where more than 20% of *O. moubata* carried spirochetes, that were once highly pathogenic for laboratory animals but were now no longer capable of infecting them. The same phenomenon was also reported for certain strains of *Borrelia sogdiana* after eight successive generations in *O. tholozani* (5).

Although transovarial transmission of tick-borne relapsing fever spirochetes is common, venereal transmission from male to female ticks during copulation appears to be rare. According to Wagner-Jevseenko (57), *B. duttonii* could occasionally be demonstrated in the fluid of spermatophores, but in only 2 of 96 females who mated with infected males did the sexually transferred spirochetes produce infections. In contrast, Gaber et al. (24), studying venereal transfer of *B. crocidurae* in Egyptian *O. erraticus,* concluded that polygamous males contribute significantly to the spread of the spirochete in this tick vector. After the first and second gonotropic cycles, spirochetes were observed in 23 and 37% of female ticks, respectively. Although the mechanism(s) of transmission was not clear, the authors speculated that infected fluids from the males' genital accessory glands or infected saliva secreted during copulation or both played a role.

The Behavior of *Borrelia theileri* in Its Tick Vectors

Borrelia theileri, the etiological agent of bovine borreliosis, is prevalent in Africa, India, Indonesia, Australia, and South America, where it is associated with ixodid ticks, including *Boophilus microplus, Boophilus annulatus, Boophilus decoloratus,* and *Rhipicephalus evertsi.* An even wider geographical distribution has been suggested by the demonstration of *B. theileri* or similar spirochetes in the blood of cattle in North America and Europe (56).

Little information is available on the relationship of this spirochete to its tick vectors, in which it produces a systemic infection, with the ovary and central ganglion most consistently infected. Transovarial transmission with filial infection rates as high as 80% does occur, but the larval ticks seem to be incapable of transmitting the spirochetes. (The reader is reminded that ticks of the genus *Boophilus* are one-host ticks, i.e., all stages and molts occur on the same host animal.) According to Smith et al. (54), who studied the development of *B. theileri* in the progeny of field-collected *B. microplus* from Mexico, spirochetes in ovarially infected larvae, 22 hours after placement on a calf, were few in number but increased in feeding nymphs and adults, especially after repletion. The authors also noted a massive growth of organisms in hemocytes and the release of spirochetes from these cells into the hemolymph, where intensive multiplication resulted in masses of spirochetes. These heavy spirochetal infections of hemolymph, ovary, and other tissue did not alter the ticks' feeding or reproductive habits.

The Behavior of the Lyme Disease Spirochete *Borrelia burgdorferi* in Its Tick Vectors

Unlike the spirochetes of relapsing fever and bovine borreliosis that leave the midgut of their vectors shortly after ingestion, *B. burgdorferi* spirochetes, in most of their tick vectors, remain primarily in the midgut, where they aggregate near the microvillar brushborder and in the intercellular spaces of the gut epithelium. From there, they may penetrate the gut wall during or after engorgement to initiate mild systemic infections, particularly in tissues of hypodermis, ovary, and central ganglion and within muscle tissue associated with Malpighian tubules and tracheae (14).

Although our 1981 collection of flat, adult *I. dammini* from Shelter Island, New York yielded ticks with midgut infections only—77 (61%) of 126 ticks—subsequent examination of 151 additional unfed adult ticks from the same region revealed 102 with midgut infections. Of these, four (3.9%) were systemically infected (14). Similar results were obtained with the European tick vector *I. ricinus*. Of 112 infected flat adults collected from vegetation in western Switzerland, 106 (91%) had midgut infections only. The remaining six had spirochetes throughout their tissues (12). A somewhat larger percentage of ticks with systemic infections was recorded for the western black-legged tick *I. pacificus,* from northern California and southwestern Oregon. Of 1678 ticks examined, 25 (1.5%) had spirochetes in their midgut. Of these, eight (32%) were also systemically infected (15).

The percentage of ticks with spirochetes throughout their tissues increased when examinations were done after the ticks had been allowed to feed. Thus, of 46 midgut-infected *I. dammini* females evaluated six to eight weeks after repletion, 9 (19.5%) had spirochetes throughout their tissues. Similarly, of 25 engorged females that had failed to oviposit for more than three months after feeding, 23 (92.9%) had midgut infections, and 18 (78.2%) of these 23 had systemic infections (14).

Thus, it appears that "gut penetration" by *B. burgdorferi* is closely associated with histological changes in the gut epithelium during or after feeding or both. Preliminary transmission and scanning electron microscopy studies suggest that *B. burgdorferi* spirochetes aggregate within clefts or pitlike structures between epithelial cells and through them become closely associated with the basement membrane (Figures 5.1 and 5.2). Actual penetration of this membrane has not yet been observed.

According to Ribeiro et al. (50), penetration of the gut epithelium and

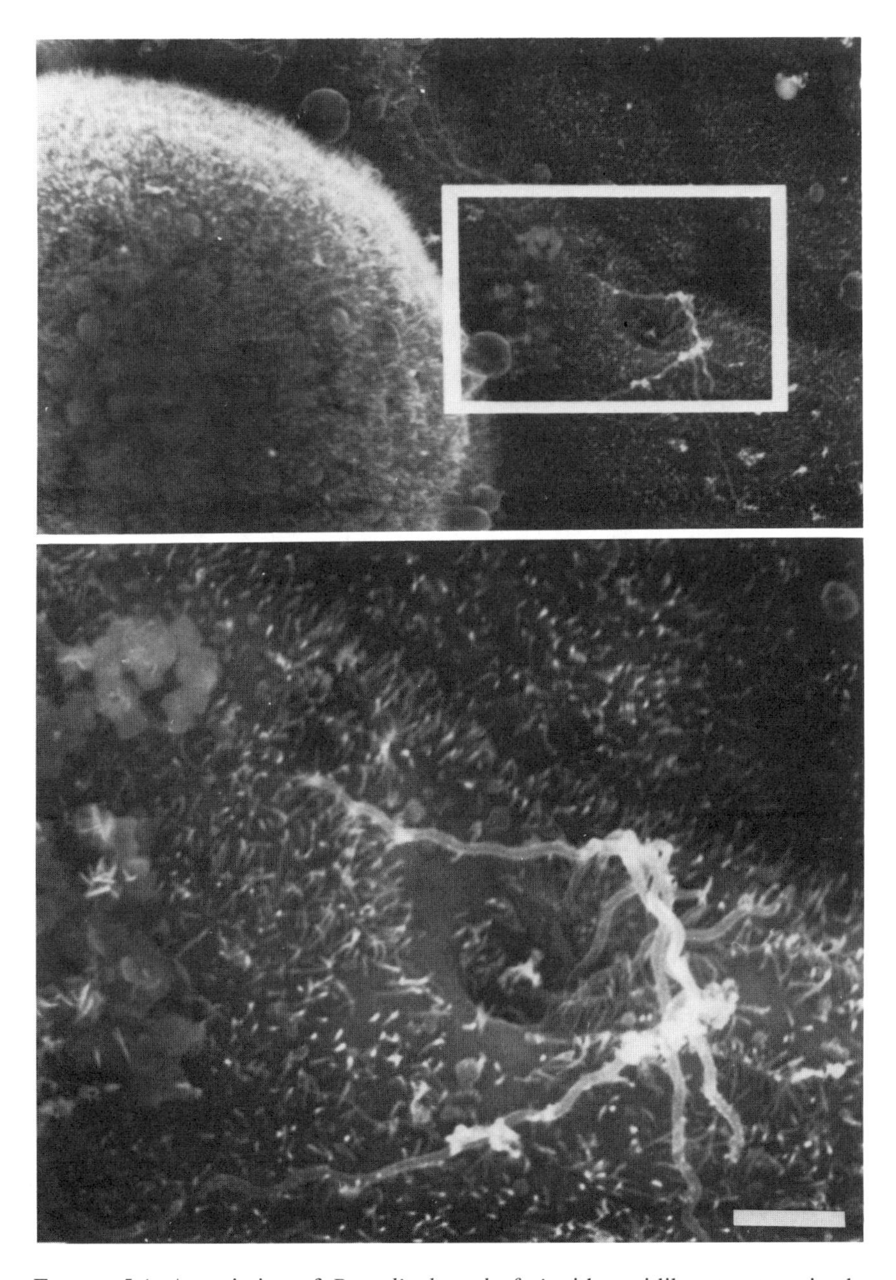

FIGURE 5.1. Association of *Borrelia burgdorferi* with a pitlike structure in the midgut epithelium of an infected *Ixodes dammini* (scanning electron micrograph; bar, 1.0 μ). (Reprinted with permission from *Reviews of Infectious Diseases*.)

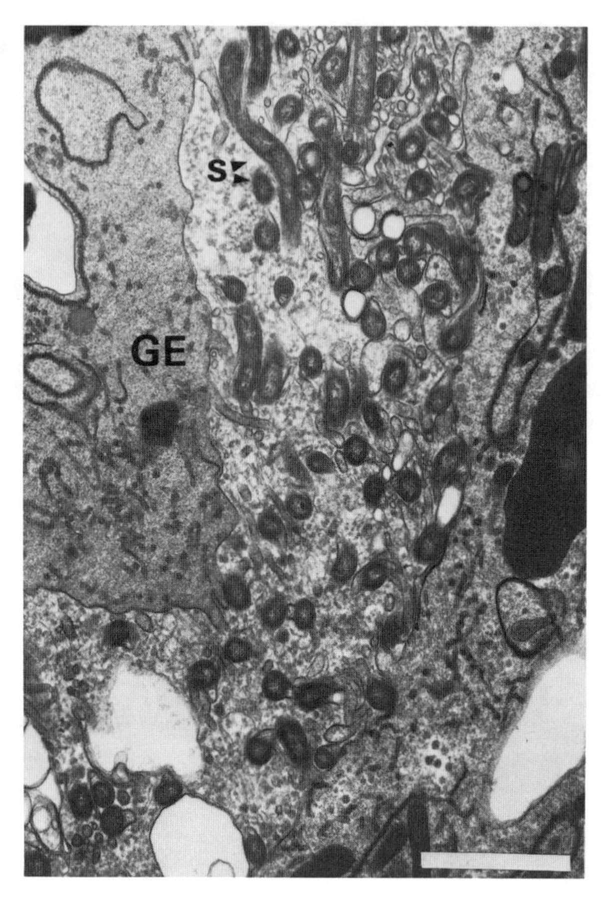

FIGURE 5.2. *Borrelia burgdorferi*-filled pit in the midgut eipthelium of an infected *Ixodes dammini*. s (arrows), typical serpentine profile of spirochete; DE, gut epithelial cell (transmission electron micrograph; bar, 1 μ). (Reprinted with permission from *Reviews of Infectious Diseases*.)

spirochete dissemination takes place during the first few days the tick is attached. By the fifth day, 20 of 35 midgut-infected *I. dammini* had spirochetes in their hemolymph. Based on their findings of sequential histological studies, Benach et al. (6) drew a similar conclusion and suggested that *B. burgdorferi* multiplies in midgut tissues during early feeding (days 1–3) and penetrates the gut wall during midfeeding (days 5–7), to enter the hemocoel. From there, the spirochetes disseminate via hemolymph to other tissues, particularly the central ganglion.

In systemically infected *I. dammini,* the Lyme disease spirochete, although present in the midgut, usually only mildly infects other tissues. It is rarely found in the hemolymph, which, unlike that of the relapsing fever and bovine borreliosis vectors, does not appear to be an optimal medium for its development. Massive spirochetal infections have been found in ovarian tissues of engorged *I. dammini* females that failed to oviposit or that laid small batches of eggs only (14). Electron microscopic evaluation of such tissues showed intensive spirochetal invasion of supporting tissue, as well as of developing oocytes. As illustrated in Figure 5.3, numerous spirochetes may be present between the vitelline and oocyte membranes, where they denude the oogonial surface and interfere with the transport of chitin, which is necessary for the formation of the oocyte shell; eggs so infected fail to mature and prevent ticks from ovipositing.

Nevertheless, demonstration of *B. burgdorferi* in field-collected, unfed larval *I. dammini* suggests that mild spirochetal infections of oocytes do permit the eggs to develop (7, 46). Recently, infection rates of 3.3 to 15.0% were reported in the progeny of five female ticks removed from a deer, and of 25 and 27% in the progeny of two female ticks from a dog (42).

Experimentally infected *I. dammini* were also found to pass spirochetes via eggs, but decreasing infection rates in the larvae and the absence of spirochetes in nymphs suggested a gradual die-off of spirochetes; the number of ovarially transferred spriochetes may have been too low to initiate permanent systemic infections (42).

The passage of *B. burgdorferi* via eggs had been established also for

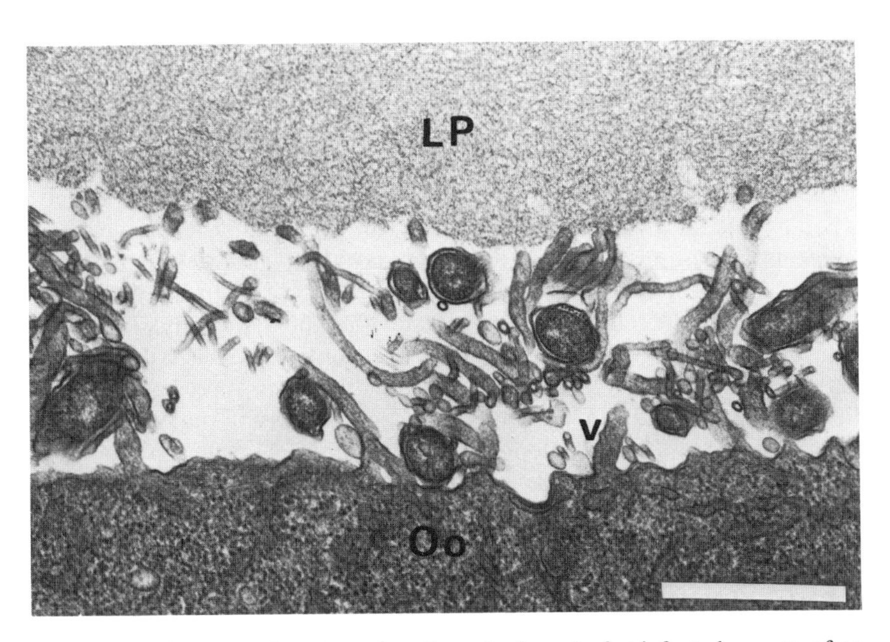

FIGURE 5.3. Cross section through a *Borrelia burgdorferi*-infected oocyte of an *Ixodes dammini* female with numerous spirochetes between the lamina propria (LP) and a oogonial cell (Oo). Note the destruction of the villar processes (v) (transmission electron micrograph; bar, 2.5 μ).

two naturally infected *I. ricinus* females from Switzerland. One female produced 100% infected eggs, the other only 60%; the F_1 larvae and nymphs were not examined (12).

So far, transovarial and subsequent transstadial passage of *B. burgdorferi* has been recorded only in *I. pacificus* (36). One of three naturally infected females produced 100% infected larvae that maintained the spirochetes transstadially; and four or five F_1 females passed them via eggs to as many as 97% of the F_2 progeny.

Of particular interest was the greatly reduced staining reaction of the ovarially passed spirochetes when they were treated with anti-*B. burgdorferi* FITC-labeled polyclonal antibody. Also, none of the spirochetes in the F_1 generation, reacted with a monoclonal antibody (H5332), which supposedly is specific for *B. burgdorferi*. Spirochetes in the F_2 generation, however, reacted.

Reduced immunofluorescence has been noted also for spriochetes in systemically infected *I. dammini* (W. Burgdorfer, unpublished information). Whereas the spirochetes in midgut tissues consistently showed strong fluorescence, those in tissues of the hemocoel often showed weaker fluorescence. This observation has been interpreted as indicating a loss of antigenic properties due to unfavorable physiological conditions for spirochetes outside the midgut. Indeed, perhaps the ability of *B. burgdorferi* to develop in tissues other than the midgut depends on its pathogenicity and interaction with the various tissues and varies from strain to strain and from one species of tick to another.

The mode(s) of transmission of *B. burgdorferi* by its tick vectors has been a controversial issue because fewer than 5% of naturally or experimentally infected. *I. dammini* are systemically infected. Thus, some investigators postulate transmission as occurring via saliva only, whereas others consider regurgitation to be an additional means of transmission. Supporters of the saliva hypothesis point out, as reviewed above, that the Lyme disease spirochete undergoes intensive multiplication during early feeding, then penetrates the gut wall to invade, via the hemolymph, various tissues, including the salivary glands. Ribeiro et al. (50) demonstrated spirochetes in the saliva as early as three days after the tick becomes attached and thereafter in increasing number. Of 20 hemolymph-positive, infected adult *I. dammini*, 11 had spirochetes in their saliva four days after attachment. Similarly, of seven infected nymphal ticks of the same species, three secreted spirochetes in their saliva as early as three days after attachment. In all these instances, spirochetes were found in saliva before rapid engorgement, which suggested that transmission occurs by the salivary gland route during the late-feeding period.

Despite extensive examination, Benach et al. (6) failed to find *B. burgdorferi* in the lumen of salivary glands or in associated tissues. Nevertheless, the presence of spirochetes in tick-feeding cavities in rabbit skin suggested that transmission occurs either by salivation or by regurgitation of midgut contents. In a follow-up study by the same laboratory, Wheeler at al. (58), using monoclonal antibody (11G1), reported detecting spirochetal components in immunoblots of salivary glands of engorging ticks and concluded that *B. burgdorferi* is transmitted by *I. dammini* via saliva during the later stages of feeding.

The proponents of transmission through regurgitation point to recent publications by Hesse (32), who used a radioactive tracer (isotope ^{32}P) to demonstrate regurgitation to be a reflux of midgut contents in the argasid tick *Ornithodoros moubata,* and by Brown (8) who provided evidence that female *Amblyomma americanum* regurgitate gut material during feeding. In addition, the following observations and considerations favor this mode of transmission. In our laboratory, several successful transmission experiments were carried out with naturally or experimentally infected *I. dammini* in which no generalized distribution of spirochetes was observed. Also, if dissemination of *B. burgdorferi* occurs regularly during engorgement, a relatively large proportion of midgut-infected nymphs should develop systemic infections detectable in the adult stage. In Long Island, where Lyme disease is highly endemic, less than 5% of infected questing adult *I. dammini* had a generalized infection. Studies are now in progress to elucidate further how *B. burgdorferi* is transmitted, with an emphasis on saliva or gut material as sources of spirochetes.

The Relationship of *Borrelia burgdorferi* to Nonspecific Tick Vectors and Other Hematophagous Arthropods

Although *I. dammini,* and *I. pacificus* in the United States; *I. ricinus* in Europe; and *I. persulcatus* in European and Asian USSR, China, and Japan are recognized as the principal vectors of *B. burgdorferi,* identical spirochetes have been detected also in the lone star tick *Amblyomma americanum* (51, 52), the dog ticks *Dermacentor variabilis* (2, 3) and *Rhipicephalus sanguineus* (49), the rabbit ticks *Ixodes dentatus* (4) and *Haemaphysalis leporispalustris* (37), the rabbit dermacentor *D. parumapertus* (49), the woodrat tick *Ixodes neotomae* (37), the winter tick *D. albipictus* (3), and the black-legged tick *Ixodes scapularis* (40). So far, the only species claimed as an additional vector to humans is *A. americanum* in the Northeast (New Jersey), where all active stages of this tick are found to harbor spirochetes (51, 52).

Detection of *B. burgdorferi* in *D. variabilis* and *A. americanum* is of interest because attempts to infect these ticks experimentally have failed in at least two laboratories. At the Rocky Mountain Laboratories, we found that larval ticks of both species readily ingested spirochetes through feeding on infected rabbits. Organisms were detectable in up to 80% of freshly molted nymphs but not in nymphs that had fasted for three to six months (W. Burgdorfer, unpublished information). Similarly,

Piesman and Sinsky (47) found that the spirochetes did not survive molting processes to nymphs. These findings suggest that either the minimum dosage requirement, that is, the number of spirochetes necessary to cause persistent infections, was not met, or that these ticks are refractory to certain strains of *B. burgdorferi*. This may also be true for other species of ticks that sometimes harbor spirochetes temporarily after engorging on a spirochetemic host.

Detection of *B. burgdorferi* in *H. leporispalustris, I. neotomae,* and *D. parumapertus* and evidence of past or current spirochetal infections in black-tailed jack rabbits (*Lepus californicus californicus*) in northern California have long suggested that lagomorphs and their ticks are possibly involved in the ecology of the Lyme disease spirochete (37). More recently, spirochetes identified as *B. burgdorferi* were also isolated from 71 of 168 *I. dentatus* taken off naturally infected cottontail rabbits (*Sylvilagus floridanus*) in New York (4).

Very little information is available on the behavior of *B. burgdorferi* in rabbit ticks. All four infected ticks (2 of 174 *H. leporispalustris* and 2 of 10 *I. neotomae*) from black-tailed jack rabbits had midgut infections and at least three of them had systemic infections involving most tissues. One of the two *H. leporispalustris* oviposited eggs, of which 67% contained spirochetes (37). As yet, there is no information concerning the development of the Lyme disease spirochete in *I. dentatus*.

A potential vector in the southern and southeastern United States is the black-legged *I. scapularis,* another member of the *I. ricinus* complex. Infected experimentally, this tick has been shown (13, 47) to maintain and distribute *B. burgdorferi* as efficiently as does *I. dammini*. So far, only two specimens, both engorged females, have been found to be naturally infected. Spriochetes were also detected in two larvae that emerged from eggs deposited by each of these females (40).

The claim by some persons that they contracted Lyme spirochetosis as a result of insect rather than tick bites prompted several investigations of the role of mosquitoes, deerflies, and horseflies as vectors of *B. burgdorferi*. Surveys conducted in southcentral Connecticut from 1984 through 1987 led to the detection of spirochetes in four species of deerflies, in seven species of horseflies, and in four species of mosquitoes (39, 41). The prevalence of the infection varied, although it reached 21% for *Chrysops callidus* and about 7 to 8% for each of the four species of mosquitoes. The number of spirochetes detected in smears of foregut tissues varied from one to five to more than twenty-five. These findings led to experimental studies in which blood-sucking *Aedes canadensis* and *Aedes stimulans* were allowed to feed on Syrian hamsters (39). Although 9 of 71 and 2 of 30 mosquitoes, respectively, had *B. burgdorferi* in their head tissues, none of the hamsters became infected with spirochetes.

Three species of laboratory-reared mosquitoes (*Aedes aegypti, Aedes atropalpus, Aedes triseriatus*) and the horsefly (*Tabanus nigrovittatus*) were also infected experimentally by allowing them to feed (through a lambskin membrane) upon beef blood mixed with BSK medium containing *B. burgdorferi* (43). Of 562 mosquitoes, 134 (23.8%) ingested spirochetes that survived in the digestive system for up to six days and could be demonstrated in preparations of head or midgut tissues for fourteen days. Similarly, of 57 *T. nigrovittatus,* 28 had living spirochetes in head and anterior digestive tract tissue. Two females harbored spirochetes for two to three days.

Even though these observations indicate that these insects may not be suitable hosts for *B. burgdorferi,* mechanical transmission as a result of their intermittent feeding habits cannot be ruled out and should be evaluated further.

Borrelia burgdorferi: Subject of a Complex Development Cycle?

The use of fluorescence, scanning, and transmission electron microscopy in studies of *B. burgdorferi* development in its tick vectors has consistently revealed this organism's ability to form vesicles along its length or at its terminal (Figure 5.4). These vesicles are similar to those described for relapsing fever borreliae (see above) and were at first considered

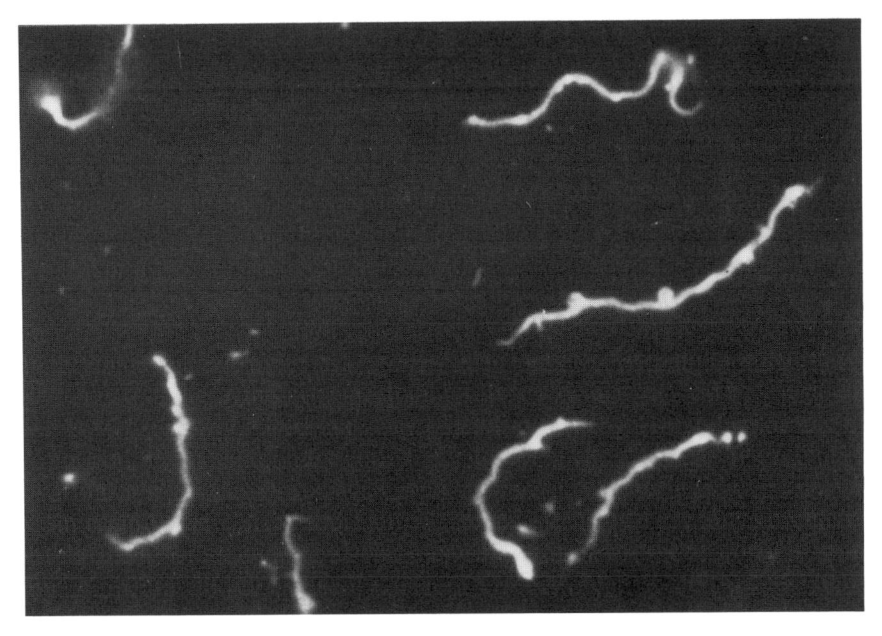

FIGURE 5.4. Fluorescent, antibody-stained *Borrelia burgdorferi* in a midgut smear of an infected *Ixodes dammini*. Note the gemmae (1330 ×).

germinative units, although later they were thought to be degenerative products.

In our experience, two types of vesicles can be identified: (a) "blebs," up to 200 nm in size, which develop as an eversion of the spirochete's outer membrane (Figure 5.5) and (b) "gemmae," up to 2.0 μ in size, with eversions not only of the outer membrane but also of the cytoplasmic cylinder (Figure 5.6). Some blebs do contain material of low electron density. Gemmae, however, usually contain one or more electron-dense chromatin bodies or granules.

Outer membrane blebbing appears to be an active process in *in vitro* and *in vivo* spirochetes. It results in a variety of structures that may be tubular (spaghetti-like) or string- or pearl-like or spherical (Figure 5.7). Gemmae are said to arise as the result of aging and/or of adverse changes

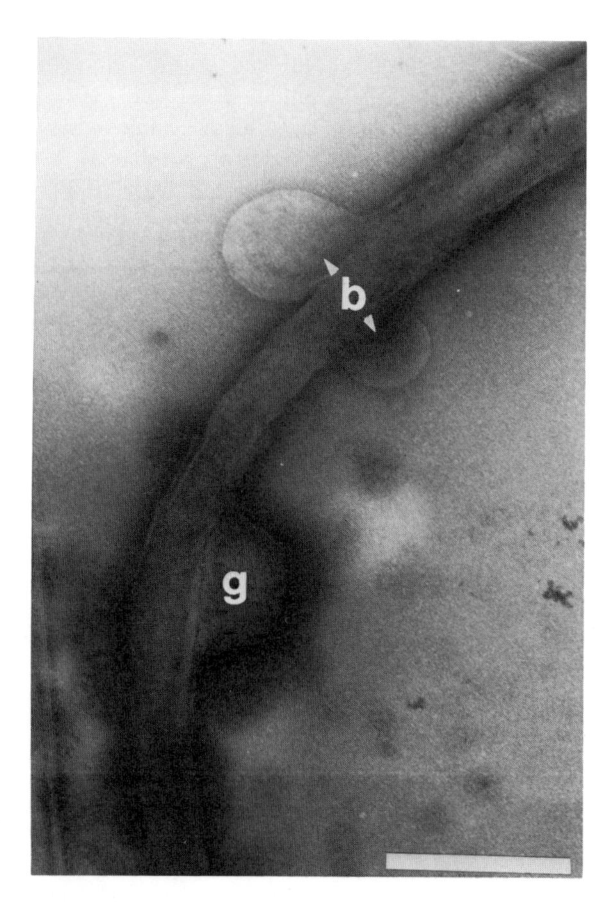

FIGURE 5.5. Negatively stained portion of *Borrelia burgdorferi* with eversion of the outer membrane (b) and the beginning of gemma (g) formation (bar, 0.5 μ).

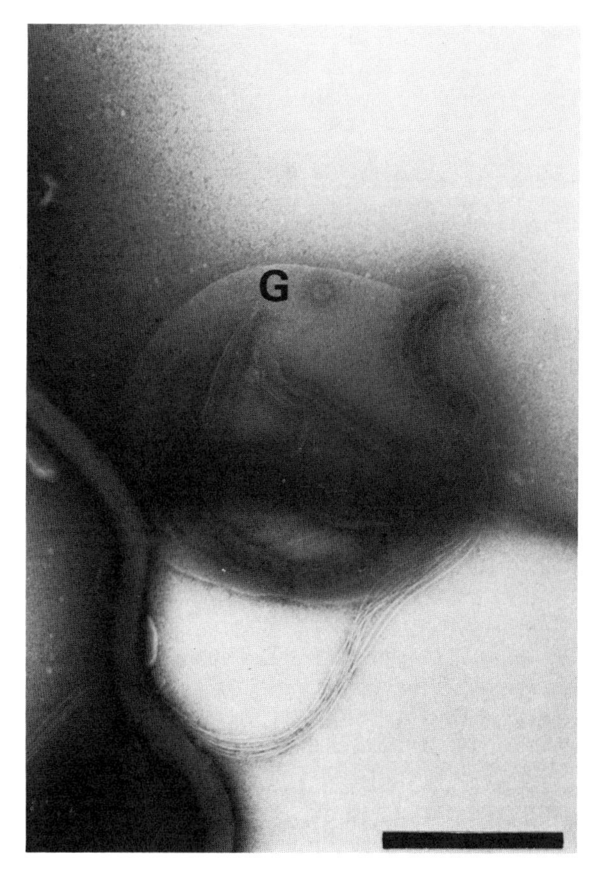

FIGURE 5.6. Negatively stained gemma (g) associated with a spirochete (bar, 1 μ).

in the spirochete's environment, such as changes in pH or depletion of metabolites (48).

In our laboratory, recent molecular investigations into the nature of these vesicles have shown that intact DNA is packed within some of them as well as within the granules of gemmae (27). Purified bleb preparations were found to contain genetic material in the form of linear and circular plasmids, which suggests that these vesicles may be involved in the exchange of genetic information. As yet, we have no evidence that the granules or chromatin bodies within gemmae give rise to a new generation of spirochetes, as has been postulated by some investigators (see above). However, we have been able to demonstrate encystment of spirochetes within elements of the outer membrane (Figure 5.8). As spirochetes begin to form gemmae, a part or, at times, the entire spirochete may coil and fold itself within its own outer membrane. A cross-sectional profile of such an entity would show a spherical unit containing several protoplasmic cylinders, as seen in Figure 5.9.

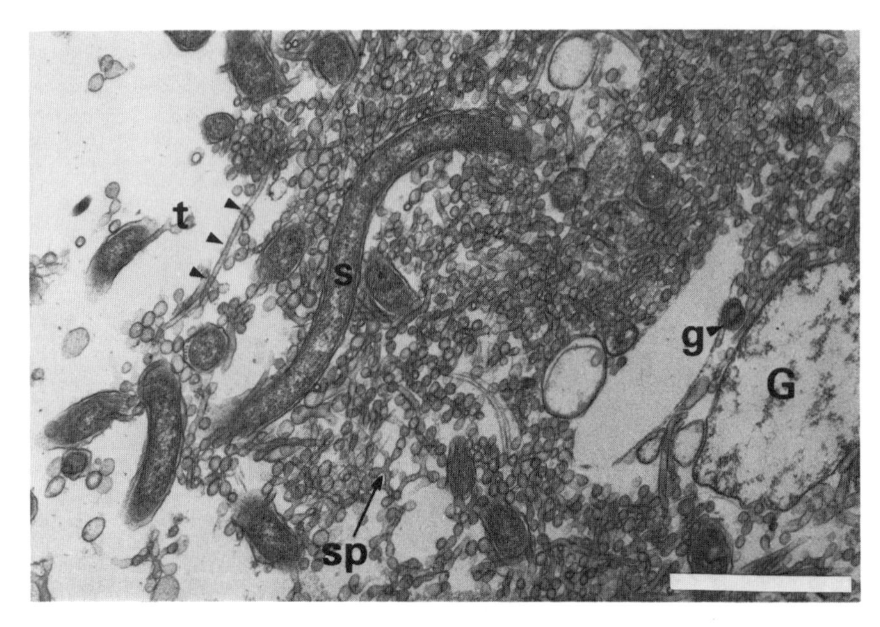

FIGURE 5.7. Extensive growth of *Borrelia burgdorferi* in the ovarial lumen of an infected *Ixodes dammini* female. Note the morphological variations of the spirochetal blebs. s, serpentine configuration of spirochete, t (arrows), tubular blebs; sp, string-of-pearl type blebs; G, empty gemma; g (arrow), granule (transmission electron micrograph; bar, 1.0 μ).

Immunochemical and molecular studies are in progress to shed additional light on the mechanism(s) of spirochetal development, which appears to be far more complex than was generally thought.

Conclusion

After spirochetes were discovered as the causative agents of louse-borne and tick-borne relapsing fevers, intensive studies on the development of these microorganisms in their vectors have shown that the ingested spirochetes leave the digestive tract to enter the hemocoel, where they undergo massive multiplication and then may invade the tissues of various organs. In the body louse, this development involves only the hemolymph and the nervous system's ganglia, whereas in the various tick vectors, it involves, with few exceptions, all organs.

With the discovery of a borrelia in ticks of the genus *Ixodes* as the etiological agent of Lyme borreliosis, a hitherto unknown tick–spirochete relationship was identified. The newly described spirochete *Borrelia burgdorferi* was found to persist only in the midgut in the majority of ticks. Thus, from 4, to 5% of *I. dammini,* for instance, had a

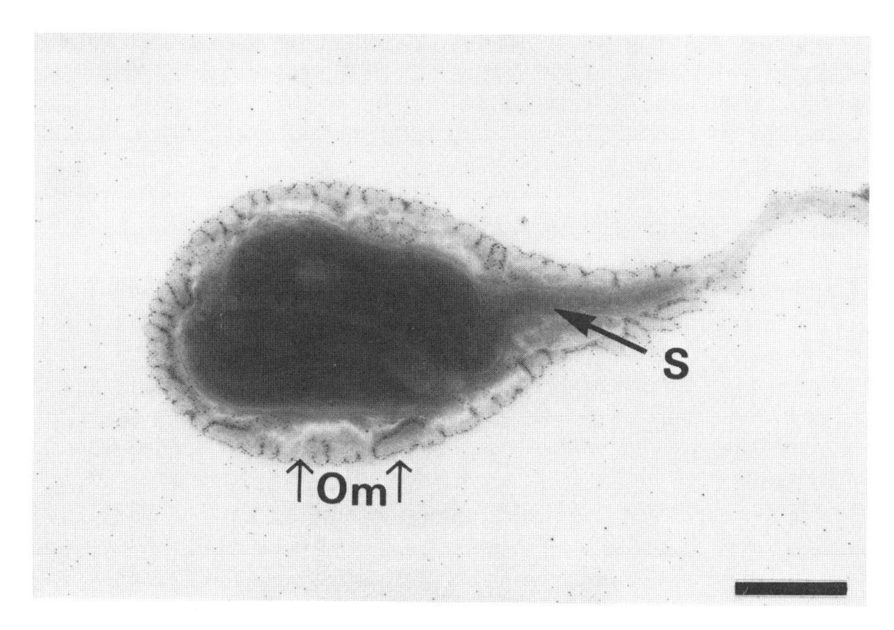

FIGURE 5.8. Negative staining of an encysted *Borrelia burgdorferi*. The spirochete folds and coils within portions of its outer membrane. S, terminus of spirochete; Om, outer membrane margin (bar, 0.2 μ).

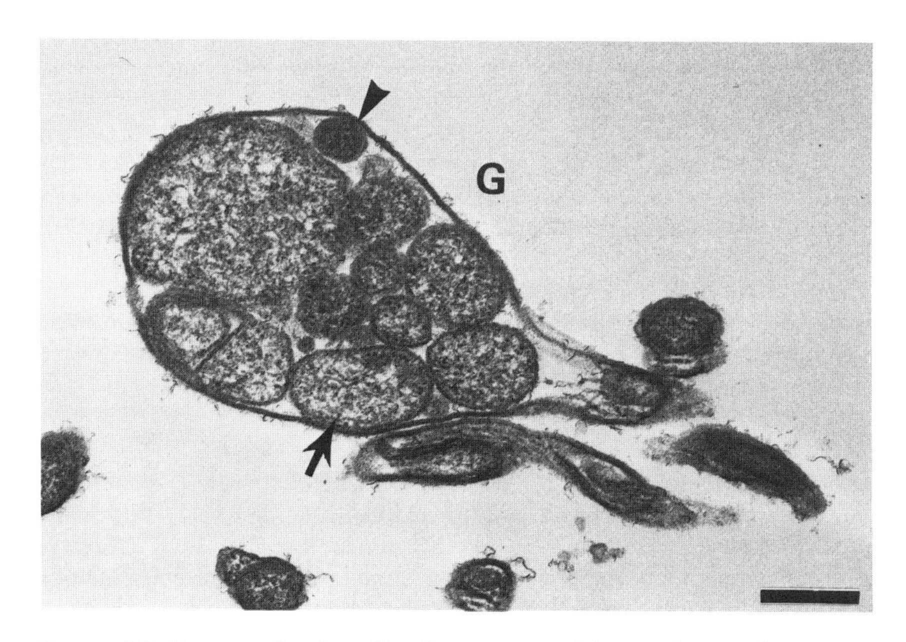

FIGURE 5.9. Cross-sectional profile of gamma containing an electron-dense granule (arrowhead) and several membrane-enclosed cytosolic masses (full arrow) (transmission electron micrograph; bar, 0.2 μ).

systemic infection with few spirochetes in the various tissues. Midgut penetration and subsequent tissue invasion are said to occur during feeding of infected ticks, and it is thought that transmission occurs via spirochete-containing saliva, although transmission via regurgitation has not as yet been ruled out.

Transovarial or vertical transmission, which is common to certain argasid tick–spirochete associations, has also been recorded for the Lyme disease borrelia, but it is rare and appears to be of no ecological significance. In fact, there is evidence that heavily infected oocytes fail to complete oogenesis.

Detection of *B. burgdorferi* in ticks not belonging to the *I. ricinus* complex suggests that other tick species might be involved in the natural history of this spirochete. Thus, there is strong evidence that the rabbit ticks *Haemaphysalis leporispalustris* and *I. dentatus* play a role in maintaining the Lyme disease spirochete among lagomorphs.

Although deerflies, horseflies, and mosquitoes are not considered vectors of *B. burgdorferi,* the presence of spirochetes in the digestive tract of these arthropods indicates they have ingested spirochetemic blood, and suggests the possibility of mechanical transmission.

Continued extensive field investigations into the ecology of *B. burgdorferi* in the United States and other countries will undoubtedly reveal not only additional vector associations of this agent but also new species of arthropod-borne borreliae.

Lastly, the development of spirochetes in their arthropod vectors and vertebrate hosts may indeed be more complex than assumed. The consistent finding of membrane-derived vesicles associated with *B. burgdorferi,* by fluorescence, scanning, and transmission electron microscopy, has rekindled interest in the nature of these vesicles, which in the past had been considered either to be degenerative products or germinative units in a complex developmental cycle. Molecular studies conducted so far have shown that these vesicles, called *blebs* and *gemmae,* contain genetic material in the form of linear and circular plasmids; they may be involved in the storage or the exchange of genetic material or both. The potential role of the chromatin bodies or granules in gemmae is the subject of ongoing research.

Acknowledgments. We thank M. D. Corwin, Laboratory of Pathobiology, Rocky Mountain Laboratories, for providing us with the scanning electron micrograph. We are also indebted to Dr. W. J. Hadlow (retired) for his valuable editorial suggestions, and to B. Kester for her secretarial assistance in preparing the manuscript.

This chapter contains in part material presented at the Roche Laboratories-sponsored Symposium "Lyme Disease and Other Spirochetal Diseases" held in Washington, D. C., February 29-March 1, 1988, and published in *Reviews of Infectious Diseases,* Vol. 11, No. 5, September-October, 1989. Permission for reproduction has been granted by Reviews of Infectious Diseases.

References

1. Aeschlimann, A., 1958, Développement embryonnaire d'*Ornithodorus moubata* (Murray) et transmission transovarienne de *Borrelia duttoni, Acta Trop.* **15:**15–64.
2. Anderson, J.F., Johnson, R.C., Magnarelli, L.A., and Hyde, F.W., 1985, Identification of endemic foci of Lyme disease: Isolation of *Borrelia burgdorferi* from feral rodents and the tick, *Dermacentor variabilis, J. Clin. Microbiol.* **22:**36–38.
3. Anderson, J.F., and Magnarelli, L.A., 1984, Avian and mammalian hosts for spirochete-infected ticks and insects in a Lyme disease focus in Connecticut, *Yale. J. Biol. Med.* **57:**627–641.
4. Anderson, J.F., Magnarelli, L.A., LeFebvre, R.B., Andreadis, T.G., Mcaninch, J.B., Perng G.Ch., and Johnson, R.C., 1989, Antigenically variable *B. burgdorferi* isolated from cottontail rabbits and *Ixodes dentatus* in rural and urban areas, *J. Clin. Microbiol.* **27:**13–20.
5. Balashov, Y.S., 1968, Transovarial transmission of the spirochete *Borrelia sogdiana* in *Ornithodoros papillipes* ticks and its effect on biological properties of the agent, *Parazitologiya* **2:**198–201 (in Russian).
6. Benach, J.L., Coleman, J.L., Skinner, R.A., and Bosler, E.M., 1987, Adult *Ixodes dammini* on rabbits: A hypothesis for the development and transmission of *Borrelia burgdorferi, J. Infect. Dis.* **155:**1300–1306.
7. Bosler, E.M., Coleman, J.L., Benach, J.L., Massey, D.A., Hanrahan, J.P., Burgdorfer, W., and Barbour, A.G., 1983, Natural distribution of the *Ixodes dammini* spirochete, *Science* **220:**321–322.
8. Brown, S.J., 1988, Evidence for regurgitation by *Amblyomma americanum. Vet. Parasitol.* **28:**335–342.
9. Burgdorfer, W., 1951, Analyse des Infektionsverlaufes bei *Ornithodorus moubata* (Murray), unter Berücksichtigung der natürlichen Übertragung von *Spirochaeta duttoni. Acta Trop.* **8:**193–262.
10. Burgdorfer, W., 1976, In Johnson, R.C. (ed): The Biology of Parasitic Spirochetes, New York, Academic Press, pp. 191–200.
11. Burgdorfer, W., Barbour, A.G., Hayes, S.F., Benach, J.J., Grunwaldt, E., and Davis, J.P., 1982, Lyme disease—a tick-borne spirochetosis? *Science* **216:**1317–1319.
12. Burgdorfer, W., Barbour, A.G., Hayes, S.F., Peter, O., and Aeschlimann, A., 1983, Erythema chronicum migrans—a tick-borne spirochetosis. *Acta Trop.* **40:**79–83.
13. Burgdorfer, W., and Gage, K.L., 1986, Susceptibility of the black-legged tick, *Ixodes scapularis,* to the Lyme disease spirochete, *Borrelia burgdorferi, Zbl. Bakt. Hyg.* **263:**15–20.
14. Burgdorfer, W., Hayes, S.F., and Benach, J.L., 1988, Development of *Borrelia burgdorferi* in Ixodid tick vectors, *Ann. NY Acad. Sci.* **539:**172–179.
15. Burgdorfer, W., Lane, R.S., Barbour, A.G., Gresbrink, R.A., and Anderson,

J.R., 1985, The western black-legged tick, *Ixodes pacificus:* A vector of *Borrelia burgdorferi, Am. J. Trop. Med. Hyg.* **34:**925–930.
16. Ciesielski, C.A., Markowitz, L.E., Horsley, R., Hightower, A.W., Russell, H., and Broome, C.V., 1988, The geographic distribution of Lyme disease in the United States, *Ann. NY Acad. Sci.* **539:**283–288.
17. Dekonenko, E.P., 1988, Clinical manifestations of tick-borne erythema in the USSR, *Ann. NY Acad. Sci.* **539:**452.
18. DeLamater, E.D., Newcomer, V.D., Haanes, M., and Wiggall, R.H., 1951, Studies on the life cycle of spirochetes. VIII. Summary and comparison of observations on various organisms, *J. Invest. Dermatol.* **16:**231–256.
19. Diab, F.M., and Soliman, Z.R., 1977, An experimental study of *Borrelia anserina* in four species of *Argas* ticks. 1. Spirochete localization and densities, *Z. Parasitenk.* **53:**201–212.
20. Dutton, J.E., and Todd, J.L., 1905, The nature of tick fever in the eastern part of the Congo Free State, *Br. Med. J.* **2:**1259–1260.
21. Dutton, J.E., and Todd, J.L., 1907, A note on the morphology of *Spirocheta duttoni, Lancet* **ii:**1523–1525.
22. Fantham, H.B., 1911, Some researches on the life-cycle of spirochaetes, *Ann. Trop. Med. Parasitol* **5:**479–496.
23. Feng, L.C., and Chung, H.L., 1936, Studies on the development of *Spirochaeta duttoni* in *Ornithodorus moubata, Chinese Med. J.* **50:**1185–1190.
24. Gaber, M.S., Khalil, G.M., and Hoogstraal, H., 1982, *Borrelia crocidurae:* Venereal transfer in Egyptian *Ornithodoros erraticus* ticks, *Exp. Parasitol.* **54:**182–184.
25. Gaber, M.S., Khalil, G.M., Hoogstraal, H., and Aboul-Nasr, A.E., 1984, *Borrelia crocidurae* localization and transmission in *Ornithodoros erraticus* and *O. savignyi, Parasitology* **88:**403–413.
26. Garnham, P.C.C., Davies, C.W., and Heisch, R.B., 1947, An epidemic of louse-borne relapsing fever in Kenya, *Trans. R. Soc. Trop. Med. Hyg.* **41:**141–170.
27. Garon, C.F., Dorward, D.W., and Corwin, M.D., 1989, Structural features of *Borrelia burgdorferi*—the Lyme disease spirochete: Silver staining for nucleic acids, *Scanning Microscopy* (in press)
28. Geigy, R., and Aeschlimann, A., 1964, Langfristige Beobachtugen über transovarielle Übertragung von *Borrelia duttoni* durch *Ornithodorus moubata, Acta Trop.* **21:**1–4.
29. Hampp, E.G., 1950, Morphologic characteristics of smaller oral treponemes and *Borrelia vincenti* as revealed by stained smear, darkfield and electron microscopic technics, *J. Am. Dent. Assoc.* **40:**1–11.
30. Hatt, P., 1929, Observations sur l'évolution des spirochètes des fièvres récurrentes chez les Ornithodores, *Arch. Inst. Pasteur Tunis.* **18,** 258–264.
31. Heisch, R.B., and Harvey, A.E.C., 1962, The development of *Spirochaeta duttoni* and *S. recurrentis* in *Pediculus humanus, Parasitology* **52:**77–88.
32. Hesse, G., 1983, Evidence of regurgitation bei *Ornithodoros moubata* (Ixodoidea: Argasidae) using radioactive tracer, *Zentralbl. Bakteriol. Mikrobiol. Hyg.* **256:**267–268.
33. Hindle, E., 1911, On the life-cycle of Spirochaeta gallinarum, *Parasitology* **4:**463–477.
34. Kleine F.K., and Krause, M., 1983, Zur Kritik angeblicher Entwicklungsfor-

men von Rückfallfieberspirochaeten in der Zecke (*Ornithodorus moubata*), *Arch. Schiffs. Trop. Hyg.* **36:**190–191.
35. Kleine, F.K., and Eckard, B., 1913, Über die Lokalisation der Spirochaeten in der Rückfallfieberzecke (*O. moubata*), *Zschr. Hyg. Infektionskrankh.* **74:**389–394.
36. Lane, R.S., and Burgdorfer, W., 1987, Transovarial and transstadial passage of *Borrelia burgdorferi* in the western black-legged tick, *Ixodes pacificus, Am. J. Trop. Med. Hyg.* **37:** 188–192.
37. Lane, R.S., and Burgdorfer, W., 1988, Spirochetes in mammals and ticks (Acari: Ixodidae) from a focus of Lyme borreliosis in California, *J. Wildl. Dis.* **24:**1–9.
38. Leishman, W., 1907, Spirochaetae of relapsing fever and tick fever, *Lancet* **ii:**806.
39. Magnarelli, L.A., and Anderson, J.F., 1988, Ticks and biting insects infected with the etiologic agent of Lyme disease, *Borrelia burgdorferi, J. Clin. Microbiol.* **26:**1482–1486.
40. Magnarelli, L.A., Anderson, J. F., Apperson, C.S., Fish, D., Johnson, R.C., and Chappell, W.A., 1986, Spirochetes in ticks and antibodies to *Borrelia burgdorferi* in white-tailed deer from Connecticut, New York State, and North Carolina, *J. Wildl. Dis.* **22:**178–188.
41. Magnarelli, L.A., Anderson, J.F., and Barbour, A.G., 1986, The etiological agent of Lyme disease in deerflies, horseflies, and mosquitoes, *J. Infect. Dis.* **154:**355–358.
42. Magnarelli, L.A., Anderson, J.F., and Fish, D., 1987, Transovarial transmission of *Borrelia burgdorferi* in *Ixodes dammini* (Acari: Ixodidae), *J. Infect. Dis.* **156:**234–236.
43. Magnarelli, L.A., Freier, J.E., and Anderson, J.F., 1987, Experimental infections of mosquitoes with *Borrelia burgdorferi,* the etiologic agent of Lyme disease, *J. Infect. Dis.* **156:**694–695.
44. Nicolle, C., Anderson, Ch., and Colas-Belcour, J., 1930, Recherches expérimentales poursuivies à l'Institut Pasteur de Tunis sur les conditions de la transmission des spirochètes récurrentes par les Ornithodores, *Arch. Inst. Pasteur Tunis.* **19:**133–227.
45. Nicolle, C., Blaizot, L., and Conseil, E., 1912, Conditions de transmission de la fièvre récurrente par le pou, *C. R. Acad. Sci. (Paris)* **155:**481–484.
46. Piesman, J., Donahue, J.G., Mather, T.N., and Spielman, A., 1986, Transovarially acquired Lyme disease spirochetes (*Borrelia burgdorferi*) in field-collected larval *Ixodes dammini* (Acari: Ixodidae), *J. Med. Entomol.* **23:**219–219.
47. Piesman, J., and Sinsky, R.J., 1988, Ability of *Ixodes scapularis, Dermacentor variabilis* and *Amblyomma americanum* (Acari: Ixodidae) to acquire, maintain, and transmit Lyme disease spirochetes (*Borrelia burgdorferi*), *J. Med. Entomol.* **25:**336–339.
48. Pillot, J., Dupouey, P., and Ryter, A., 1964, La significance des formes atypiques et la notion de cycle évolutif chez les spirochètes, *Ann. Inst. Pasteur* **107:**484–502.
49. Rawlings, J.A., 1986, Lyme disease in Texas, *Zentralbl. Bakteriol. Mikrobiol. Hyg.* **263:**483–487.
50. Ribeiro, J.M.C., Mather, T.N., Piesman, J., and Spielman, A., 1987, Dissemi-

nation and salivary delivery of Lyme disease spirochetes in vector ticks (Acari: Ixodidae), *J. Med. Entomol.* **24:**201–205.
51. Schulze, T.L., Bowlen, G.S., Bosler, E.M., Laket, M.F., Parkin, W.E., Altman, R., Ormiston, B.G., and Shisler, J.K., 1984, *Amblyomma americanum:* A potential vector of Lyme disease in New Jersey, *Science* **224:**601–603.
52. Schulze, T.L., Lakat, M.F., Parkin, W.E., Shisler, J.K., Charette, D.J., and Bosler, E.M., 1986, Comparison of rates of infection by the Lyme disease spirochete in selected populations of *Ixodes dammini* and *Amblyomma americanum* (Acari: Ixodidae), *Zentralbl. Bakteriol. Mikrobiol. Hyg.* **263:**72–78.
53. Sergent, E., and Foley, H., 1910, Recherches sur la fièvre récurrente et son mode de transmission, dans une épidémie algérienne, *Ann. Inst. Pasteur* **24:**337–375.
54. Smith, R.D., Brener, J., Osorno, M., and Ristic, M., 1978, Pathobiology of *Borrelia theileri* in the tropical cattle tick *Boophilus microplus, J. Invertebr. Pathol.* **32:**182–190.
55. Stanek, G., Pletschette, M., Flamm, H., Hirschl, A.M., Aberer, E., Kristoferitsch, W., and Schmutzhard, E., 1988, European Lyme borrelioses, *Ann. NY Acad. Sci.* **539:**274–282.
56. Van Veen Schillhorn, T.W., and Leyendekkers, G.J., 1971, *Borrelia theileri* (Laveran 1903) in cattle in the Netherlands, *Tijdschr. Diergeneesk.* **96:**1028–1031.
57. Wagner-Jevseenko, O., 1958, Fortpflanzung bei *Ornithodorus moubata* und genitale Übertragung von *Borrelia duttoni, Acta Trop.* **15:**118–168.
58. Wheeler, Ch.M., Coleman, J.L., Habicht, G.S., and Benach, J.L., 1989, Adult *Ixodes dammini* on rabbits: Development of acute inflammation in the skin and immune responses to salivary gland, midgut, and spirochetal components, *J. Infect. Dis.* **159:**265–273.
59. Wittrock, O., 1913, Beitrag zur Biologie der Spirochaeta des Rückfallfiebers, *Zschr. Hyg.* **74:**55–60.
60. Zaher, M.A., Soliman, Z.R., and Diab, F.M., 1977, An experimental study of *Borrelia anserina* in four species of *Argas* ticks. 2. Transstadial survival and transovarial transmission, *Z. Parasitenk.* **53:**213–223.

6 Distribution of Viruses and Their Nematode Vectors

Giovanni P. Martelli and Charles E. Taylor

Introduction

There are several landmarks on the pathway of our expanding knowledge of nematode transmissiom of plant viruses. The initial discovery of *Xiphinema index* as vector of grapevine fanleaf virus (GFLV) (51) stimulated the search for nematode vectors of other soil-borne viruses, and this was accompanied by research on many aspects of the biology, ecology, and taxonomy of both nematodes and viruses. Early investigations established that plant viruses specifically associate with their nematode vectors, and the mechanism of this association began to emerge when it was discovered that the virus coat protein was a key factor in the adsorption of particles at virus retention sites within the nematodes. The importance of wild hosts for both viruses and vectors, the perennation of viruses in weed seeds, and the insight into the feeding behavior of vector nematodes improved our understanding of how viruses survive and spread in nature, and a basis for their control in commercial crops.

In recent years, improved technology has provided detailed information on the characteristics of viruses of the Nepovirus and Tobravirus groups. Members of both groups have a bipartite genome made up of two functional and separately encapsidated RNA species, which may recombine under both experimental and natural conditions to give rise to pseudo-recombinant strains (48, 108).

Now that the physicochemical composition of nepoviruses and their hydrodynamic and serological properties are known, subgroups that are broadly consistent with the geographical distribution and presumed origin of the different viruses have been established. Interest in the taxonomy of the virus vectors longidorids and trichodorids continues, as the number of

---

Giovanni P. Martelli, Dipartimento di Patologia vegetale, University of Bari and Centro di Studio del CNR sui Virus e le Virosi delle Colture Mediterranee, 70126 Bari, Italy.
Charles E. Taylor, Honorary Research Associate, Scottish Crops Research Institute, Invergowrie, Dundee, DD2 5DA Scotland.

species known increases and the role of nematodes as vectors is better understood (63, 139).

There have been numerous reviews on many aspects of nematode transmission of viruses (47, 63, 68, 69, 83, 86, 122, 124, 125, 127, 128, 132, 133). In this most recent review, here, we consider the geographical distribution of the nepoviruses and tobraviruses as an approach to understanding the ecological and biological association between these viruses and their vectors.

Nematode-Transmitted Viruses as Plant Pathogens

Diseases Induced by Nepoviruses

Nepoviruses are reported to infect wild plants, annual crops, and perennial crops. The natural host range varies greatly with individual viruses, as does the severity of the diseases they induce. Some nepoviruses are pathogens of primary economic importance, since they affect and damage major crops. Other nepoviruses are restricted to a single or a few hosts; thus, they are, for the most part, only of scientific interest. Grasses and cultivated cereals do not appear to be hosts to nepoviruses, and the only gymnosperm host so far reported is *Cycas revoluta* in Japan (39, 61).

The diseases caused by nepoviruses have been reviewed repeatedly, and most recently by Murant et al. (89) and Stace-Smith and Ramsdell (122). Thus, they will only be summarized here, with a few selected examples in the following plant categories.

Vegetables

The artichoke (*Cynara scolymus*) seems to be one of the vegatables most frequently attacked by nepovirus. Three different viruses, viz., artichoke Italian latent (AILV), artichoke yellow ringspot (AYRV), and artichoke vein banding (AVBV) viruses, are named after this host (36, 73, 95). Two additional members of the group, strains of raspberry ringspot (RRV) and of tomato black ring (TBRV) viruses, have been recovered from artichoke plants in the eastern Mediterranean area (96) and in France (80), respectively.

Depending upon the cultivar and, perhaps, growing conditions, nepovirus-infected artichokes may either be symptomless (AILV, RRV) or exhibit symptoms ranging from mild chlorotic discolorations (AVBV, RRV, TBRV) to generalized yellowing and stunting (AILV), scattered yellow blotches (RRV), and intense chrome yellow rings and line patterns accompanied by necrosis and stunting (AYRV). The yield is variously affected, but no estimates of crop loss have been made (94).

Potato (*Solanum tuberosum*) has its own nepoviruses in Peru, such as potato black ringspot (PBRV), potato virus U (PVU), and arracacha virus B (AVB) (55, 56, 112), of which only PBRV induces a field syndrome characterized by necrotic spotting or generalized yellowing of the leaves (112). In Europe, potato is affected by TBRV, the cause of "bouquet" and "pseudo-aucuba" diseases, the symptoms of which are necrotic spots, rings and malformations of the leaves, and stunting. Crop losses of up to 30% have been estimated in secondarily infected plants (5).

Cucurbits (melon, watermelon, squash, cucumber) are seriously affected by tobacco ringspot virus (TobRSV) in the United States. Infected plants have mottled and malformed leaves, ringspotting, stunting and poor fruit set (122). In Europe, arabis mosaic virus (ArMV) induces mottling, ringspotting and stunting of field-grown cucumbers (84).

Five different nepoviruses, ArMV, chicory yellow mottle (CYMV), grapevine chrome mosaic (GCMV), strawberry latent ringspot (SLRV), and TBRV have been associated in Europe with diseases of celery (*Apium graveolens*), in which they cause bright yellow mottling (GCMV, CYMV) or chlorotic mottling, distortion and crinkling of the leaves, stunting, and necrosis (ArMV, SLRV, TBRV) (72, 84, 85, 93).

Interestingly, of the three nepoviruses named after tomato, that is, TBRV, tomato ringspot (TomRSV), and tomato top necrosis (TTNV) viruses, none seems to be of economic importance to this crop.

Small Fruits

The consensus is that grapevine (*Vitis vinifera*), raspberry (*Rubus idaeus*), and strawberry (*Fragaria* × *ananassa*) are affected by nepoviruses more than any other small fruit species.

No less than twelve different nepoviruses have been found associated with two major grapevine disorders known as "grapevine degeneration" and "grapevine decline." These viruses are ArMV, AILV, blueberry leaf mottle (BBLMV), grapevine Bulgarian latent (GBLV), GCMV, GFLV, peach rosette mosaic (PRMV), RRV, SLRV, TobRSV, TBRV, and TomRSV.

Grapevine degeneration is typically caused by GFLV in the Mediterranean basin and all other viticultural areas, except for Central Europe and the Balkans where several other nepoviruses (ArMV, GCMV, RRV, SLRV, and TBRV), either alone or in association with GFLV, are able to induce a comparable disease. Grapevine degeneration consists of three distinct syndromes, characterized either by deformations of the leaves, shoots, and canes; chlorotic mottling; reduced vigor and poor fruit setting (fanleaf); or by bright yellow discolorations of the foliage (yellow mosaic) or chrome yellow flecks along the main veins, which sometimes spread into the interveinal tissues (vein banding) (10). The crop may be drastically affected, with average losses up to or above 60% (110).

Grapevine decline typically occurs in *V. vinifera* and *Vitis labrusca* grown in the northern United States and Canada. It shows leaf and cane symptoms comparable to those of fanleaf, but affected vines die more frequently, especially when they are European cultivars. Tomato ringspot virus is the main cause of grapevine decline, together with three additional American nepoviruses [BBLMV, PRMV, and TobRSV (71)].

Raspberry has been reported to be a host to seven different nepoviruses: ArMV, cherry leafroll (CLRV), cherry rasp-leaf (CRLV), RRV, SLRV, TBRV, and TomRSV, all of which are pathogenic, except for CRLV whose infections are latent (122).

The field syndromes induced in raspberry by nepoviruses vary with the agent, or association of agents, and the cultivar. There are cultivars that are resistant or immune to individual viruses (ArMV, RRV, SLRV, TBRV) or some of their isolates (ArMV). Susceptible cultivars react to viral infections with a variety of foliage changes (chlorotic mottling, vein yellowing, yellow speckling, yellow or chlorotic ringspotting, curling), reduced vigor, stunting, and reduction and deformation of fruit (89).

Although, compared with raspberry, fewer nepoviruses (ArMV, RRV, SLRV, TBRV, and TomRSV) have been found to infect strawberry plants, their effects on this crop are equally destructive. Except for TomRSV, which rarely causes a natural infection in strawberry (26), all other viruses are major pathogens, especially in Great Britain and Central Europe, where mixed infections (e.g., ArMV and SLRV; RRV and TBRV) are common.

Symptoms consist of chlorotic spots, rings, and/or yellow blotches of the leaves, which may also be twisted, cupped, or crinkled. With mixed infections, the symptoms are usually more severe; the plants are stunted and often die (89).

Fruit Trees

Nepovirus-induced diseases of pome fruit trees are rare, apple (*Malus sylvestris*) being the only species known to be affected. Two disorders of apple have been described, both in North America: "flat apple" and "union necrosis and decline," which are induced by CLRV and TomRSV, respectively (123).

Flat apple derives its name from the flattened appearance of the fruit borne by diseased plants. The affected plants become progressively weaker, stunted, and densely bushy.

In apple decline, the infected trees develop a necrosis of the woody cylinder at the graft union, possibly due to the hypersensitive reaction of the scion to the virus, which leads to a progressive decline.

Several nepoviruses cause diseases of economic importance in stone fruits.

Cherry (*Prunus avium*) is reported to be susceptible to six different nepoviruses two of which (CLRV and CRLV) are named after it. A typical, nepovirus-induced disease of cherry is rasp-leaf, which is characterized by enations on the underside of the leaves. Rasp-leaf syndromes are reported from western North America and Europe, but the causal agents differ. Rasp-leaf is caused by a single virus (CRLV) in North America (121), whereas in Europe it originates from mixed infections of nepoviruses (ArMV or RRV) and viruses of the prunus necrotic ringspot type (29).

Stem pitting and decline is another serious disease of cherry caused by TomRSV. Affected trees have reduced vigor and yield, with extensive pitting of the vascular cylinder. A similar disease, induced by the same agent (TomRSV), affects peach (*Prunus persica*) in North America. With this disease, known as yellow bud mosaic, stem pitting is accompanied by different patterns of yellow discoloration and severe distortion of the leaves (119).

Other economically important diseases of peach caused by nepoviruses are peach rosette mosaic, caused by PRMV in North America (58), and peach willow leaf rosette, caused by SLRV in Europe (28). In both disorders, there are various degrees of mottling and distortion of the leaves and a progressive decline of the tree.

A high incidence of CLRV infections has been found in English walnut (*Juglans regia*) in Europe and North America. The effect of the virus differs dramatically, depending on whether the host is on its own roots or is grafted to rootstocks of species other than *J. regia,* or their hybrids (e.g., Paradox = *J. regia* × *Juglans hindisii*). Self-rooted English walnuts are tolerant of infection; most plants are symptomless or, at most, show chlorotic ringspots and line patterns, or an occasional bright yellow blotching (117). Grafted walnuts, however, go into a severe decline, a condition known as "black line," which, in several areas, constitutes a limiting factor to walnut production. This disease depends upon CLRV hypersensitivity of walnut rootstocks, the tissues of which, when invaded by the virus, necrotize to give rise to a black line of dead cells at the graft union (89).

Finally, the olive (*Olea europaea*) should be briefly mentioned as a natural host of nepoviruses. A long debate, started in 1938 on whether olive was affected by virus diseases, ended some 10 years ago with the detection of virus particles in developing pollen grains (91). So far, seven different viruses have been recovered, mostly from symptomless olive trees, by sap inoculation (70). Of these viruses, four are nepoviruses, viz., olive latent ringspot virus (OLRV), ArMV, CLRV, and SLRV. None of them, however, causes a specific disease, except for SLRV, which has been found to be associated with striking malformations of leaves and fruits of cv. Ascolana tenera (67).

Diseases Induced by Tobraviruses

Among the tobraviruses, tobacco rattle virus (TRV) has the widest natural host range. This range includes herbaceous weeds and wild, woody perennials, as well as annual and perennial crops (44–46).

In naturally infected plants, TRV tends to remain localized in the roots—the initial site of infection. In certain hosts, however, the virus moves to the above-ground parts, as in the case of pepper (*Capsicum annuum*), in which TRV induces bright yellow ring and line patterns in the leaves and yellow blotching, puckering, and malformation of the fruit (25).

Limited TRV systemic infection also occurs in potato plants originating from tubers affected by spraing disease—a severe disorder characterized by areas of corky tissues in the tuber flesh and mottled foliage (21).

Pea early-browning virus (PEBV) has a natural host range restricted to Leguminosae. Pea (*Pisum sativum*), French bean (*Phaseolus vulgaris*), broad bean (*Vicia faba*), and alfalfa (*Medicago sativa*) are the only plants from which the virus has been recovered. The infection is usually systemic and the symptoms shown by the foliage range from mild mottling and deformation to yellow chevrons and bands to extended necrosis (46).

Pepper ringspot virus (PRV) has been reported only from Brazil, where it systemically invades crops like tomato, pepper, and artichoke, to produce various patterns of bright yellow rings, lines, and bands on the foliage (46).

Causal Agents

Nepoviruses

The nepovirus group is one of the most rapidly expanding taxonomic groups of plant viruses. Its initial membership of eight, inclusive of definitive and possible members (47), had already grown to twenty-six in 1982 (79), and currently numbers thirty-four (Table 6.1). Of these, only eleven have a recognized nematode vector (see also Table 6.4). The rest owe their present taxonomic assignment to the possession of specific biological characteristics (i.e., host range responses, transmission through seeds) and physicochemical and other properties such as the type of intracellular behavior that conform those typical of the group.

Although all nepoviruses have isometric particles ~30 nm in diameter and a bipartite genome with two functional RNA species (42), wide differences exist in the physicochemical and hydrodynamic properties of individual members. Their serological properties, geographical distribution, vectors, and means of natural spread also differ. Such differences may be used to subdivide the group into smaller coherent clusters.

Grouping on a Molecular Basis

Nepoviruses differ with respect to their physicochemical properties and hydrodynamic behavior. Normally, these viruses contain three types of particles, corresponding to centrifugal component T, empty protein shells; M, nucleoproteins containing one molecule of the smaller genomic RNA (RNA-2); and B, nucleoproteins containing one molecule of the larger RNA (RNA-1). However, a few members (six in all), contain a different type of particle that encapsidates two molecules of RNA-2. This determines whether the B component yields one (homogeneity) or two (heterogeneity) buoyant density classes when centrifuged at equilibrium.

Nepoviruses with a homogeneous B component also exhibit clear-cut differences in the molecular weight of RNA-2, which influences the sedimentation behavior and, hence, the sedimentation coefficient of the M particles.

Finally, the protein coat of eight members of the group, rather than having a single polypeptide with a molecular weight of ~55,000 daltons, is made up of two or three smaller polypeptides (e.g., AVBV) with a different molecular weight (21,000–29,000 daltons and 42,000 to 47,000 daltons, respectively).

Taking these differences in physical properties into account, Martelli and coworkers (68, 74) divided the Nepovirus group into four distinct clusters; Murant and Taylor (87) however, divided the group into three clusters. These subdivisions were questioned by Francki et al. (33), since the properties of some nepoviruses are incompletely known and since many of the published values of particle sedimentation and RNA molecular weight may not be correct. They (33) proposed instead two subgroups, based on whether RNA-1 and RNA-2 differed significantly in size.

These two ways of subgrouping both have their merits. The scheme shown in Table 6.1 delineates the complexity of the group as a whole; it separates definitive from tentative members, except for TTNV, which is not fully characterized; and it is consistent with serological clustering, in that related viruses fall into the same subgroup. In no case is there serological cross-reactivity between members of different clusters.

The presence of two kinds of coat proteins seems to constitute the major single criterion whereby some nepoviruses are still regarded as tentative members of the group. In fact, except for lucerne Australian symptomless virus (LASV) and rubus Chinese seed-borne virus (RCSV), in which nucleoproteins apparenly sediment as a single component (4, 99), all other tentative nepoviruses, including SLRV (35), have two nucleoprotein centrifugal components, as do the definitive members of the group.

The importance of the difference in protein coat composition in the separation of definitive and tentative nepoviruses has been questioned by Francki et al. (33). They pointed out that the smaller polypeptides

TABLE 6.1. Grouping of nepoviruses according to their physicochemical properties.

Protein coat with one polypeptide			Protein coat with two polypeptides
B component homogeneous (one molecule of RNA-1)		B component heterogeneous (one molecule of RNA-1 or two molecules of RNA-2)	Arracacha virus B (AVB) Artichoke vein banding virus (AVBV) Cherry raspleaf virus (CRLV) Lucerne Australian symtomless virus (LASV) Lucerne Australian latent virus (LALV) Rubus Chinese seed-borne virus (RCSBV) Satsuma dwarf virus (SDV) Strawberry latent ringspot virus (SLRV)
M component with $S_{20,w}$ up to 100;MW of RNA-2, $\sim 1.5 \times 10^6$	M component with $S_{20,w}$ above 110; MW of RNA-2, above 2×10^6	Arabis mosaic virus (ArMV) Grapevine fanleaf virus (GFLV) Olive latent ringspot virus (OLRV) Potato black ringspot virus (PBRV) Raspberry ringspot virus (RRV) Tobacco ringspot virus (TobRSV)	

Arracacha virus A (AVA)	Artichoke yellow ringspot virus (AYRV)
Artichoke Italian latent virus (AILV)	Blueberry leaf mottle virus (BBLMV)
Cocoa necrosis virus (CNV)	Cassava green mottle virus (CGMV)
Crimson clover latent virus (CCLV)	Cherry leafroll virus (CLRV)
Cycas necrotic stunt virus (CNSV)	Chicory yellow mottle virus (CYMV)
Grapevine chrome mosaic virus (GCMV)	Grapevine Bulgarian latent virus (GBLV)
Mulberry ringspot virus (MRV)	Hibiscus latent ringspot virus (HLRV)
Tomato black ring virus (TBRV)	Lucerne Australian latent virus (LALV)
Tomato top necrosis virus (TTNV)	Myrobalan latent ringspot virus (MyLRV)
	Peach rosette mosaic virus (PRMV)
	Potato virus U (PVU)
	Tomato ringspot virus (TomRSV)

detected in protein coat preparations of viruses like satsuma dwarf (SDV) may just be dimers and trimers of the true coat protein subunit, which is estimated to have a molecular weight of 14,500 daltons, that is, a value comparable to that calculated for the smallest polypeptides observed in dissociated virus protein preparations of TomRSV, TobRSV, and OLRV. This polypeptide is thought to be the basic unit of the tetrameric coat protein (55,000–60,000 daltons) typical of the group (23, 24, 115).

Different views are held by other workers (99), who consider some of the tentative nepoviruses with more than one coat polypeptide to be sufficiently distinct from the rest to warrant classification either as a true taxonomic subgroup of nepoviruses or as a new group with SLRV as the type member; they propose the name Slaterivirus for this new group.

This controversy results from the fact that it is not known whether the two polypeptides originate from (a) a *Comovirus*-like translational strategy of RNA-2, whereby a single large polyprotein precursor, produced in vivo, cleaves by internal proteolysis to form two smaller capsid proteins (for a review, see ref. 38) or (b) the originally single, large subunit, simply cleaves during chemical dissociation of the protein coat for electrophoretic analysis.

Serological Grouping

The taxonomy of nepoviruses, that is, the establishment of individual "species" within the group, is largely based on serology, which, as indicated in the preceeding section, is in turn linked with their physicochemical properties.

Most of the nepoviruses (22 out of 34 definitive and possible members) are serologically distinct and are apparently not related to any other member of the group. Their identification as separate entities is, therefore, unambiguous.

Serological stability seems to be highest with viruses infecting a single host or a narrow range of hosts. A primary example of this is GFLV, the populations of which, regardless of their geographical origin, their host (species or cultivars of *Vitis*), and the type of symptomatological responses they induce in host plants, exhibit a remarkable serological uniformity (69). A naturally occurring serological variant of GFLV was only recently discovered in Tunisia (116) after a long search.

A possible explanation offered for the striking serological uniformity of GFLV—which may be applicable to comparable cases with other viruses—is the low selection pressure to which the virus has been subjected in nature because of its strict adaptation to a single host (*V. vinifera* in particular) (69).

Nepoviruses with a wide natural host range apparently vary much more serologically and often give rise to distinct "species." Here, the distinction between close and distant relationship is arbitrary, and, therefore, it may be hard to decide whether virus isolates are best considered different viruses or different strains of the same virus.

A conservative approach has been used with CLRV, for which many serologically distinguishable variants have been regarded as strains of the type virus, rather than as different viruses, even though American and European strains form two distinct clusters (54). However, the eucharis mottle isolate of TobRSV has been considered either a distant serological variant of the type virus (86) or a separate entity worthy of its own name (120, 122).

A comparable situation exists with the RRV "strain" recoverd from artichoke in the eastern Mediterranean area. This virus, of which minor serological variants from Greece and Turkey are known, differs from Scottish and English serotypes of RRV by a serological differentiation index of 3 to 6 (96), that is, a value equal to or above that separating GFLV from ArMV, or TBRV from GCMV or cocoa necrosis virus (CNV). When the geographical origin of the virus and the fact that the host it infects is a typical Mediterranean species are considered, it would seem appropriate to regard the artichoke strain of RRV as a distinct nepovirus. This possibility appears to be strongly supported by recent information indicating that although the Greek isolate has only a 9% sequence homology with the English serotype of RRV, it shares 73% of its sequence with the Turkish isolate (105).

An intriguing aspect of serological clustering of nepoviruses is that it is largely consistent with the geographical distribution of the viruses (Table 6.2 and 6.3) and hence, with their possible centers of origin. For instance,

TABLE 6.2. Grouping of nepoviruses according to serological relatedness.

Serological clusters	Geographical origin
1. Arabis mosaic virus (ArMV) (type and hop strain)	Europe
Grapevine fanleaf virus (GFLV)	Mediterranean–Near East
2. Tomato black ring virus (TBRV) (type and beet ringspot strain)	Europe
Grapevine chrome mosaic virus (GCMV)	Europe
Cocoa necrosis virus (CNV)	Africa
3. Raspberry ringspot virus (RRV) (Scottish, English, grapevine, and cherry strains)	Europe
Artichoke strain	Mediterranean–Near East
4. Strawberry latent ringspot virus (SLRV)	Europe
Rubus Chinese seed-borne virus (RCSBV)	Far East
5. Blueberry leaf mottle virus (BBLMV) (blueberry and grapevine strain)	North America
Grapevine Bulgarian latent virus (GBLV)	Europe
6. Tobacco ringspot virus (TobRSV) (type strain)	North America
Eucharis mottle strain	South America
Potato black ringspot virus (PBRV)	South America

TABLE 6.3. Grouping of nepoviruses according to presumed geographical origin.

Presumed origin and viruses	Natural host range[a]	Present distribution[a]
1. *Europe*		
Arabis mosaic virus (ArMV)	Very wide (fruits, vegetables, ornamentals)	Very wide
Cherry leafroll virus (CLRV)	Very wide (fruits, shrubs)	Very wide
Crimson clover latent virus (CCLV)	Crimson clover	Wide
Grapevine Bulgarian latent virus (GBLV)	Grapevine	Wide
Grapevine chrome mosaic virus (GCMV)	Grapevine, celery	Wide
Raspberry ringspot virus (RRV)	Very wide (especially small fruits)	Wide
Strawberry latent ringspot virus (SLRV)	Very wide (fruits, vegetables, ornamentals)	Wide
Tomato black ring virus (TBRV)	Very wide (fruits, vegetables, ornamentals)	Wide
2. *Mediterranean—Near East*		
Artichoke Italian latent virus (AILV)	Narrow (vegetables, grapevine)	Wide
Artichoke vein banding virus (AVBV)	Artichoke	Restricted
Artichoke yellow ringspot virus (AYRV)	Narrow (vegetables)	Restricted
Chicory yellow mottle virus (CYMV)	Narrow (vegetables)	Restricted
Grapevine fanleaf virus (GFLV)	Grapevine	Ubiquitous
Myrobalan latent ringspot virus (MyLRV)	Narrow (fruits)	Restricted
Olive latent ringspot virus (OLRV)	Olive	Restricted
Raspberry ringspot virus, artichoke strain	Artichoke	Restricted
3. *North America*		
Blueberry leaf mottle virus (BBLMV)	Blueberry, grapevine	Restricted
Cherry rasp leaf virus (CRLV)	Narrow (fruits)	Restricted
Peach rosette mosaic virus (PRMV)	Narrow (fruits)	Restricted
Tobacco ringspot virus (TobRSV)	Very wide (fruits, vegetables, ornamentals)	Wide
Tomato ringspot virus (TomRSV)	Very wide, (fruits, vegetables, ornamentals)	Very wide
Tomato top necrosis virus (TTNV)	Tomato	Restricted
4. *South America*		
Arracacha virus A (AVA)	Arracacha	Restricted
Arracacha virus B (AVB)	Arracacha, ocra, potato	Restricted
Potato black ringspot virus (PBRV)	Potato	Restricted

TABLE 6.3. *Continued*

Presumed origin and viruses	Natural host range[a]	Present distribution[a]
Potato virus U (PVU)	Potato	Restricted
Tobacco ringspot virus, eucharis strain	Eucharis	Restricted
5. *Africa*		
Cocoa necrosis virus (CNV)	Cocoa	Restricted
Hibiscus latent ringspot virus (HLRV)	Hibiscus	Restricted
6. *Australia*		
Cassava green mottle virus (CGMV)	Cassava	Restricted
Lucerne Australian latent virus (LALV)	Alfalfa, white clover	Restricted
Lucerne Australian symptomless virus (LASV)	Alfalfa	Restricted
7. *Far East*		
Cycas necrotic stunt virus (CNSV)	Cycas	Restricted
Mulberry ringspot virus (MRSV)	Mulberry	Restricted
Rubus Chinese seed-borne virus (RCSBV)	*Rubus*	Restricted
Satsuma dwarf virus (SDV)	Satsuma mandarin	Restricted

[a] Ubiquitous, occuring in all major areas of cultivation of the host plant; very wide, recorded from many countries in two or more continents; wide, recorded from many countries in the same continent; restricted, recorded from a single or two adjacent countries.

North Americam TobRSV has serological counterparts in the southern part of the continent (TobRSV eucharis strain and PBRV), and, similarly, European ArMV, RRV, and TBRV have serologically related "species" further south: RRV artichoke strain and GFLV in the Mediterranean area, and CNV in Africa. The viruses within each of these clusters are serologically interrelated, which indicates evolution from a common ancestor. The fact that these viruses occur in physically contiguous, though separated regions, is therefore in line with the likelihood that they have developed in these regions.

Geographical Origin and Distribution

There is a consensus that nepoviruses are primarily pathogens of wild plants and thus depend for their survival and spread in natural environments (as opposed to man-made agricultural environments) on dissemination by nematode vectors and host plant seeds that they may infect (see reviews, refs. 40 and 86). It follows that these viruses have little natural mobility, so they tend to be localized in specific territorial enclaves in which they become firmly established.

It is therefore conceivable that the geographical distribution of nepoviruses broadly corresponds to their areas of origin or differentiation, in which their hosts, primary and alternative (usually crop plants), and vectors are readily available.

The notion that nepoviruses may have differential geographical origins, first put forward with reference to viruses infecting grapevines in Europe and North America (68, 69) and recently extended to other nematode-borne viruses of the American continent (122), seems tenable and is consistent with the distribution of vectors. Therefore it seems reasonable to hypothesize a presumed geographical origin of currently recognized nepoviruses, as shown in Table 6.3. From this table, it is evident that nepoviruses that generally infect a wide range of hosts have a much wider distribution, especially if the hosts are vegetatively propagated perennial crops, than have viruses with a few or a single host. Such viruses, as would be expected, have a restricted distribution.

A remarkable exception to the latter is GFLV, which although a highly specialized pathogen has the widest geographical distribution of the nepoviruses. Uncontrolled marketing of infected budwood and rooted cuttings have greatly facilitated the spread of GFLV and its major vector, *X. index,* to virtually all the viticultural areas of the world. This also applies to records of European nepoviruses such as SLRV, TBRV and ArMV, from grapevines in eastern Mediterranean regions (Turkey and Israel) and Japan (9); from grapevines, cherry, rhubarb and parsley in North America (122), as well as the American TomRSV, from a shrub in Australia (24). No plausible explanation is presently available for the records of ArMV from a native shrub in the United States (122) and of TobRSV from soybean in the People's Republic of China (144).

Another widely distributed virus, CLRV, is recorded from cultivated and native plant species in Europe and North America. Simple dissemination through infected propagative material may not account for the widespread occurrence of CLRV outside Europe, its presumed area of origin, since there is serological evidence that strains of CLRV may have independently arisen in Europe and North America. This is compatible with the notion that, in nature, CLRV spreads by air-borne pollen rather than by nematodes (54, 122) (Table 6.4).

Tobraviruses

Tobraviruses constitute the other recognized taxonomic group of plant viruses transmitted by nematodes (*Trichodorus* and *Paratrichodorus*). These viruses have rigid, rod-shaped particles that vary in length: very short, ~45 nm.; short (S), 50 to 110 nm; and long (L), 185 to 200 nm. Their

TABLE 6.4. Grouping of nepoviruses according to means of natural spread.

Virus	Seed transmission in naturally and/or artificially infected hosts	Vector
1. *Transmitted by nematodes*		
Arabis mosaic virus (ArMV)	Yes	
All strains		*Xiphinema diversicaudatum*
Artichoke Italian latent virus (AILV)	Not detected	
Italian strain		*Longidorus apulus*
Greek strain		*Longidorus fasciatus*
Cherry rasp leaf virus (CRLV)	Yes	*Xiphinema americanum*
Grapevine fanleaf virus (GFLV)	Yes	*Xiphinema index*, *Xiphinema italiae*
Mulberry ringspot virus (MRSV)	Yes	*Longidorus martini*
Peach rosette mosaic virus (PRMV)	Yes	*Xiphinema americanum*, *Longidorus diadecturus*, *Longidorus elongatus*
Raspberry ringspot virus (RRV)	Yes	
Scottish strain		*Longidorus elongatus*
English strain		*Longidorus macrosoma*
Strawberry latent ringspot virus (SLRV)	Yes	
All strains		*Xiphinema diversicaudatum*
Tobacco ringspot virus (TobRSV)	Yes	
All strains		*Xiphinema americanum*
Tomato black ring virus (TBRV)	Yes	
Type strain		*Longidorus attenuatus*
Beet ringspot strain		*Longidorus elongatus*
Tomato ringspot virus (TomRSV)	Yes	
Type strain		*Xiphinema americanum*, *Xiphinema rivesi*
Grapevine yellow vein strain		*Xiphinema californicum*
2. *Transmitted by pollen to mother plants, no vector found*		
Cherry leafroll virus (CLRV)	Yes	
Blueberry leaf mottle virus (BBLMV)	Yes	
Artichoke yellow ringspot virus (AYRV)	Yes	
3. *Vector unknown*		
Arracacha virus A (AVA)	Yes	
Arracacha virus B (AVB)	Yes	

TABLE 6.4. *Continued*

Virus	Seed transmission in naturally and/or artificially infected hosts	Vector
Artichoke vein banding virus (AVBV)	Not tested	
Cassava green mottle virus (CGMV)	Not tested	
Chicory yellow mottle virus (CYMV)	Yes	
Cocoa necrosis virus (CNV)	Yes	
Crimson clover latent virus (CCLV)	Yes	
Cycas necrotic stunt virus (CNSV)	Yes	
Grapevine Bugarian latent virus (GBLV)	Not tested	
Grapevine chrome mosaic virus (GCMV)	Not tested	
Hibiscus latent ringspot virus (HLRV)	Not detected	
Lucerne Australian latent virus (LALV)	Yes	
Lucerne Australian symptomless virus (LASV)	Yes	
Myrobalan latent ringspot virus (MyLRV)	Not tested	
Olive latent ringspot virus (OLRV)	Not tested	
Potato black ringspot virus (PBRV)	Not detected	
Potato virus U (PVU)	Yes	
Rubus Chinese seed-borne virus (RCSBV)	Yes	
Satsuma dwarf virus (SDV)	Yes	
Tomato top necrosis virus (TTNV)	Not tested	

bipartite genome has two functional RNA species, their S particles encapsidate one molecule of the smaller RNA (RNA-2), and their L particles contain one molecule of the larger RNA (RNA-1). Coat protein subunits are of a single type and have a molecular weight of 21,000-23,000 daltons.

In contrast with nepoviruses, tobraviruses have increased very little in number: The two members constituting the group when it was first established (47) are now three (46) (Table 6.5).

The classification of tobraviruses is based on molecular hybridization, that is, the extent of sequence homology between RNA-1 species, rather

TABLE 6.5. Tobravirus group: Members, host range, vectors, and geographical distribution.

Virus	Natural Host Range	Seed transmission	Vectors	Geographical distribution
Tobacco rattle virus (TRV)	Very wide (vegtables, ornamentals, woody perennials, shrubs)	Yes	*Trichodorus cylindricus, T. hooperi, T. primitivus, T. similis, T. viruliferus; Paratrichodorus allius, P. anemones, P. christiei, P. minor, P. nanus, P. pachydermus, P. porosus, P. teres, P. tunisiensis*	Very wide (Europe, Mediterranean, North America, Japan, New Zealand)
Pea early-browning virus (PEBV)	Narrow (legumes only)	Yes	*Trichodorus primitivus, T. viruliferus; Paratrichodorus anemones, P. pachydermus, P. teres*	Wide (Europe, Mediterranean)
Pepper ringspot virus (PRV)	Narrow (vegetables)	Yes	*Paratrichodorus christiei*	Restricted (Brazil)

than on serology. This led to the recognition of three distinct viruses, each with its own separate gene pool: TRV, PEBV, and PRV (106).

Isolates belonging to any given virus or "species" have strongly conserved RNA-1 genes, whereas their RNA-2 genes vary. Therefore, serology is, at most, only useful for separating strains within each "species." For example, broad bean yellow band virus (BBYBV) was originally considered to be a possible new member of the group because it is not apparently related serologically to the English or Dutch strains of PEBV (111). However, it was later demonstrated that its RNA-1 has substantial sequence homology with PEBV RNA-1, and, therefore, despite its lack of serological relatedness, it was synonymized with PEBV as a new serotype (107).

The use of serology to identify certain tobraviruses can be misleading. In fact, since sequences of a gene pool of a tobravirus "species" may be captured in nature by a gene pool of a different "species", new pseudo-recombinants arise in which the RNA-2 (i.e., the part of the genome responsible for serological specificity as it codes for the coat protein) of a given virus becomes dependent for its replication on the RNA-1 of another virus, conferring upon it the serological characteristics of the former virus (108).

The geographical distribution of tobraviruses seems to differentiate, to a certain extent, individual members of the group from one another, thus justifying the concept of the existence of gene pools.

Vectors

Many species of nematodes ingest viruses when they feed on the roots of virus-infected plants, but it is now well established that the natural transmission of nepoviruses is only by longidorid nematodes, and of tobraviruses by trichodorid nematodes. However, of the 157 species of *Xiphinema* and 82 species of Longidorus described to date (early 1989), relatively few have been implicated as vectors and, indeed, not all nepoviruses require nematode vectors for their survival and dissemination (Table 6.4). So far, at least 14 of about 50 described species of *Trichodorus* and *Paratrichodorus* are vectors of the tobraviruses TRV and PEBV, but each species may only transmit a particular strain. A third tobravirus, PRV, has been described from Brazil and *P. christiei* has been implicated as a vector (22, 113) (Table 6.5).

Distribution of *Xiphinema* and *Longidorus*

Xiphinema and *Longidorus* have been reported from most parts of the world where nematode surveys have been undertaken. Individual species mostly occur as discrete populations in a particular region, and analyses of their distribution have been made in an attempt to deduce their phylogenetic relationships. From a comparison of several morphological characteristics of different longidorid genera, Coomans (27) concluded that *Xiphinema* originated in Gondwanaland and, before the break-up of Pangaea, the genus spread to Laurasia. The main speciation occurred in Africa, from where the majority of species have been described, with South American regarded as another important speciation area.

Longidorus, with *Paralongidorus* and *Longidoroides*, is considered to have originated in Southeast Africa and India when these areas were still united, and a later spread to Laurasia was accompanied, and followed, by a main speciation of Longidorus in the holarctic region, especially Europe.

In their analysis of the European longidorid fauna, Topham and Alphey (135) relate the relative impoverishment of species in the northern regions to Quaternary glaciation and attribute the highly diverse fauna of the eastern Mediterranean countries of Israel, Italy, and Malta to Miocene plate tectonic activity in that area. The distribution of longidorid species in the Americas also provides evidence of the effect of changing latitude on species richness and diversity.

Much of the present distribution of longidorid species can, in broad terms, be related to paleoecology (16, 30, 78, 98, 135), but in relatively recent times, many species, especially those associated with crop plants, have been disseminated from their centers of origin by man's activities. Examples include *X. index*, the vector of GFLV, which has been distributed throughout Europe and the areas of the world where grapevines are grown from its center of origin in ancient Persia (50, 82). *Xiphinema rivesi*, a vector of TomRSV in the eastern United States, has probably been exported to Europe where it occurs in scattered localities particularly in western France, but so far without association with the virus. Among *Longidorus* species, there is evidence that *L. elongatus* and *L. vineacola* have been introduced into the Scottish Western Islands with garden planting material (7, 16) and that *L. apulus* has been distributed in Apulia (southern Italy) on soil adhering to artichoke sprouts used for propagation (109).

Species that have been widely dispersed survive in new biotopes because of their genetic adaptability. With time, many of the geographically separated populations may change sufficiently in their taxonomic characteristics to be considered new species. Certainly there is much evidence of morphometric variation within widely dispersed species of *Xiphinema* and *Longidorus*, and this has caused problems not only in terms of taxonomy but also in the identification of their role in virus transmission.

Brown and Topham (17) found that populations of *Xiphinema diversicaudatum* from different countries were distinguishable morphometrically, as well as by certain aspects of their biological behavior, including their reproductive ability and their ability to transmit virus. However, although populations could be grouped morphometrically, the differences were not considered to be sufficient to establish new species. Morphometric differences between dispersed populations have also been noted in *Xiphinema coxi, L. elongatus, L. profundorum,* and *L. vineacola,* and these species, together with *X. diversicaudatum,* may be regarded as species complexes in which the biological characteristics of the populations also differ to some degree. *Xiphinema americanum* was recognized as a species complex by Lima (62), who concluded that it comprised seven parthenogenetic species, four of which he described as new. Other workers (52, 123) supported this view, although they thought the demarcation of these species were problematical and unsatisfactory. However, Lamberti and Bleve-Zacheo (62) divided *X. americanum sensu lato* into six groups of species, totaling 25 in all, with 15 of them new. They thought that *X. americanum sensu stricto* is restricted in its geographical distribution to the eastern part of North America, and they designated *Xiphinema californicum* a new species to define the morphologically distinct group of the western seaboard of the United States. Apart from some outstanding queries, records of *X. americanum* in European countries have been assigned to *Xiphinema pachtaicum* of *Xiphinema brevicolle,* neither of which species has been shown to be a vector in field situations(16).

Because of the taxonomic reconstruction of *X. americanum* (62) many of the records of its association with TomRSV or TobRSV in North America need to be reconsidered. *Xiphinema americanum senus stricto* remains as the vector of some strains of TomRSV, and so far it is the only recognized vector of TobRSV, although the geographical distribution of the virus is not entirely coincident with that of the vector; however, *X. californicum* is established as the vector of California-type strains of TomRSV (53) and is presumed to be the vector of CRLV (62, 90). Similarly, *X. rivesi* is the vector of strains of TomRSV in eastern Canada and Pennsylvania (USA) (34), and *X. utahense* and *X. occidum* are also considered to be potential vectors of some strains of TomRSV (62, 144).

Records of virus transmission of *X. americanum* or derived species outside North America may be authentic but at most are associated with outlier populations of the nematode that have been dispersed, through man's agency, from the center of origin of both the vector and the virus in North America.

The other species in North America that have been associated with the transmission of nepoviruses in field situations are *X. index,* which was introduced from Europe, and which is unique in its association with GFLV, and *Longidorus diadecturus,* which is a vector of PRMV in Ontario (Canada), with *X. americanum* also being recorded as a less efficient vector of the virus (1, 2). In a recent paper (2a), an Ontario population of *L. elongatus* was also recorded as a vector of PRMV but only at a low transmission level.

Longidorus diadecturus and *Longidorus martini*, the vector of MRSV in Japan (143), are unknown in Europe; they are distinct from other *Longidorus* species in certain morphological details (11).

Distribution of *Trichodorus* and *Paratrichodorus*

Trichodorid nematodes are widespread in North America and Europe and have been recorded from many parts of the world, including some relatively isolated islands. Most species appear to be locally distributed so that different species are present in different landmasses, although within a geographical region some species may be more widespread than others. In a survey of trichodorids in Europe (3), *Paratrichodorus pachydermus* and *Trichodorus primitivis* were found to occur in most of the northern countries, whereas *Paratrichodorus tunisiensis* has so far been found only in Italy, and *Trichodorus hooperi* only in the southwest of England. Trichodorid species described from Africa, India, and Japan also appear to be localized, but surveys in those regions have been insufficient to establish the extent of their geographical distribution.

Although groupings of trichodorids species have been recognized (31, 65), their taxonomy does not indicate evolutionary directions and centers of origin. However, the abundance of species in Europe, and the usual occurrence of several species in a single soil sample (64, 90), suggests that active speciation is occurring and new biotopes are being invaded.

Some species are cosmopolitan in their distribution, for example, *Paratrichodorus minor* and *Paratrichodorus porosus* (66), and may have been distributed by man, although trichodorids are susceptible to such mechanical injury as occurs in the rough handling of soil samples (8), and are unlikely to survive casual transportation from one region to another. However, they may be successfully dispersed in flood or irrigation waters (6, 114), and their ability to reproduce rapidly allows them to invade and exploit new environments quickly (59).

Most of the records of tobravirus transmission by trichodorid species are from Europe and North America (63, 127, 140), but this reflects the research interest in these regions, and, in due course, more virus–vector associations may be expected to be identified in other parts of the world. Currently, there are isolated records of TRV transmissions from Japan and New Zealand, and of PRV from Brazil.

Isolates of TRV from North America differ from those from Europe and are transmitted by different species, which supports the view that geographical separation is associated with differentiation of virus and vector. So far, PEBV has been found only in Europe, but several strains have been indentified and at least five trichodorid species have been implicated as vectors.

Virus–Vector Associations

Much of the accumulated experimental evidence of nematode transmission of plant viruses indicates that there is a high degree of specificity in the association between virus and vector. Thus, although serologically unrelated nepoviruses may share a common vector species (e.g., *X. diversicaudatum* transmits ArMV and SLRV), strains of a virus that are serologically distinct are transmitted by different, although closely related species of the same nematode genus. For example, the Scottish strains of RRV and the unrelated TBRV are transmitted by *L. elongatus,* but the English strains of these viruses have *Longidorus macrosoma* and *Longidorus attenatus,* respectively, as vectors. Further, the division of *X. americanum* species complex into several discrete species (62) now supposes the association of serologically distinct strains of TomRSV with different vector species (63).

Evidence of the specific association between tobraviruses and trichoroid vectors is less clear. American and European isolates of TRV are serologically distinguishable (see earlier section) and, in nature, are associated with different species of *Trichodorus* and *Paratrichodorus;* indeed, only species of the latter genus have been recorded as vectors in North America.

In comparing the transmission of TRV by nine species of *Trichodorus* and *Paratrichodorus* from the Netherlands, van Hoof (141) found that transmission occurred only when the nematode and virus came from the same locality. A high level of specificity is also apparent with the transmission of PEBV, different isolates of which are transmitted by several trichodorid species (37, 142). Other evidence suggests that specificity of transmission is not well developed. In Britain, the spinach yellow mottle strain of TRV was transmitted by a mixed population of *Trichodorus* and *Paratrichodorus* species (60); in Belgium, five trichoroid species transmitted TRV, infecting a potato crop (92), although in this case it was not recorded whether a single virus strain was involved.

A recent study in eastern Scotland (19) showed that a close relationship may be established between different species of trichodorid nematodes and serologically distinct isolates of TRV. At two field sites, *P. pachydermus* transmitted the majority of the several isolates of TRV (PRN serotype) that were present, but *Trichodorus cylindricus* transmitted isolates of a previously uncharacterized serotype.

For some years, it has been recognized that if the vector status of a nematode is to be established with any certainty, several criteria must be met in experimental work (75, 127, 136). These include (a) the virus must be available to the nematode; (b) test conditions must be suitable for transmission to occur; (c) the possibility of virus contamination of the bait plants must be avoided. To these criteria, Trudgill et al. (139) added (d) the virus and nematode must be correctly identified; (e) bait plant tissues must be shown to be infected with the virus being tested; and (f) the nematode being tested must be shown to be the only possible vector in that experiment. Test procedures have now been developed to meet these criteria, and they are sufficiently sensitive to detect small differences in efficiency of transmission between different species of longidorids and trichodorids (19, 138). Further, the refinements incorporated into the procedures have led to the conclusion that some of the anomalous results previously obtained in laboratory experiments and unsupported by field evidence (127, 132) might have been due to contamination (77) and that two-thirds of the published results of virus–vector associations are invalid (139).

Efficient extraction of nematodes from the soil is a prerequisite for virus transmission tests. In a comparison of methods used to extract virus–vector nematodes, Brown and Boag (18a) concluded that a decanting and sieving technique, with 200-g soil samples, is the most satisfactory procedure for longidorid and trichodorid nematodes.

Variation in Transmission

Although the authenticity of many of the virus–vector associations is well established, the different results of transmission tests obtained by different workers suggest that vector species differ in their efficiency of transmission; but because of the different experimental conditions, this has been difficult to substantiate. However, when precise test procedures were used, it was demonstrated that *X. diversicaudatum* is an efficient vector of ArMV, whereas *L. elongatus* and *L. macrosoma* are inefficient vectors of RRV (Scottish strain) and RRV (English strain), respectively (134, 138).

Recent experiments have also shown that vector populations that are widely separated geographically may differ in their efficiency of transmission of the virus with which they are normally associated. Comparing populations of *X. diversicaudatum* from 10 countries as vectors of ArMV and SLRV, Brown (12, 14) found that those from France, Italy, and Spain rarely transmitted the viruses, whereas populations from other countries were all efficient vectors (Table 6.6). These populations had been exposed to the British strains of the viruses, but when the Italian population was exposed to an Italian strain of SLRV, the efficiency of transmission did not improve. In another test (18), *X. diversicaudatum* from Scotland and Italy was exposed to two Italian strains and the type (British) strain of SLRV. The Scottish population readily transmitted the type strain of the virus, but did not transmit the Italian strains; the Italian population transmitted all three virus strains but at a very low frequency. Immunosorbent electron microscopy (100) of the nematodes demonstrated that they had ingested the viruses to which they had been exposed, but electron microscopy of sections of the odontophore, the site of virus

TABLE 6.6. Transmission of the type strains of arabis mosaic (ArMV) and strawberry latent ringspot (SLRV) viruses by 10 populations of *Xiphinema diversicaudatum*.[a]

	Percentage number of transmissions[b]	
Nematode Population	ArMV	SLRV
Bulgaria	100	48
England	96	60
New Zealand	96	60
Norway	96	40
Scotland	92	68
Switzerland	96	56
United States	48	36
France	10	10
Italy	4	2
Spain	0	15

[a] Compiled from Brown (12, 14) and Brown and Trudgill (18).
[b] Using groups of two nematodes per test pot; 25 replicates of each test.

retention in *Xiphinema* vectors, revealed that few or no virus particles were present in those populations that failed to transmit. Although no significant morphometric differences were apparent between the populations of *X. diversicaudatum* from the 10 countries (17), the marked difference in transmission efficiency of the Italian, and possibly also the French and Spanish, populations could be considered to indicate putative new species. However, it is interesting to note that the Scottish and Italian *X. diversicaudatum* were capable of cross-breeding and that the resulting progeny were intermediate between the parents in efficiency of transmission (13).

Differences in efficiency of transmission have also been shown to occur among *Longidorus* vectors. A Scottish population of *L. elongatus* transmitted the type strains of TBRV and RRV more frequently than the English population, and neither population transmitted the German potato bouquet strain of TBRV, which is a distinct serotype and is considered to have as a vector *L. attenuatus* (15).

In laboratory tests, potato bouquet and two other isolates of TBRV were transmitted less frequently by an English population of *L. attenuatus* then were several English isolates of the virus, including an isolate associated with celery yellow vein disease (20). This, and similar evidence for other vectors, supports the contention that local populations of a vector species are most efficient at transmitting local virus isolates, and, thus, geographical separation tends to lead to high levels of specificity between virus and vector.

Nematode–Virus Interactions

Electron microscopy of thin sections of nematode vectors has identified the virus retention sites within each of the vector genera. In *Longidorus*

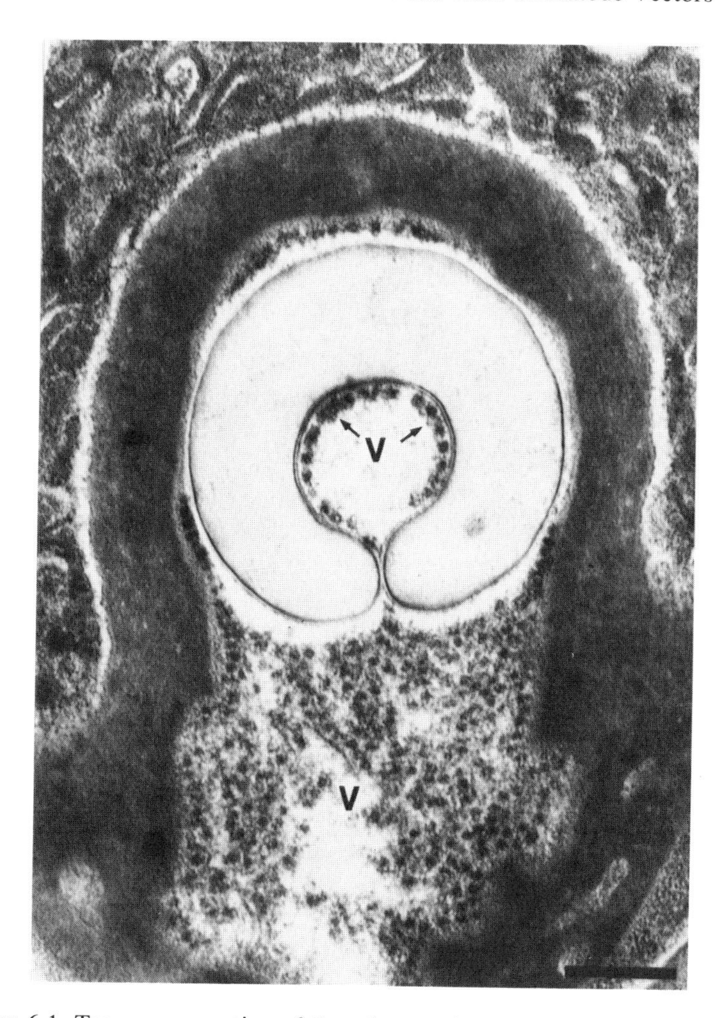

FIGURE 6.1. Transverse section of the odontostyle and guiding sheath of *Longidorus elongatus* reared on a plant infected with raspberry ringspot virus. Particles of the virus (V) are present in association with the inner surface of the odontostyle and between the odontostyle and the guiding sheath (bar, 200 nm). Courtesy of W.M. Robertson.

species, virus particles are adsorbed to the inner surface of the odontostyle (129, 132, 138), and in *L. elongatus,* particles of RRV and TBRV may also be located between the odontostyle and the guiding sheath (129) (Figure 6.1). In *X. diversicaudatum* carrying ArMV or SLRV (130), *X. americanum* carrying TomRSV (76), or *X. index* carrying GFLV (97, 130), virus particles are specifically associated with the cuticular lining of the odontophore, the slender esophagus, and the esophageal pump; the maximum concentration of particles usually occurs in the anterior region of the odontophore.

In trichodorid vectors, TRV particles have been found to be retained in association with the lining of the food canal from the anterior region of the esophastome to the esophagointestinal valve (131) but not attached to the onchiostyle. The tubular particles may be attached by their sides or their ends: the long particles tend to line up parallel to the long axis of the food canal, whereas the short particles tend to adhere by their ends (104, 132).

Experiments with pseudo-recombinant isolates of RRV and TBRV have indicated that the specific association of a nepovirus with its vector is determined by the RNA-2 of the virus genome, which carries the coat protein cistron (41, 43, 49). Thus, association between virus and vector appears to depend on some feature of the protein coat that interacts specifically with the retention site within the nematode.

The tobraviruses also have RNA genomes in two pieces. Pseudo-recombinants of TRV strains have been produced (46), and although they have not been used in transmission experiments with trichodorid vectors, it seems likely that the mechanism of specific association between virus and vector is similar to that of nepoviruses.

The mechanism whereby virus particles are adsorbed specifically at the retention site within the nematode vector has been a subject for speculation for some time (127, 133). Recent investigations indicate that specific recognition between virus and vector may involve the interaction of complementary molecules at their point of contact, as occurs in a variety of host–pathogen systems (118). In *X. diversicaudatum,* a discontinuous layer of carbohydrates lines the odontophore and esophagous, and ArMV and SLRV particles attach only to the carbohydrate zones (101, 102, 103) (Figure 6.2). In *P. pachydermus,* a vector of TRV, the total lining of the wall of the esophagous also stains for carbohydrates (102). Thus, virus retention in *Xiphinema* and trichodorid vectors may involve an interaction between carbohydrate moieties on the food canal wall and complementary lectin-like molecules on the protein coat of the virus.

Carbohydrates have not been detected on the guiding sheath or the odontostyle in *L. elongatus* (101). However, by labeling the odontostyle with cationized ferritin, a strong negative charge was shown to be present on the exterior surface of the adontostyle and on the wall of the lumen, and this may account for the retention of positively charged virus particles (101).

Dissociation of virus particles from the retention site is thought to occur when the pH of the lumen is changed by the passage of secretions from the esophageal glands during the initial stages of feeding (127, 133). In Longidorus vectors, specificity and efficiency of transmission may be determined in some cases, if not in all, by the mechanism of dissociation of the virus particles from the retention site (127). For example, when *L. macrosoma* was exposed to the English and Scottish strains of RRV, the former was transmitted, but not the latter, as expected, although virus particles were found to be adsorbed to the inner surface of the odonto-

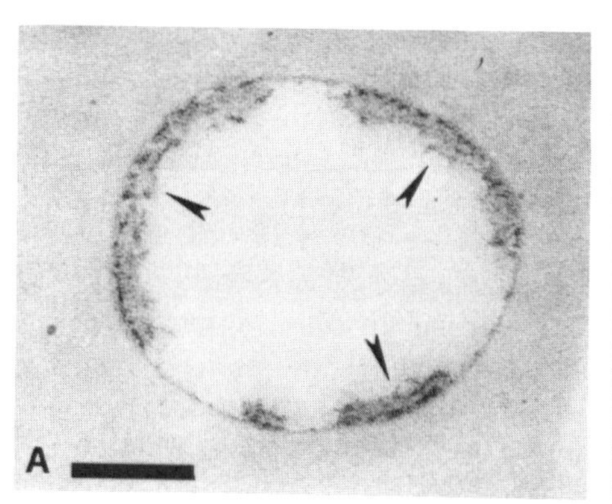

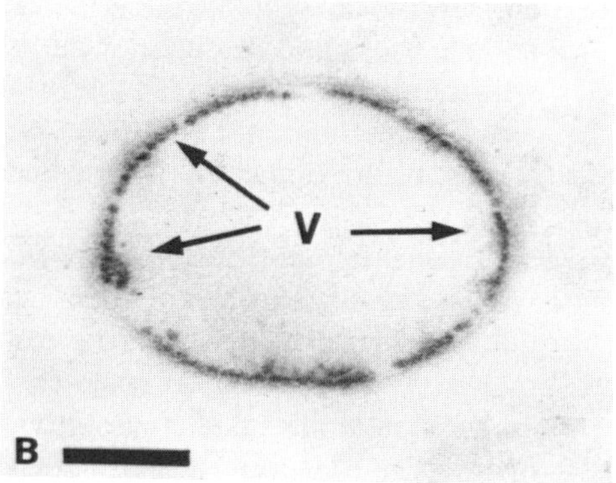

FIGURE 6.2. (A) Transverse section of the lumen of the odontophore of *Xiphinema diversicaudatum* reared on a plant infected with arabis mosaic virus. The section is stained to show the thin discontinuous carbohydrate layer in association with cloud-like areas (arrows). The unstained virus particles are within the cloud-like areas. (B) Transverse section of the lumen of the odontophore of *X. diversicaudatum* reared on a plant infected with arabis mosaic virus and stained with uranyl acetate and lead citrate to show virus particles (V) (bars, 200nm). Courtesy of W.M. Robertson.

style in both sets of nematodes (132, 137). In terms of surface charge density, this was considered to indicate that the change in pH brought about by the esophageal gland secretions altered the surface charge of the particles of the English strain and resulted in detachment and transmission but that it did not have a similar effect on the virus particles of the Scottish strain (127).

The assumed difference in the mechanism of transmission in *Longidorus* vectors, compared with *Xiphinema* and trichodorid vectors, also reflects other observed differences between the two groups of nematodes. Viruses are retained for only a few weeks, at most, in *Longidorus* vectors and transmission is inefficient; in *Xiphinema* and trichodorid vectors, however, associated viruses may be retained for several months and transmission is efficient.

Discussion

The survival and dissemination of plant viruses depends on their effective transmission and their access to suitable host plants for their multiplication. Tobraviruses and nepoviruses are dispersed over short distances by their nematode vectors, that is, to the extent of the area occupied by the local population, but dissemination to new sites may occur through the distribution of infected week seeds or pollen. Thus, nematodes may not appear to be essential for the maintenance and spread of tobraviruses and nepoviruses, and, indeed, some of the nepoviruses are not associated with nematode vectors (Table 6.4). However, the interaction between virus, nematode, and host plant is a dynamic process, which, at various stages, offers opportunities for genetic selection to the advantage of the virus.

The nematode-transmitted viruses are essentially parasites of wild plants and usually infect them without causing obvious symptoms of infection. However, when crop plants are infected, symptoms are almost invariably severe and, in some cases, may cause the death of the plants, as RRV in some raspberry cultivars and TomRSV and CLRV in union necrosis in apple and walnut, respectively. Similarly, most of the longidorid and trichodorid vectors have wide host ranges among wild plants with which they are compatible enough as not to cause excessive injury by their feeding; but when the roots of crop plants are attacked, they are often severely galled and the growth of the plant is affected (127).

The feeding apparatus of *Longidorus* and *Xiphinema* nematodes is a long tubular odontostyle with which they pierce the young roots and feed on the cell contents. Feeding commences with the penetration of a column of cells near the root apex, and each cell may be fed on progressively until the tip of the odontostyle is located at the feeding site some five to seven cells distant from the rhizodermis. Secretions from the esophageal glands induce a hypertrophic reaction in the root cells around the feeding site, and the coenocyte or cisternum that is formed, depending on the species, provides a rich and readily accessible food source for the nematode. The high metabolic activity of the cells presumably is also conducive to the multiplication and translocation of viruses within the root tissues.

In trichodorid nematodes, the feeding apparatus is in the form of a solid tooth, or onchiostyle, which is used to penetrate the rhizodermal cells of the root tip. Secretions from the esophageal glands are injected into the cell soon after the wall has been penetrated and the contents of the cell are usually ingested within a few seconds, after which the nematode moves to another rhizodermal cell. It is thought that virus particles are also injected into the cell with the secretions of the esophageal glands, and that the virus quickly passes to adjacent cells, or possibly that virus transmission occurs on those occasions when the nematode fails to ingest the total contents of the cell on which it is feeding.

Seemingly, nematodes are passive carriers of the viruses with which they are associated, in the sense that the viruses do not invade the nematode tissues or have any obvious effect on their biological behavior. Nevertheless, the vectors may affect the transmission of the viruses. For instance, vector species vary from being efficient to highly inefficient in their ability to transmit virus. Populations of *X. diversicaudatum* from Britain are efficient vectors of ArMV and SLRV, whereas Italian populations only infrequently transmit these viruses. Thus, the same viruses may be disseminated at different rates in different regions. Also, the adsorption of virus particles at the site of retention within the vectors involves the interaction of the virus coat protein with particular features of the cuticular lining of the nematode food canal, and this in itself may have a selective effect on the virus.

Because of the requirement of compatibility between virus and vector to ensure the survival of the virus, it might be expected that the coat protein would be a relatively invariant property of each virus. There is, however, increasing evidence that, in field situations, several minor antigenic variants of a virus may be present together. Such variants have often been revealed when cultivars that are considered to be immune to a particular virus are planted and become infected.

Variants of RRV and ArMV that broke the resistance of raspberry cultivars in Scotland were revealed in this way (88, 134), and, more recently, further isolates of both viruses were found to infect raspberry cultivars that had been shown to be immune to the viruses in graft-inoculation tests (57). Crops constitute a large monoculture area that can apply a selection pressure to the viruses, but the new isolates may have characteristics that are unfavorable in an ecological environment outside of the crop. Such characteristics include poor seed transmissibility and lack of virulence (40), which are conferred by RNA-1, and thus are not believed to be influenced by nematode transmission.

In the above examples, the viruses have wide host ranges that include wild herbaceous species, and the genetic variation displayed by these viruses contrasts with the lack of variation of GFLV, which in nature has so far been found only in association with *Vitis* spp.

The genome of nepoviruses is bipartite, with RNA-1 and RNA-2 located in separate particles, but because these viruses have a similar coat protein, it is assumed that each type is equally ingested and transported by the nematode vector. However, although there is good circumstantial evidence for linking transmission with the properties of the virus coat protein, and hence linking it with the antigenic characteristics of the virus, there remains the possibility that the regions in the coat protein that are important for the attachment and release of particles at retention sites in the nematode vector are not involved in the immunological reaction (20). Further, some properties determined by RNA-1 may affect transmissibility by the vector, as suggested to explain the poorer transmissibility of a pseudorecombinant isolate of TBRV compared to that of the parental source of its RNA-2 (43).

The RNA-1 and RNA-2 of tobraviruses are located in particles of different lengths, and these are readily visualized by electron microscopy of thin sections of the vector species. In the limited number of observations so far made, short and long particles of TRV are randomly distributed at the site of retention in the nematode. Different isolates of the virus, however, cannot be identified in thin sections of the nematodes and the experimental evidence is inconclusive about specific transmission by trichodorid vectors.

The RNA-1 of tobraviruses is strongly conserved, and the RNA-2 is variable; in nepoviruses, however, both parts of the genome diverge more or less in parellel (32). The RNA-2 nucleotide sequence seems to differ markedly between isolates. Harrison and Robinson (46) suggest that the variation in the tobravirus particle protein indicates that there is no selection pressure for its conservation, and hence it does not play a key role in determining vector transmissibility and specificity. A wide range of naturally occurring strains of both tobraviruses and nepoviruses has been found, but the processes by which these variants are produced remain a matter of speculation.

Nepoviruses and tobraviruses have two complementary methods of dispersion that ensure their survival in a particular location and their distribution to new areas. The nematode vectors are usually static populations, and the spread of virus is slow, but this slow spread is compensated for by the long period of retention of the viruses in the vectors, which, in the case of *Xiphinema* and trichodorid nematodes, may ensure survival between plantings of susceptible crops or through periods when plants are absent in natural situations. Infection of the seeds of weed hosts provides a means of perennation of the viruses over long periods of time and a means of spread to new sites. Weed seed infection is more prevalent among *Longidorus* vectors, which retain viruses for only a few weeks, compared with retention for several months in *Xiphinema* and trichodorid nematodes.

In the past, dispersal of viruses to new areas was probably attained solely through infected week seeds, and continued existence of viruses would depend on their coming into contact with suitable vector species. In more recent times, man has been responsible for the distribution of virus and vector with commercial vegetative material, examples of which are *X. diversicaudatum* and ArMV and *X. index* and GFLV. However, only the less vulnerable of the nematode vector species have been widely distributed, and most species remain in relatively limited ares. Thus, it is not surprising that surveys continue to record more and more species in the genera associated with virus transmission. If nematode vectors apply some selection pressure on the viruses they carry, then it seems likely that there will be a continuing genetic drift of both viruses and vectors to establish new associations.

References

1. Allen, W.R., Van Schagen, J.G., and Eveleigh, E.S:, 1982, Transmission of peach rosette mosaic virus to peach, grape and cucumber by *Longidorus diadecturus* obtained from diseased orchards in Ontario, Can. J. Pl. Pathol. **4:**16–18.

2. Allen, W.R., Van Schagen, J.G., and Ebsary, B.A., 1984, Comparative transmission of the peach rosette mosaic virus by Ontario populations of *Longidorus diadecturus* and *Xiphinema americanum* (Nematoda:Longidoridae), *Can. J. Pl. Pathol.* **6**:29–32.

2a. Allen, W.R., and Ebsary, B.A., 1988, Transmission of raspberry ringspot, tomato black ring and peach rosette mosaic viruses by an Ontario population of *Longidorus elongatus, Can. J. Plant. Pathol.* **10**:1–5.

3. Alphey, T.J.W., and Taylor, C.E., 1987, European Atlas of Longidoridae and Trichodoridae, *Scott. Crop Res. Inst., Dundee, Scotland,* 35 p.

4. Barbara, D.J., Ashby, S.C., and McNamara, D,G., 1985, Host range, purification and some properties of rubus Chinese seed-borne virus, *Ann. Appl. Biol.* **107**:45–55.

5. Beemster, A.B.R., and de Bokx, J.A., 1987, Survey of properties and symptoms, in de Bokx, J.A., and Van der Want J.P.H. (eds): Viruses of Potato and Seed-Potato Production, Pudoc, Wageningen, pp. 84–113.

6. Boag, B., 1986, Detection, survival and dispersal of soil vectors, in Garret, R., Ruesink, B., and G. McLean (eds): Virus Epidemics: Monitoring, Modelling and Predicting Outbreaks, New York, Academic Press, pp. 119–145.

7. Boag, B., and Brown, D.J.F., 1987, The occurrence of *Longidorus vineacola* in Scotland with notes on its distribution in Europe, *Nematol. Medit.* **15**:51–57.

8. Bor, N.A., and Kuiper, K., 1966, Gevoilighed van *Trichodorus teres* en *Trichodorus pachydermus* voor uitvendige invloeden, *Meded. Rijks Landouwetensch. Gent.* **31**:609–616.

9. Bovey, R., and Martelli, G.P., 1986, The viroses and virus-like disease of the grapevine. A bibliographic report 1979–1984, *Vitis* **25**:227–275.

10. Bovey, R., Gartel, W., Hewitt, W.B. Martelli, G.P., and Vuittenez, A, 1980, Virus and virus-like diseases of grapevine, Lausanne, Payot, 181 p.

11. Brown, D.J.F., 1983, The distribution of Longidoridae (Nematoda) in Europe and variation in the morphology, biology and virus transmission of *Xiphinema diversicaudatum* (Micol.) Thorne, 1939 Ph.D. Thesis, Open University, Milton Keynes, England, 303 p.

12. Brown, D.J.F., 1985., The transmission of two strains of strawberry latent ringspot virus by populations of *Xiphinema diversicaudatum* (Nematoda: Dorylamoidea), *Nematol. Medit.* **13**:217–223.

13. Brown, D.J.F., 1986a, Transmission of virus by the progeny of crosses between *Xiphinema diversicaudatum* (Nematoda:Dorylamoidea) from Italy and Scotland, *Rev. Nematol.* **9**:71–74.

14. Brown, D.J.F., 1986b, The transmission of two strains of arabis mosaic virus from England by populations of *Xiphinema diversicaudatum* (Nematoda: Dorylamoidea) from ten countries, *Rev. Nematol.* **9**:83–87.

15. Brown, D.J.F., and Taylor, C.E., 1981, Variazioni nella trasmissione di virus tra popolazioni di nematodi vettori Longidoridae, *Atti Giornate Nematol., Soc. ital. Nematol., Firenze 1981,* pp. 191–204.

16. Brown, D.J.F., and Taylor, C.E., 1987, Comments on the occurrence and geographical distribution of longidorid nematodes in Europe and the Mediterranean region, *Nematol. Medit.* **15**:333–373.

17. Brown, D.J.F., and Topham, P.B., 1985, Morphometric variability between

populations of *Xiphinema diversicaudatum* (Nematoda:Dorylamoidea), *Rev. Nematol.* **8:**15–26.

18. Brown, D.J.F., and Trugdill, D.L., 1983, Differential transmissibility of arabis mosaic and strawberry latent ringspot viruses by three populations of *Xiphinema diversicaudatum* (Nematoda:Dorylamoidea), *Rev. Nematol.* **6:**229–238.

18a. Brown, D.J.F., and Boag, B., 1988, An examination of methods used to extract virus–vector nematodes (Nematoda:Longidoridae and Trichodoridae) from soil samples, *Nematol. Medit.* **16:**93–99.

19. Brown, D.J.F., Ploeg, A.T., and Robinson, D.J., 1988, Transmission of viruses by trichodorid nematodes, *Ann. Rpt. Scott. Crop Res. Inst. 1987:* 130–131.

20. Brown, D.J.F., Murant, A.F., and Trudgill, D.L., 1988, Transmission of serotypes of tomato black ring virus (TBRV) by *Longidorus attenuatus, Ann. Rpt. Scott. Crop. Res. Inst. 1987:*131–132.

21. Cadman, C.H., 1959, Potato stem-mottle disease in Scotland, *Eur. Potato J.* **2:**165–175.

22. Chagas, C.M., and Silberschmidt, K., 1972, Virus da faixa amarela da alcachofra no Estado de Sao Paulo, *O Biologico,***38:**35–40.

23. Chu, P.W.G., and Francki, R.I.B., 1979, The chemical subunit of tobacco ringspot virus coat protein, *Virology* **93:**398–412.

24. Chu, P.W.G., Francki, R.I.B., and Hatta, T., 1983, Some properties of tomato ringspot virus isolated from *Penta lanceolata* in South Australia, *Pl. Pathol.* **32:**353–356.

25. Conti, M., and Masenga, V., 1977, Identification and prevalence of pepper viruses in Northern Italy, *Phytopathol. Z.* **90:**212–222.

26. Converse, R.H., 1981, Infection of cultivated strawberries by tomato ringspot virus, *Phytopathology* **71:**1149–1152.

27. Coomans, A., 1985, A phylogenetic approach to the classification of the Longidoridae (Nematoda:Dorylaimida), *Agricult. Ecosystems Environ.* **12:**335–354.

28. Corte, A., 1968, D. Soil-borne viruses associated with a peach disease occurring in North Italy, *Tagunsgber. Dal.* **97:**187–194.

29. Cropley, R., 1961, Viruses causing rasp-leaf and similar diseases in sweet cherry, *Ann. Appl. Biol.* **49:**530–538.

30. Dalmasso, A., 1970, Influence directe de quelques facteurs écologiques sur l'activité biologique et la distribution des espèces françaises de la famille Longidoridae (Nematoda:Dorylaimida), *Ann. Zool. Ecol. Anim.* **2:**163–200.

31. DeWaele, D., Mancini, G., Roca, F., and Lamberti, F., 1982, *Trichodorus taylori* sp n. (Nematoda:Dorylaimida) from Italy, *Nematol. Medit.* **10:**27–37.

32. Dodd, S.M., and Robinson, D.J., 1984, Nucleotide sequence homologies among RNA species of strains of tomato black ring virus and other nepoviruses, *J. Gen. Virol.* **65:**1731–1740.

33. Francki, R.I.B., Milne, R.G., and Hatta, T., 1985, *Atlas of Plant Viruses* II, Boca Raton, FL, CRC Press, pp. 23–38.

34. Forer, L.B., Hill, N., and Powell, C.A., 1981, *Xiphinema rivesi,* a new tomato ringspot vector, *Phytopathology* **71:**874.

35. Gallitelli, D., Savino, V., and Martelli, G.P., 1982, The middle component of strawberry latent ringspot virus, *J. Gen. Virol.* **55:**169–172.

36. Gallitelli, D, Martelli, G.P., and Rana G.L., 1984, Artichoke vein banding virus, CMI/AAB Descriptions of Plant Viruses, No. 285, Commonwelath Mycological Institute/Association of Applied Biologists, Kew, Surrey, England.
37. Gibbs, A.J., and Harrison, B.D., 1964, A form of pea early browning virus found in Britain, *Ann. Appl. Biol.* **54:**1–11.
38. Goldbach, R., and Van Kammen, A., 1985, Structure, replication and expression of the bipartite genome of cowpea mosaic virus, in Davies J.W. (ed): Molecular Plant Virology, Boca Raton, FL, CRC Press, pp. 83–120.
39. Hanada, K., Kusunoki, M., and Iwaki, M., 1986, Properties of virus particles, nucleic acid and coat protein of cycas necrotic stunt virus, *Ann. Phytophathol. Soc. Japan* **52:**422–427.
40. Harrison, B.D., 1977, Ecology and control of viruses with soil-inhabiting vectors, *Ann. Rev. Phytopathol.* **15:**331–360.
41. Harrison, B.D., and Hanada, K., 1976, Competitiveness between genotypes of raspberry ringspot virus is mainly determined by RNA-1, *J. Gen. Virol.* **31:**455–457.
42. Harrison, B.D., and Murant, A.F., 1977, Nepovirus group, CMI/AAB Descriptions of Plant Viruses No. 185, Commonwealth Mycological Institute/Association of Applied Biologists, Kew, Surrey, England.
43. Harrison, B.D., and Murant A.F., 1977, Nematode transmissibility of pseudo-recombinant isolates of tomato black ring virus *Ann. Appl. Biol.,* **86:**209–212.
44. Harrison, B.D., and Robinson, D.J., 1978, The tobraviruses, *Adv. Virus Res.* **23:**25–77.
45. Harrison B.D., and Robinson, 1981, Tobraviruses, in Kuzstak, E. (ed): Handbook of Plant Viruse Infections and Comparative Diagnosis, Amsterdam, Elsevier/North Holland, pp. 516–540.
46. Harrison, B.D., and Robinson, D.J., 1986, Tobraviruses in Van Regenmortel, M.H.V., and Fraenkel-Konrat, H. (eds): The Plant Viruses, Volume 2, New York, Plenum Press, pp. 339–369.
47. Harrison, B.D., Finch, J.T., Gibbs, A.J., Hollings, M., Shepherd, R.J., Valenta, V., and Wetter, C., 1971, Sixteen groups of plant viruses, *Virology* **45:**356–363.
48. Harrison, B.D., Murant, A.F., and Mayo, M.A., 1972, Two properties of raspberry ringspot virus determined by its smaller RNA, *J. Gen. Virol.* **17:**137–141.
49. Harrison, B.D., Murant, A.F., Mayo, M.A., and Roberts, I.M., 1974, Distribution of determinants for symptom production, host range and nematode transmissibility between two RNA components of raspberry ringspot virus, *J. Gen. Virol.* **22:**233–247.
50. Hewitt, W.B., 1968, Viruses and virus disease of grapevine, *Rev. Appl. Mycol.* **47:**433–455.
51. Hewitt, W.B., Raski, D.J., and Goheen, A.C., 1958, Nematode vector of soil-borne fanleaf virus of grapevine, *Phytopathology* **48:**586–595.
52. Heyns, J., 1974, The genus *Xiphinema* in South Africa. I. *X. americanum* group (Nematoda:Dorylaimida), *Phytophylactica* **6:**157–164.
53. Hoy, J.W., Mircetich, S.M., and Lownsbery, B.F., 1984, Differential

transmission of prunus tomato ringspot virus strains by *Xiphinema californicum*, *Phytopathology* **74**:332–335.

54. Jones, A.T., 1985, Cherry leafroll virus, AAB Descriptions of Plant Viruses, No. 306, Association of Applied Biologists, Warwick, England.
55. Jones, R.A.C., and Kenten, R.H., 1983, Arracacha virus B, CMI/AAB Descriptions of Plant Viruses, No. 270, Commonwealth Mycological Institute/Association of Applied Biologists, Kew, Surrey, England.
56. Jones, R.A.C., Fribourg, C.E., and Koenig, R., 1983, A previously undescribed nepovirus isolated from potato in Peru, *Phytopathology* **73**:195–198.
57. Jones, A.T., Mitchell, M.J., and Brown, D.J.F., 1989, Infectibility of some new raspberry cultivars with arabis mosaic and raspberry ringspot viruses and further evidence for variation in British isolates of these two nepoviruses. *Ann. appl. Biol.* **114** (in press).
58. Klos, E.J., 1976, Rosette mosaic, in Virus Diseases and Noninfectious Disorders of Stone fruits in North America, Agric. Handbook 437, USDA, Washington D.C., pp. 135–138.
59. Kuiper, K., and Loof, P.A.A., 1962, *Trichodorus flavensis* n. sp. (Nematoda:Enoplida). A plant nematode from new polder soil, *Versl. Pl. Ziekt. Dienst. Wageningen* **136**:193–200.
60. Kurppa, A, Jones, T.A., Harrison, B.D., and Bailiss, K.W., 1981, Properties of spinach yellow mottle, a distinctive strain of tobaco rattle virus, *Ann. Appl. Biol.* **98**:243–254.
61. Kusunoki, M., Hanada, K., Iwaki, M., Chang, M.U., Doi, Y., and Yora, K., 1986, Cycas necrotis stunt virus, a new member of nepoviruses found in *Cycas revoluta*. Host range, purification, serology and some other properties, *Ann. Phytopathol. Soc. Japan* **52**:302–311.
62. Lamberti, F., and Bleve-Zacheo, T., 1979, Studies on *Xiphinema americanum sensu lato* with descriptions of fifteen new species (Nematoda:Longidoridae), *Nematol. Medit.* **7**:51–106.
63. Lamberti, F., and Roca, F., 1987, Present status of nematodes as vectors of plant viruses, in Veech, J., and Dickson, D.W. (eds): Vistas on Nematology, Soc. Nematol., Hattsville, pp. 321–328.
64. Lima, M.B., 1965, Studies on species of the genus *Xiphinema* and other nematodes, Ph. D. Thesis, University of London, 163 p.
65. Loof, P.A.A., 1973, Taxonomy of the *Trichodorus-aequalis* complex (Diphtherophorina), *Nematologica* **19**:49–61.
66. Loof, P.A.A., 1975, Taxonomy of Trichodoridae, in Lamberti, F., Taylor, C.E., and Seinhorst, J.W. (eds): Nematode Vectors of Plant Viruses, New York, Plenum Press, pp. 103–127.
67. Marte, M., Gadani, F., Savino V., and Rugini, E., 1986, Strawberry latent ringspot virus associated with a new disease of olive in Central Italy, *Pl. Disease* **70**:171–172.
68. Martelli, G.P., 1975, Some features of nematode-borne viruses and their relationships with the host plants, in Lamberti, F., Taylor, C.E., and Seinhorst, J.W. (eds): Nematode Vectors of Plant Viruses, New York, Plenum Press, pp. 223–252.
69. Martelli, G.P., 1978, Nematode-borne viruses of grapevine, their epidemiology and control, *Nematol. Medit.* **6**:1–27.

70. Martelli, G.P., and Gallitelli, D., 1985, Virosi dell'olivo, *Italia Agric.* **122(2):**150–156.
71. Martelli, G.P., and Prota, U., 1985, Virosi della vite, *Italia Agric.* **122(2):**201–228.
72. Martelli, G.P., and Quacquarelli, A., 1972, Grapevine chrome mosaic virus, CMI/AAB Descriptions of Plant Viruses, No. 103, Commonwealth Mycological Institute/Association of Applied Biologists, Kew, Surrey, England.
73. Martelli, G.P., Rana, G.L., and Savino, V., 1977, Artichoke Italian latent virus, CMI/AAB Descriptions of Plant Viruses, No. 176, Commonwealth Mycological Institute/Association of Applied Biologists, Kew, Surrey, England.
74. Martelli, G.P., Quacquerelli, A., Gallitelli, D., Savino, V., and Piazzolla, P., 1978, A tentative grouping of nepoviruses, *Phytopath. Medit.* **17:**147–147.
75. McElroy, F.D., 1977, Nematodes as vectors of plant viruses. A current review, *Proc. Am. Phytopathol. Soc.* **4:**1–10.
76. McGuire, J.M., Kim, K.S., and Douthit, L.M., 1970, Tobacco ringspot in the nematode *Xiphinema americanum, Virology* **42:**212–216.
77. McNamara, D.G., 1978, Studies on the ability of the nematode *Xiphinema divesicaudatum* (Micol.) to transmit raspberry ringspot virus and to survive in plant-free soil, Ph.D. Thesis, University of Reading, England, 280 p.
78. McNamara, D.G., and Flegg, J.J.M., 1981, The distribution of virus–vector nematodes in Great Britain in relation to past and present vegetation, in Thresh, J.M. (ed): Pests, Pathogens and Vegetation, London, Pitman, pp. 225–235, 977.
79. Matthews, R.E.F., 1982, Classification and nomenclature of viruses, *Intervirology* **17:**1–199.
80. Migliori, A., Marzin, H., and Rana, G.L., 1984, Mise en evidence du tomato black ring virus (TBRV) chez l'artichaut en France, *Agronomie* **4:**683–686.
81. Mircetich, S.M., Sandborn, R.R., and Ramos, D.E., 1980, Natural spread, graft transmission and possible etiology of walnut blackline disease, *Phytopathology* **70:**962–968.
82. Mojtahedi, H., Sturhan, D., Akhiani, A., and Barooti, S., 1980, *Xiphinema* species in Iranian vineyards, *Nematol. Medit.* **8:**165–170.
83. Murant, A.F., 1970, The importance of wild plants in the ecology of nematode-transmitted viruses, *Outlook Agric.* **6:**114–121.
84. Murant, A.F., 1970, Arabis mosaic virus, CMI/AAB Descriptions of Plant Viruses, No. 16, Commonwelath Mycological Institute/Association of Applied Biologists, Kew, Surrey, England.
85. Murant, A.F., 1974, Strawberry latent ringspot virus, CMI/AAB Descriptions of Plant Viruses, No. 126, Commonwealth Mycological Institute/Association of Applied Biologists, Kew, Surrey, England.
86. Murant, A.F., 1981, Nepoviruses, in Kurstak, E. (ed): Handbook of Plant Virus Infections and Comparative Diagnosis, Amsterdam, Elsevier/North Holland, pp. 197–238.
87. Murant, A.F., and Taylor, M., 1978 Estimates of molecular weights of nepovirus RNA species by polyacrylamide gel electropphoresis under denaturing conditions, *J. Gen. Virol.* **41:**53–61.
88. Murant, A.F., Taylor, C.E., and Chambers J., 1968, Properties, relationships and transmission of a strain of raspberry ringspot virus infecting

rasberry cultivars immune to the common Scottish strain, *Ann. Appl. Biol.* **61:**175–186.

89. Murant, A.F., Jones, A.T., Martelli, G.P., and Stace-Smith, R., 1990, Nepoviruses: Diseases and virus identification, in Harrison, B.D., and Murant, A.F. (eds): The Plant Viruses, Volume 5, New York, Plenum Press, (in press).
90. Nyland, G., Lownsbery, B.F., Loew, S.K., and Mitchell, J.F., 1969, The transmission of cherry rasp-leaf virus by *Xiphinema americanum, Phytopathology* **59:**1111–1112.
91. Pacini, E., and Cresti, M., 1977, Viral particles in developing pollen grains of *Olea europaea, Planta* **137:**1–4.
92. Pelmaeker, M. de, and Coomans, A., 1986, Virusvectornematoden in Aadeppelvelden, *Parasitica* **42:**11–16.
93. Quacquarelli, A., Martelli, G.P., and Vovlas, C., 1974; Chicory yellow mottle virus, CMI/AAB Descriptions of Plant Viruses, No. 132, Commonwealth Mycological Institute/Association of Applied Biologists, Kew, Surrey, England.
94. Rana, G.L., and Martelli, G.P., 1983, Virosi del carciofo, *Italia Agric.* **120(1):**27–38.
95. Rana, G.L., Kyriakopoulou, P.E., and Martelli, G.P., 1983, Artichoke yellow ringspot virus, CMI/AAB Descriptions of Plant Viruses, No. 271, Commonwealth Mycological Institute/Association of Applied Biologists, Kew, Surrey, England.
96. Rana, G.L., Castrovilli, S., Gallitelli, D., and Kyriakopoulou, P.E., 1985, Studies on two serologically distinct raspberry ringspot virus strains from artichoke, *Phytopathol. Z.* **112:**222–228.
97. Raski, D.J., Maggenti, A.R., and Jones, N.O., 1973, Location of grapevine fanleaf and yellow mosaic virus particles in *Xiphinema index, J. Nematol.* **5:**208–211.
98. Rau, J., 1975, Das Vorkommen Virusubertragender Nematoden in ungerstorten Biotopen Niedersachsens, Diss. Techn. Univ. Hannover, 169 p.
99. Remah, A., Jones, A.T., and Mitchell, M.J., 1986, Purification and properties of lucerne Australian symptomless virus, a new virus infecting lucerne in Australia, *Ann. Appl. Biol.* **109:**307–315.
100. Roberts, I.M., and Brown, D.J.F., 1980, Detection of six nepoviruses in their nematode vectors by immunosorbent electron microscopy, *Ann. Appl. Biol.* **96:**187–192.
101. Robertson, W.M., 1987, Possible mechanisms of virus retention in virus vector nematodes, *Ann. Rpt. Scott, Crop Res. Inst. 1986*:127.
102. Robertson, W.M., and Henry, C.E., 1986a, A possible role of carbohydrates in the retention of nematode-transmitted viruses, *Ann. Rpt. Scott. Crop Res. Inst. 1985*:113.
103. Robertson, W.M., and Henry, C.E., 1986b, An association of carbohydrates with particles of arabis mosaic virus retained within *Xiphinema diversicaudatum, Ann. Appl. Biol.* **109:**299–305.
104. Robertson, W.M., and Wyss, U., 1983, Feeding processes of virus-transmitting nematodes, in Harris, K.F., (ed): Current Topics in Vector Research, New York, Springer-Verlag, pp. 271–295.
105. Robinson, D.J., and Clark, J., 1987, Genome sequence homology among

strains of raspberry ringspot nepovirus, *Ann. Rpt. Scott. Crop Res. Inst. 1986*:172–173.

106. Robinson, D.J., and Harrison, B.D., 1985a, Unequal variation in the two genome parts of tobraviruses with evidence for the existence of three separate viruses, *J. Gen. Virol.* **66:**171–176.
107. Robinson, D.J., and Harrison, B.D., 1985b, Evidence that broad bean yellow band virus is a new serotype of pea-early browning virus, *J. Gen. Virol.* **66:**2003–2009.
108. Robinson, D.J., Hamilton, W.D.O., Harrison, B.D., and Baulcombe, D.C., 1987, Two anomalous tobravirus isolates. Evidence for RNA recombination in nature, *J. Gen. Virol.* **68:**2551–2561.
109. Roca, F., Martelli, G.P., Lamberti, F., and Rana, G.L., 1975, Distribution of *Longidorus attenuatus* Hooper in Apulian artichoke fields and its relationship with artichoke Italian latent virus, *Nematol. Medit.* **3:**91–101.
110. Rudel, M., 1985, Grapevine damage induced by particular virus–vector combinations, *Phytopatholo. Medit.* **24:**183–185.
111. Russo, M., Gallitelli, D., Vovlas, C., and Savino, V., 1984, Properties of broad bean yellow band virus, a possible new tobravirus, *Ann. Appl. Biol.* **105:**223–230.
112. Salazar, L.F., and Harrison, B.D., 1979, Potato black ringspot virus, CMI/AAB Descriptions of Plant Viruses, No. 206, Commonwealth Mycological Institute/Association of Applied Biologists, Kew, Surrey, England.
113. Salomao, T.A., 1973, Soil transmission of artichoke yellow band virus, *Atti II Congr. Internaz. Studi Carciofo, Bari 1973*:831–854.
114. Sauer, M.R., 1968, Nematodes in an irrigated vineyard, *Nematologica,* **14:**457–458.
115. Savino, V., Gallitelli, D., and Barba, M., 1983, Olive latent ringspot virus, a newly recognized virus infecting olive in Italy, *Ann. Appl. Biol.* **1203:**243–253.
116. Savino, V., Cherif, C., and Martelli, G.P., 1985, A natural serological variant of grapevine fanleaf virus, *Phytopathol. Medit.* **24:**29–34.
117. Savino, V., Quacquarelli, A., Gallitelli, D., Piazzolla, P., and Martelli, G.P., 1977, Il virus dell'accartocciamento fogliare del ciliegio nel noce. I. Identificazione e caratterizzazione, *Phytopathol. Medit.* **16:**96–102.
118. Sequeira, L., 1978, Lectins and their role in host-pathogen specificity, *Ann. Rev. Phytopathol.* **16:**453–481.
119. Smith, S.H., and Traylor, J.A., 1969, Stem pitting of yellow bud mosaic virus-infected peaches, *Pl. Dis. Rep.* **53:**666–667.
120. Stace-Smith, R., 1985, Tobacco ringspot virus, AAB Descriptions of Plant Viruses, No. 309, Association of Applied Biologists, Warwick, England.
121. Stace-Smith, R., and Hansen, A.J., 1976, Cherry rasp-leaf virus, CMI/AAB Descriptions of Plant Viruses, No. 159, Commonwealth Mycological Institute/Association of Applied Biologists, Kew, Surrey, England.
122. Stace-Smith, R., and Ramsdell, D.C., 1987, Nepoviruses of the Americas, in Harris, K.F. (ed): *Current Topics in Vector Research,* New York, Springer-Verlag, pp. 131–166.
123. Tarjan, A.C., 1969, Variation within the *Xiphinema americanum* group (Nematoda:Longidoridae), *Nematologica* **15:**241–252.
124. Taylor, C.E., 1971, Nematodes as vectors of plant viruses, in Zuckerman,

B.M., Mai, W.F., and Rhode, R.A. (eds): Plant Parasitic Nematodes/ Volume II, New York, Academic Press, pp. 185–211.
125. Taylor, C.E., 1978, Plant-parasitic Dorylaimida: Biology and virus transmission, in Southey, J.F. (ed): Plant Nematology, London, HMSO, pp. 232–243.
126. Taylor, C.E., 1980, Nematodes, in Harris, K.F., and Maramorosch, K. (ed): Vectors of Plant Pathogens, New York, Academic Press, pp. 375–416.
127. Taylor, C.E., and Brown, D.J.F., 1981, Nematode-virus interactions, in Zuckerman, B.M., Rhode, R.A. (ed): Plant Parasitic Nematodes Volume III, Academic Press, New York, pp. 281–301.
128. Taylor, C.E., and Cadman, C.H., 1969, Nematode vectors, in Maramorosch, K. (ed): Viruses, Vectors and Vegetation, New York, Interscience, pp. 55–94.
129. Taylor, C.E., and Robertson, W.M., 1969, The location of raspberry ringspot and tomato black ring viruses in the nematode vector *Longidorus elongatus* (de Man), *Ann. Appl. Biol.* **64:**43–48.
130. Taylor C.E., and Robertson, M.W., 1970a, Sites of virus retention in the alimentary tract of the nematode vectors *Xiphinema diversicaudatum* (Micol.) and *X. index* (Thorne et Allen), *Ann. Appl. Biol.* **66:**373–380.
131. Taylor, C.E., and Robertson, W.M., 1970b, The location of tobacco rattle virus in the nematode vector *Trichodorus pachydermus* Seinhorst, *J. Gen. Virol.* **6:**179–182.
132. Taylor, C.E., and Robertson, W.M., 1975, Acquisition, retention and transmission of viruses by nematodes, in Lamberti, F., Taylor C.E., and Seinhorst, J.W. (eds): Nematode Vectors of Plant Viruses, New York, New York, pp. 253–276.
133. Taylor, C.E., and Robertson, M.W., 1977, Virus vector relationships and mechanics of transmission, *Proc. Am. Phytopathol. Soc.* **4:**20–29.
134. Taylor, C.E., Thomas, P.R., and Converse, R.H., 1966, An outbreak of arabis mosaic virus and *Xiphinema diversicaudatum* (Micolestsky) in Scotland, *Pl. Pathol.* **15:**170–174.
135. Topham, P.B., and Alphey, T.J.W., 1985, Faunistic analysis of Longidorid nematodes in Europe, *J. Biogeogr.* **12:**165–174.
136. Trudgill, D.L., and Brown, D.J.F., 1978a, Frequency of transmission of some nematode-borne viruses, in Scott, P.R., and Bainbridge, A. (eds): Plant Disease Epidemiology Oxford, Blackwell, pp. 281–289.
137. Trudgill, D.L., and Brown, D.J.F., 1978b, Ingestion, retention and transmission of two strains of raspberry ringspot virus by *Longidorus macrosoma, J. Nematol.* **10:**85–89.
138. Trudgill, D.L., Brown, D.J.F., and Robertson, W.M., 1981, A comparison of the four British vector species of *Longidorus* and *Xiphinema, Ann. Appl. Biol.* **99:**63–70.
139. Trudgill, D.L., Brown, D.J.F., and McNamara, D.G., 1983, Methods and criteria for assessing the transmission of plant viruses by longidorid nematodes, *Revue Nematol.* **6:**133–141.
140. van Hoof, H.A., 1962, *Trichodorus pachydermus* and *T. teres,* vectors of the early browning virus of peas. *Tijdschr. Plantenziekten* **68:**391.
141. van Hoof, H.A., 1968, Transmission of tobacco rattle virus by *Trichodorus* species, *Nematologica* **14:**20–24.

142. Vrain, T.C., and Yorston, J.M., 1987, Plant parasitic nematodes in orchards of the Okanagan valley of British Columbia, Canada, *P. Disease* **71:**85–86.
143. Yagita, H., and Komuro, Y., 1972, Transmissiom of mulberry ringspot virus by *Longidorus martini* Merny, *Ann. Phytopathol. Soc. Japan* **38:**275–283.
144. Zhigang, Xu, Polston, J., and Goodman, R.M., 1986, Identification of soybean mosaic, southern bean mosaic and tobacco ringspot viruses from soybean in the People's Republic of China, *Ann. Appl. Biol.* **108:**51–57.

7
Detecting Plant Viruses in Their Vectors

Roger T. Plumb

Introduction

Until the advent of such techniques as electron microscopy and serology, the principal way of determining whether a suspected virus vector contained a plant-infecting virus was to allow it access to a plant host known to be susceptible to the virus. In other words, a bioassay of the vector for the appropriate virus was performed. An alternative approach was to use extracts of whole suspected vectors to inoculate susceptible test species mechanically. The latter method was obviously of limited use, and its usefulness was greatly affected by low virus concentrations in potential vectors and inactivation of virus in vector extracts.

Our fascination with the interaction of plant virus and vector remains, and new techniques allow some of the existing problems to be overcome. There are three principal reasons to detect viruses in vectors: to investigate modes and mechanisms of transmission and sites of specificity; to assess the presence of virus-carrying vectors rapidly; and to facilitate epidemiological studies and disease control.

In this chapter, the various methods that have been, or are currently being, used to further these aims are described in relation to the different vector groups. Broadly, the recent methods fall into four categories.

1. Electron microscopy. *Directly* observing extracts of whole putative vectors or separate organs or body fluids and ultrathin sections of vector tissues or *indirectly* visualizing viruses by immunological tagging using fluorescence, ferritin, or other labels, either in situ or in extracts of vectors using immunospecific electron microscopy (ISEM).
2. Nucleic acid hybridization. Using cDNA probes to identify the presence of viruses in extracts of whole vectors.
3. Serology. Testing extracts of whole vectors, usually using variations of the enzyme-linked immunosorbent assay (ELISA).

Roger T. Plumb, Institute of Arable Crops Research, Rothamsted Experimental Station, Harpenden, Herts., AL5 2JQ, England.

4. Alternative methods. Tagging viruses or virus components with radioactive isotopes.

These methods are not mutually exclusive; thus, the labeling techniques, especially those used to locate virus in vectors, are usually linked with electron microscopy or, if only a "present" or "absent" answer is needed, light or fluorescence microscopy may be adequate.

The extent of the work done on different vector groups varies greatly. This reflects both the interest in and ease of working with the particular vectors as well as the complexity of the virus vector interaction. Thus, more work has been done on aphids and plant and leafhoppers than on other groups presumably because they are the largest vector groups and because they, and, especially aphid vectors, have a wide range of relationships with viruses, from the ephemeral nonpersistent relationship to the circulative/propagative relationship.

Aphids

Plant virologists were concerned for many years, and to some extent still are concerned, with the apparent lack of involvement of a vector or vectors, other than man, in the spread of tobacco mosaic virus (TMV). Thus, Ossiannilson (55) dissected out the guts of *Aulacorthum solani* and *Myzus persicae* fed on TMV-infected plants and detected TMV, but the visual evidence for the presence of particles was not strong. The relevance of these observations to the putative transmission of TMV was not clarified by the discovery of TMV-like particles in the leafhopper *Eupteryx atropunctata*. These earliest results also illustrated a problem that has continued to complicate the interpretation of the results when the presence of virus in a vector was being assessed, irrespective of the method used. The mere presence of virus implies no functional relationship with the vector in which the virus is found. The distinction between vectors that are viruliferous, that is, that contain or carry a virus, and those that are infective (inoculative), that is, that can infect a plant with the virus they carry, is critical.

When virus–vector interactions that were known to result in transmission were investigated, the shortcomings and inadequacies of the available methods became clear. Kikumoto and Matsui (43), using electron microscopy, investigated the acquisition of TMV, potato virus X (PVX), and potato virus Y (PVY) by *M. persicae*. Only PVY is transmitted by this aphid but the evidence obtained, by dissecting aphids and examining them for the presence of virus particles, was that TMV and PVX had been acquired but that PVY had not. The conclusion, in the face of the knowledge of PVY transmission by *M. persicae,* was that because PVY was present at a lower concentration than the other viruses in infected plants, not enough of the virus was imbibed during feeding to be detected.

Ultrathin sections of aphids examined after they had had access to tobacco severe etch virus (86), anthriscus yellows virus (AYV) (50), or PVY (47) demonstrated that particles of the severe etch virus were apparently linked to the 20 μm distal part of the maxillary food canal, that PVY was present in the anterior alimentary tract, and that particles of AYV were apparently embedded in a matrix in the anterior part of the foregut and midgut. For PVY, fixing with formaldehyde increased the number of particles detected. However, Lopez-Abella et al. (47) questioned the usefulness of thin sections in detecting ingested virus, but in their work, ELISA of whole aphids or parts of aphids was found to be ineffective in detecting the virus present.

These results did, however, support the concept that nonpersistently and semipersistently transmitted viruses, at least, appeared to be carried on sites close to the point of acquisition—the stylets or immediate foregut.

When electron microscopy was first used to investigate transmission of virus, the possibility that viruses were carried externally was still a live issue. The favored sites were the stylets, on which scanning electron microscopy (SEM) could detect a variety of ridges and furrows. In the earliest SEM studies, the choice of materials was again *M. persicae* and TMV, but it, additionally, included southern bean mosaic virus (SBMV), another highly infectious agent that reaches large concentrations in plants but that is not transmitted by aphids (21). Both TMV and SBMV could be seen, although the images were poor, but they could not be found on stylets. When visualization of particles was enhanced with an immunospecific latex label, more pea seed-borne mosaic virus (PSbMV) carried on an efficient vector biotype of *Macrosiphum euphorbiae* could be distinguished, compared with that on an inefficient vector biotype (46). The latex-labeled virus was detected mostly on the inner surface of the mandibular stylets, 50 μm from the tip. No labeled virus was seen on the mandibular ridges, maxillary projections, food duct, or common food chamber. It was also shown that, after a subsequent inoculation feed, most of the label had been lost.

The development of quick, sensitive immunological methods for detecting plant viruses, such as ELISA (16), and serological (or immunological) specific electron microscopy [S(I)SEM] (23, 69), revived interest in trying to detect and localize virus in their vectors (61). Because of the ephemeral association of viruses transmitted nonpersistently by aphids, it was thought that detecting them in vectors would be difficult. However, Gera et al. (31), using strains of cucumber mosaic virus with different efficiencies of transmission by *Aphis gossypii*, detected virus in single homogenized aphids. The absorbance values were greater after a 90-second than a

20-minute acquisiton feed, and an estimate of 0.01 to 0.1 ng virus/aphid was made; the ability to detect virus was lost after a short probe. All the observations were made on wingless aphids, and the authors cautioned that, for epidemiological application the results would need to be confirmed on winged aphids.

Studies with PVY were less successful (12). Although the virus could be detected in groups of *M. persicae* and there was a positive correlation between detection and transmission by equivalent aphids, ELISA always underestimated the rate of transmission and did not detect the virus in *Aphis craccivora, A. citricola,* or *A. gossypii,* all vectors of PVY, but did in some groups of *Acyrthosiphon pisum,* which is not a vector. Similar difficulties were encountered with the semipersistently transmitted citrus tristeza virus (11), which could be detected only in large groups of aphids; nor could vector and nonvector species be distinguished here.

Problems in detecting nonpersistently transmitted virus in aphids were also encountered by Berger et al. (7). Using a dot blot immunobinding assay, they detected as little as 0.5 pg of purified PVY and tobacco vein mottling virus. Although such sensitivity should have permitted the detection of virus in *M. persicae,* no virus was detected even when 100 aphids were extracted and clarified by low-speed centrifugation. Aphids fed on healthy plants or healthy plant extracts reacted as strongly as those exposed to virus, due to their endogenous alkaline phosphatase, the enzyme used in the ELISA test, giving high ELISA readings. Subsequently, when ^{131}I was used to detect PVY and tobacco etch virus (TEV), freeze sections showed that when helper component (HC) (63) was present, the radioactive label was associated with the maxillary stylets and anterior parts of the alimentary canal, as well as the gut of the vector aphids. Where HC was not present, the label accumulated only in the gut. This observation supported a binding role for HC, but its presence had no effect on the amount of virus taken up. Further work with ^{131}I measured the volume of material taken up as 1 to 300 pl when purified virus was acquired through thin membranes and the number of virus particles ranged from 10 to 4000 (64), but the number of particles acquired did not correlate with ability to transmit the virus. Of particular concern relative to these epidemiological studies was that the amount of virus required for transmission was so small as to be below the sensitivity of current detection methods.

The detection of persistently transmitted viruses in aphids has proved rather more straightforward than the detection of nonpersistent viruses, although thresholds of detection and sensitivity, relative to the minimum amount of virus an aphid needs to acquire before it can transmit virus and the mechanism of the passage of viruses through aphids, are far from resolved.

The presence of pea enation mosaic virus (PEMV) in *Acyrthosiphon pisum* was detected by electron microscopy (39, 71) in smears, by an immunofluorescence method (48) and by ELISA (24) used to monitor acquisition and infection. The virus was detectable in nymphs and adults; the quantities measured ranged up to concentrations of 40 ng/ml from nymphs to 200 μg/ml from adults. With prolonged feeding (up to 16 hours), the virus titer increased, and most virus was detected in the gut. No detectable virus remained after a six-day infection feed, and the transmission efficiency of the individual aphids was only poorly related to the amount of virus they contained.

Immunospecific electron microscopy (ISEM) readily detected the potato leaf roll virus (PLRV) in groups and single *M. persicae* and *Macrosiphum euphorbiae* (83) and the *Sitobion avenae* specific isolate of barley yellow dwarf was also detected in *S. avenae* and non-vectors (60). For PLRV, the number of detectable particles in each species ranged from six-to ten-fold. Refinements of ELISA to minimize nonspecific reactions that often make virus detection in vectors difficult allowed as little as 0.01 ng of PLRV/aphid to be detected (83), but again there were large differences in the virus concentration in individual aphids. Clarke et al. (17) could only detect PLRV in groups of 30 or more *M. persicae* fed on infected *Physalis floridana*. Tamada and Harrison (83) detected PLRV in 67 *M. persicae* after a three-day infection feed, but only 28 of these aphids were transmitters. There was no evidence that PLRV multiplied in its vectors and when they were transferred to a host immune to the virus, PLRV concentration in them declined, quickly at first, possibly as the virus in the gut passed through the aphid and was lost, and then more slowly, perhaps because it was limited by the rate of movement of the virus from the hemocoel through the salivary gland membrane.

The possible route and mechanism by which luteoviruses reach the salivary glands have been postulated by Gildow and Rochow (36) and Gildow (34, 35) for barley yellow dwarf virus (BYDV). Particles were often associated with the hindgut apical plasmalemma in ultrathin sections of the infective aphid species *Rhopalosiphum padi* and appeared to pass through the hindgut cell cytoplasm via coated vesicles. When antiserum homologous to the strain of virus was injected into the hemocoel, clumps of particles were subsequently detected on the basal plasmalemma.

Once they are present in the hemocoel, particles have access to many aphid organs although they have been seen associated only with the accessory salivary glands, often in the tubular vesicles. Gildow (35) used immunocytochemical methods to identify the isolate of BYDV present by ferritin labeling. Such methods have much wider application to studies of viruses and their localization within vectors. These results obtained in fine structural studies of the movement of BYDV through its vectors supported results obtained by ultrathin sectioning of *M. persicae* that had had access to either PLRV or to beet western yellows virus (BWYV) (33).

Virus detection in vectors in epidemiological studies has not replaced direct or biological tests of infectivity. This is partly because of the difficulties of interpreting the results in a biologically significant way. Some of these difficulties are as follows:

1. Is the method too sensitive, will it detect the presence of virus in quantities less than that needed for transmission?
2. Is the method too insensitive, are infective vectors being missed?
3. Is the presence of a virus evidence for its transmission?
4. Can the test insect's biological fitness as a vector be assumed (i.e., will it feed on a plant susceptible to the virus it carries)?

The relative importance of these questions will differ for each host–virus–vector combination. One such combination in which many of these problems are seen involves the BYDV. In much of north and west Europe, BYDV is endemic, but damaging epidemics are sporadic, partly because the severity of the winter varies enough so that the relative survival value of sexual and asexual morphs changes. Current forecasting schemes for autumn (September, October) integrate vector infectivity, monitored through the use of feeding tests, and potential vector populations, in which insects are caught in suction traps (66). Measurement of infectivity takes two to three weeks, and a quicker measurement would be of great value; a direct aphid vector test would eliminate the need for the time- and labor-intensive biological procedure. Denéchère et al. (22) have demonstrated that ELISA could be used to detect BYDV in groups of *Rhopalosiphum padi*. However, to determine whether such a system is practicable for single aphids, and how the results relate to the biological test, Torrance et al. (92) caught cereal aphids alive in suction traps during the period 1982 to 1984 and assessed their infectivity, either by feeding tests or a modified ELISA with a fluorogenic substrate (MUP-ELISA) (91). As had other workers, they found a wide range in the readings of virus concentration. For *R. padi* there was a fourfold range; for *Sitobion avenae* a fivefold range, for aphids that had access to infected plants.

In the only autumn (1982) when BYDV was found to occur in a relatively large proportion of migrant aphids, biological and immunological assessments gave similar results: 5.1% infective by feeding tests, 8.3% by MUP-ELISA. In autumn 1983 and autumn 1984, there was much less infectivity, which was reflected by both assessment methods, but there was a greater disparity between the results. Because much of the autumn migration of cereal aphids is of sexual forms that are biologically "unfit" vectors of BYDV, this agreement was both surprising and encouraging. Much less encouraging was the discrepancy between feeding tests and MUP-ELISA in the spring and summer, when sexual morphs are not a complicating factor. Ten times as many aphids contained BYDV, as detected by MUP-ELISA, than when the virus was detected biologically through transmission to test plants.

The use in epidemiological studies of immunological methods to detect persistently transmitted virus seems to have a brighter future than for the more ephemeral virus–vector association. Enzyme amplification methods can increase the sensitivity (90), and the use of monoclonal antibodies can overcome some of the problems of detecting persistently transmitted viruses in nonvector species. Thus, for BYDV, the *Sitobion avenae*-transmitted isolate can be acquired by *R. padi,* but rarely, if ever, is the virus transmitted. So its presence in this aphid can be discounted for epidemiological work; indeed, it would be more efficient to use monoclonal antibodies only to those viruses a vector is known to transmit. A weighting factor could be introduced, based on the relative efficiency of transmission, to assess the importance in field studies of a positive immunological result. Also, BYDV and BWYV can be detected by ISEM in aphids when polyclonal (65) and monoclonal antibodies (25) are used.

Viruses, such as the sowthistle yellow vein virus that multiply in their aphid vectors can be readily seen in ultrathin sections (56); lettuce necrotic yellows virus (LNYV) has been detected in *Hyperomyzus lactucae* by ELISA (15). However, although the method is quick, there were discrepancies between detectable virus and transmitted virus. The aphids in which no LNYV was detected did not transmit virus, but some aphids that contained virus failed to infect. Also, LNYV and *H. lactucae* were studied by electron microscopy (51), which was of especial interest because this was the first time multiplication of LNYV had been observed in its vector. Coated and uncoated particles were found throughout the vector, but none was seen in the gut lumen or in the embryos developing within infective aphids.

Complementary DNA probes have also been used to detect a luteovirus, subterranean clover red leaf virus (SCLRV) in *Aulacorthum solani,* in which the virus content was calculated as 850 pg/aphid. However, the nonvector *M. persicae* also contained trace amounts of virus. The extent of hybridization and the degree of transmission, however, were not related (42).

Plant and Leafhoppers

Interest in the location of plant viruses in hopper vectors predates investigations of other vector groups by many years. This is possibly because their comparatively large size made them easier to dissect than aphids. Storey (81) used a novel injection method to investigate the relationship of maize streak virus and *Cicadulina mbila*. By extracting various vector tissues and injecting the extracts into nonviruliferous hoppers, he demonstrated the association of virus with the vector's gut. Bennett and Wallace (6) examined the relationship between the curly top virus and *Eutettix tenellus* by feeding nonviruliferous vector on extracts of tissue from infective vectors. They concluded that blood was the main source of virus and detected a slow decline in virus content after initial acquisition.

The development of serological methods revived interest in virus–vector relations, and the combination most examined was that between wound tumor virus and *Agallia constricta*. Whitcomb and Black (94) used precipitin ring tests to detect soluble antigen and demonstrated multiplication of the virus in the vector. They also tried to relate positive serological reactions to transmission and found generally good agreement, although some nonreacting insects did transmit virus. However, they found a sufficiently consistent antigen titer in insects to believe that they could determine the proportion of vectors that were infective in a batch test. Another method used fluorescent antibodies associated with a smear test (75, 76). Again, there was a good correlation between smear results when the two methods were compared. Fluorescent antibodies were also used to follow the progression of virus through the vector (74). Virus was detected in many tissues but was first detected four days after acquisition in the intestine, from which it spread to all other tissues, 17 days after acquisition. Complementary electron microscopy studies demonstrated WTV particles in the fat body of *A. constricta* (72) and in extracts of vector tissues (29).

Another virus vector combination studied in detail was that of rice dwarf virus and *Nephotettix cincticeps*. Fukushi et al. (28) and Fukushi and Shikata (27) first demonstrated virus in ultrathin sections of vector tissue. Sinha (73) determined the presence of potato yellow dwarf virus in *A. constricta* by injecting extracts of infective vectors into nonviruliferous hoppers that were then tested for transmission.

Two different hopper-transmitted viruses are characterized by their aggregation in vector tissues. Particles of a rhabdovirus, the barley yellow striate mosaic virus, frequently formed membrane-bound aggregates in the salivary glands of *Laodelphax striatellus* (18); reovirus-like particles associated with rice ragged stunt are embedded in viroplasm-like inclusions in salivary gland, nerve tissue, muscles, fat bodies, and the foregut of *Nilaparvata lugens* and are also abundant in crystalline arrays in salivary glands and fat bodies (41). The transmission of maize chlorotic dwarf virus (MCDV) by its vector, *Graminella nigrifrons,* does not fit the usual pattern and can be described as semipersistent or noncirculative. Fine structural observations showed that MCDV accumulated in the foregut and was not associated with the stylets (13, 38, 39).

Recent immunological detection methods have been based on ELISA or ISEM. Oat blue dwarf virus was detected by ELISA in *Macrosteles fascifrons* (4); it appeared, however, that more than 80% were not infective. Thus, the use of such a method would be of little value in epidemiological studies. There was much better correlation of ELISA and infectivity with beet curly top and *Circulifer tenellus* (49), and the method was used to test groups of 10 hoppers from the field to assess virus risk. The maize rayado fino virus can also be detected in salivary glands, the digestive tract, and hemolymph of its vector *Dalbulus maidis* (30, 67), and there is evidence from ELISA that the virus multiplies in the vector (37).

Not surprisingly, several hopper-borne viruses have been tested for detection immunologically. These include rice ragged stunt virus in *Nilaparvata lugens* (40), rice gall dwarf and rice dwarf viruses in *Nephotettix nigropictus,* rice grassy stunt in *N. lugens,* and rice stripe in *Laodelphax striatellus* (52). Three methods were compared to detect Fiji disease virus in *Perkinsiella saccharicida:* ISEM, ELISA, and immuno-osmophoretic assay (26). In general, the results agreed, but viral antigens could be detected in twice as many hoppers as transmitted the virus. An ELISA of maize rough dwarf virus in *L. striatellus* detected virus in 15–20% of vectors that did not transmit virus, and virus concentrations varied widely in individual hoppers but without correlating with transmitters and nontransmitters (10). As with many other tests based on immunological assay, a substantial proportion, 15 to 30%, of insects that were positive in ELISA did not transmit, and the authors thus urged caution in the use of such methods in epidemiological studies.

A serological method was applied to a very different maize virus, maize stripe virus, to test for the presence of a capsid and a noncapsid protein in plants and vectors; only the capsid protein was found in vectors. The significance of this finding is uncertain, but it may be related to the transmission process, although the accompanying evidence that the virus multiplied in the vector suggests otherwise.

Nucleic acid hybridization has been little used so far to detect viruses in hoppers, although doubtless this method as a laboratory and field tool will see more use in the future. Boulton et al. (8) and Boulton and Markham (9) used a spot hybridization method to detect maize streak virus in several hopper species. The known vector *Cicadulina mbila* contained much virus DNA after a seven-day acquisition feed, but the quantities of virus DNA in individual insects had a 100-fold range. Other hoppers with access to infected plants all contained some virus DNA but much less than in *C. mbila.*

Nematodes

Viruses can be detected in nematodes by the "slash test." In this test the nematodes are chopped up in a small drop of water, which is then used as inoculum for test plants. However, this method does not detect all viruses in vector nematodes (68). Electron microscopy has been used to visualize particles of raspberry ringspot and tomato black ring viruses in the lumen of the buccal cavity and in the space between the stylet and the guiding sheath of *Longidorus elongatus* (84), as well as particles of artichoke

Italian latent virus (87). Arabis mosaic virus was also seen to be adsorbed to the cuticle lining the lumen of the odontophore, the anterior esophagus, and the esophageal bulb of *Xiphinema diversicaudatum,* but no particles of raspberry ringspot or grapevine fanleaf viruses, which this vector does not transmit, were found (85). The tubular virus tobacco rattle virus was also found throughout the length of the lumen of the pharynx and esophagus of its vector nematode *Trichodorus pachydermus.*

Immunosorbent electron microscopy has also been used to visualize polyhedral nematode-transmitted viruses in *Xiphinema* spp. and *Longidorus* spp. but although ISEM is quicker and more sensitive than infectivity assays, it did not always detect virus when other methods did (68).

Mites

The eriophyid mites are vectors for several viruses and other virus-like agents, and most of these viruses are associated with members of the Gramineae (58). Interest in the location of plant viruses in mites is almost as great for viruses in nonvectors as it is for known virus–vector interactions. Nearly all this work has used ultrathin sectioning.

To demonstrate interactions between virus and vector when transmission is known, much work has been done with *Eriophyes (Aceria) tulipae* and wheat streak mosaic virus (WSMV) (62). Here, WSMV antigens were detected in vectors, and ultrathin sections showed massive accumulations of virus, particularly in the gut. Virus was detectable after a two-hour acquisition access feed and the number of particles present increased with feeding time (82). In this early work, WSMV was not found in the salivary glands (79), and there was little evidence of any particular mechanism of transmission. However, more recent studies (59) detected virus in salivary glands, which suggests a circulative relationship. No WSMV invades mite cells, and there is no evidence that the virus multiplies in vectors. The particles of onion mosaic virus have also been detected in extracts of *E. tulipae* and in ultrathin sections of mites from infected plants, but the infected tissues were not identified.

Tobacco mosaic virus, potato virus X, brome mosaic virus, tobacco ringspot virus, onion yellow dwarf virus, and tomato bushy stunt virus are not transmitted by mites but all were detected and found to be infectious in the alimentary canal of *Tetranychus urticae.* A detailed search for location of TMV, however, failed to find it on the stylets or in the pharynx (53, 54). Therefore, there seems to be no biological interaction between a vector and the viruses that it carries passively, which illustrates, yet again, the problems associated with the practical use of methods to detect virus in vectors.

Brome mosaic virus in *E. tulipae* has been the subject of studies by

Paliwal (57) and Stein-Margolina (78), and Paliwal (58) has also investigated the presence of the barley stripe mosaic virus, which is pollen- and seed-borne in cereals, in eriophyid mites that do not transmit it.

Fungi

Two groups of soil-borne fungi are known to transmit plant viruses, the Plasmodiophoromycetes and the Chytridiomycetes (5), although much of the evidence for this transmission is by constant association in the field rather than by demonstration in the laboratory. However, *Olpidium brassicae,* a chytrid fungus, and its transmission and relationship with tobacco necrosis virus (TNV) has been examined in detail (88). When zoospores of *O. brassicae* were mixed with suspensions of virus and examined by electron microscopy, virus particles were seen to be adsorbed to the surface membranes of the zoospores from vector-competent strains but not from nonvector strains. When there was a gradation in vector efficiency, it was reflected in the number of adsorbed virus particles (89). Thus, *O. cucurbitacearum* adsorbed cucumber necrosis virus (CNV) particles, but not TNV. The specificity of this attachment appears to be mediated through the virus coat protein, since CNV RNA was not transmitted upon exposure to zoospores of *O. radicale* (80). Thin sections of zoospores showed that virus particles entered the cytoplasm by endocytosis. Virus attached to the plasmalemma of zoospores became detached before cysts were formed, but retraction of the flagellum upon encystment was another means by which virus was taken up.

There is less information on the presence of virus in the plasmodiophorid vectors, of which *Polymyxa* spp. are the most common examples. Adams (2) has reviewed our current knowledge of viruses in vectors. Serological tests have demonstrated beet necrotic yellow vein virus (BNYVV) in resting spores of *Polymyxa betae* (1) and barley yellow mosaic virus in zoospores and resting spores of *P. graminis* (3); it was estimated that 50 to 75 particles were present in each zoospore. Ultrathin sectioning has only demonstrated the presence of BNYVV in *P. betae* (1). Nevertheless, as with many of the other virus vector interactions discussed above, the presence of virus in or on a vector is not necessarily conclusive proof of the ability of the vector to transmit that virus.

Beetles

The presence of infectious virus from macerated vectors of cowpea mosaic virus, and the regurgitant produced by these vectors was demonstrated by mechanical inoculation of indicator plants by Dale (20); Gergerich and Scott (32) and Walters (93) have concentrated on this aspect of the relations of viruses to their bettle vectors. Ultrastructural studies of bean leaf beetles injected with southern bean mosaic virus (SBMV) or the cowpea strain of TMV showed that less than 1% of beetle hemocytes contained SBMV, whereas aggregated TMV was present in 30% (44). Both viruses were present in hemolymph and regurgitant (70), but only SBMV was transmitted. Inoculation methods that determine virus concentration demonstrated that the concentration of virus in regurgitants could equal that in plant sap (45) but that, in *Ceratoma trifurcata,* the ability to recover virus from hemolymph declined with time from acquisition, although the virus could be recovered from all areas of the intestinal tract (77).

Other Vector Groups

No doubt partly because of the difficulty of working with whiteflies and thrips, very few studies have been performed to determine the presence of viruses in their tissues. Tomato spotted wilt virus was detected in individual thrips (*Frankliniella occidentalis* and *Frankliniella schultzei*) that had had access to virus as nymphs but were tested as adults (14). However, there was not good agreement between proportion of virus found and transmission rate.

Nucleic acid hybridization has been used to detect tomato yellow leaf curl in its whitefly vector, *Bemisia tabaci,* after crushing and clarifying the tissues of the vector (19).

Concluding Remarks

In some ways, it is surprising that the quick, very sensitive diagnostic tests now available to determine the presence of plant viruses in their vectors have not been used more. The advantage of crushing a putative vector in the well of an ELISA plate or on a nylon membrane for analysis seems clear, when the alternative is a laborious bioassay. However, a comparison of the former with the latter, and the results obtained with each, has led to reservations about further study. Rarely has there been perfect agreement between presence of virus and transmission of virus, although if the test were used only for virus–vector combinations known, through other work, to be effective, spurious results should decrease. Nevertheless, all remote tests of infectivity give a continuum of results and a line has to be drawn, arbitrarily or statistically, to separate vectors from nonvectors. Results that fall into this uncertain or "gray" area may be few, but they may also be critical. This contrasts with the clear-cut presence of infection or absence of infection as determined by the bioassay.

In the future, the more refined methods that use specific sequence-defined DNA probes or monoclonal antibodies, alone or linked to a cytochemical marker, or colloidal gold, may increase their usefulness in defining sites of activity and recognizing vectors and and observing virus–vector interactions. Subsequent refinements may make these methods very useful in epidemiological vector studies. The fascination of how vectors and viruses interact will ensure continued interest in this area.

References

1. Abe, H., and Tamada, T., 1986, Association of beet necrotic yellow vein virus with isolates of *Polymyxa betae*. Keskin, *Ann. Phytopathol. Soc. Japan* **52:**235–247.
2. Adams, M.J., 1988, Evidence for virus transmission by plasmodiophorid vectors, In Cooper J.I., and Asher M.J.C. (eds): Viruses with Fungal Vectors, Wellesbourne, Warwick, England, Association of Applied Biologists, pp. 203–211.
3. Adams, M.J., Swaby, A.G., and Jones, P., 1988, Confirmation of the transmission of barley yellow mosaic virus (BaYMV) by the fungus *Polymyxa graminis, Ann. Appl. Biol.* **122:**133–141.
4. Azar, M., and Banttari, D., 1981, Enzyme-linked immunosorbent assay versus transmission assay for detection of oat blue dwarf virus in aster leafhoppers, *Phytopathology* **71:**856.
5. Barr, D.J.S., 1988, Zoosporic plant parasites as fungal vectors of viruses: taxonomy and life cycles of species involved, in Cooper J.I. and Asher, M.J.C. (eds): Viruses with Fungal Vectors, Wellesbourne, Warwick, England, Association of Applied Biologists, pp. 123–137.
6. Bennett, C.W., and Wallace, H.E., 1938, Relation of the curly top virus to the vector *Eutettix tenellus, J. Agricult. Res.* **56:**31–50.
7. Berger, P.H., Thornbury, D.W., and Pirone, T.P., 1985, Detection of picogram quantities of potyviruses using a dot blot immunobinding assay. *J. Virol. Methods* **12:**31–39.
8. Boulton, M.I., Markham, P.G., and Davies, J.W., 1984, Nucleic acid hybridization techniques for the detection of plant pathogens in insect vectors, *1984 Brit. Crop. Prot. Cong.—Pests and Disease* **1:**181–186.
9. Boulton, M.I., and Markham, P.G., 1986, The use of squash blotting to detect plant pathogens in insect vectors, in Jones R.A.C., and Torrance, L. (eds): Developments and Applications in Virus Testing, Wellesbourne, Warwick, England, Association of Applied Biologists, pp. 55–69.
10. Caciagli, P., Roggero, P., and Luisoni, E., 1985, Detection of maize rough dwarf virus by enzyme-linked immunosorbent assay in plant hosts and in the planthopper vector, *Ann. Appl. Biol.* **107:**463–471.
11. Cambra, M., Hermoso de Mendoza, H., Moreno, P., and Navarro, L., 1981, Detection of citrus tristeza virus (CTV) in aphids by enzyme-linked immunosorbent assay, *Abst. Meet. Virus Dis. Epid. Oxford,* Wellesbourne, Warwick, England, Association of Applied Biologists, pp. 71–72.
12. Carlebach, R., Raccah, B., and Loebenstein, G., 1982, Detection of potato virus Y in the aphid *Myzus persicae* by enzyme-linked immuno-sorbent assay (ELISA), *Ann. Appl. Biol.* **101:**511–516.

13. Childress, S.A., and Harris, K.F., 1989, Localization of virus-like particles in the foreguts of viruliferous *Graminella nigrifrons* leafhoppers carrying the semi-persistent maize chlorotic dwarf virus, *J. Gen. Virol.* **70:**247–251.
14. Cho, J.J., Mau, R.F.L., Hamaski, R.T., and Gonsalves, D., 1988, Detection of tomato spotted wilt virus in individual thrips by enzyme-linked immunosorbent assay, *Phytopathology* **78:**1348–1352.
15. Chu, P.W.G., and Francki, R.I.B., 1982, Detection of lettuce necrotic yellows virus by an enzyme-linked immunosorbent assay in plant hosts and the insect vector, *Ann. Appl. Biol.* **100:**149–156.
16. Clark, M.F., and Adams, A.N., 1977, Characteristics of the microplate method of enzyme-linked immunosorbent assay for the detection of plant viruses, *J. Gen. Virol.* **34:**475–483.
17. Clarke, R.G., Converse, R.H., and Kojima, M., 1980, Enzyme-linked immunosorbent assay to detect potato leafroll virus in potato tubers and viruliferous aphids. *Pl. Dis.* **64:**43–45.
18. Conti, M., and Plumb, R.T., 1977, Barley yellow striate mosaic virus in the salivary glands of its planthopper vector *Laodelphax striatellus* Fallén, *J. Gen. Virol.* **34:**107–114.
19. Czosnek, H., Ber, R., Navot, N., Zamir, D., Antignus, Y., and Cohen, S., 1988, Detection of tomato yellow leaf curl virus in lysates of plants and insects by hybridization with a viral DNA probe, *Pl. Dis.* **72:**949–951.
20. Dale, W.T., 1953, The transmission of plant viruses by biting insects, with particular reference to cowpea mosaic, *Ann. Appl. Biol.* **40:**384–392.
21. de Zoeten, G.A., 1968, Application of scanning microscopy in the study of virus transmission of aphids, *J. Virol.* **2:**745–751.
22. Denéchère, M., Cante, F., and Lapierre, H., 1979, Détection immunoenzymatique du virus de la jaunisse nanisante de l'orge dans son vecteur *Rhopalosiphum padi* (L.), *Ann. Phytopathol.* **11:**567–574.
23. Derrick, K.S., 1973, Quantitative assay for plant viruses using serologically specific electron microscopy, *Virology* **56:**652–653.
24. Fargette, D., Jenniskens, M.J., and Peters, D., 1982, Acquisition and transmission of pea enation mosaic virus by the individual pea aphid, *Phytopathology* **72:**1386–1390.
25. Forde, S.M.D., 1989, Strain differentiation of barley yellow dwarf virus isolates using specific monoclonal antibodies in immuno-sorbent electron microscopy, *J. Virol. Methods* **23:**313–320.
26. Francki, R.I.B., Ryan, C.C., Hatta, R., Rohozinski, J., and Grivell, C.J., 1986, Serological detection of Fiji disease virus antigens in the planthopper *Perkinsiella saccharicida* and its inefficient ability to transmit the virus, *Pl. Pathol.* **35:**324–328.
27. Fukushi, T., and Shikata, E., 1963, Localization of rice dwarf in its insect vector, *Virology* **21:**503–505.
28. Fukushi, T., Shikata, E., and Kimura, I., 1962, Some morphological characters of rice dwarf virus, *Virology* **18:**192–205.
29. Gamez, R., and Black, L.M., 1968, Particle counts on wound tumor virus during its peak concentration in leafhoppers, *Virology* **34:**444–451.
30. Gamez, R., Rivera, C., and Kitajima, E.W., 1981, The biological cycle of maize rayado fino virus in its insect vector *Dalbulus maidis, Proc. Int. Congr. Virol.* 5th Strasbourg, p. 213 (Abstr.).

31. Gera A., Loebenstein, G., and Raccah, B., 1978, Detection of cucumber mosaic virus in viruliferous aphids by enzyme-linked immunosorbent assay, *Virology* **86:**542–545.
32. Gergerich, R.C., and Scott, H.A., 1988, The enzymatic function of ribonuclease determines plant virus transmission by leaf-feeding beetles, *Phytopathology* **78:**270–272.
33. Gildow, F.E., 1982, Coated-vesicle transport of luteoviruses through salivary glands of *Myzus persicae, Phytopathology* **72:**1289–1296.
34. Gildow, F.E., 1985, Transcellular transport of barley yellow dwarf virus into the hemocoel of the aphid vector, *Rhopalosiphum padi, Phytopathology* **75:**292–297.
35. Gildow, F.E., 1987, Virus-membrane interactions involved in circulative transmission of luteoviruses by aphids, in Harris, K.F., (ed): Current Topics in Vector Research Volume 4, New York, Springer-Verlag, pp. 94–120.
36. Gildow, F.E., and Rochow, W.F., 1980, Role of accessory salivary glands in aphid transmission of barley yellow dwarf virus, *Virology* **104:**97–108.
37. Gingery, R.E., Gordon, D.T., and Nault, L.R., 1982, Purification and properties of an isolate of maize rayado fino virus from the United States, *Phytopathology* **72:**1313–1318.
38. Harris, K.F., 1981, Role of virus–vector interactions and vector feeding behavior in noncirculative transmission by leafhoppers, *Proc. Int. Cong. Virol. 5th, Strasbourg,* 1981, p. 213. (Abstr.).
39. Harris, K.F., Bath, J.E., Thottappilly, G., and Hooper, G.R., 1975, Fate of pea enation mosaic virus in PEMV-injected pea aphids, *Virology* **65:**148–162.
40. Hibino, H., and Kimura, I., 1983, Detection of rice ragged stunt virus in insect vectors by enzyme-linked immunosorbent assay, *Phytopathology* **72:**656–659.
41. Hibino, H., Saleh, N., and Roecham, M. 1979, Reovirus-like particles associated with rice ragged stunt diseased rice and insect vector cells, *Ann. Phytopathol. Soc. Japan* **45:**228–229.
42. Jayasena, K.W., Randles, J.W., and Barnett, O.W., 1984, Synthesis of a complementary DNA probe specific for detecting subterranean clover red leaf virus in plants and aphids, *J. Gen. Virol.* **65:**109–117.
43. Kikumoto, T., and Matsui, C., 1962, Electron microscopy of plant viruses in aphid midguts, *Virology* **16:**509–510.
44. Kim, K.S., Scott, H.A., and Robinson, M.D., 1977, Ultrastructural responses of bean leaf beetle hemocytes to beetle-transmitted and non-transmitted plant viruses, *Proc. Am. Phytopathol. Soc.* **4:**130.
45. Kopek, J.A., and Scott, H.A., 1983, Southern bean mosaic virus in Mexican bean beetle and bean leaf beetle regurgitants, *J. Gen. Virol.* **64:**1601–1605.
46. Lim, W.L., de Zoeten, G.A., and Hagedorn, D.J., 1977, Scanning electron-microscopic evidence for attachment of a nonpersistently transmitted virus to its vector's stylets, *Virology* **79:**121–128.
47. Lopez-Abella, D., Pirone, T.P., Mernaugh, R.E., and Johnson, M.C., 1981. Effect of fixation and helper component on the detection of potato virus Y in alimentary tract extracts of *Myzus persicae,* Phytopathology **71:**807–809.
48. Matisova, J., and Valenta, V., 1975, Versuche zum nachweis des Enationenvirus der Erbse im Vektor, *Acyrthosiphon pisum,* mit Hilfe der immunofluoreszenze-Methode, *Tag. Berl. Landwirtsch. Wiss, DDR, Berlin* **134:** 91–98.

49. Mumford, D.L., 1982, Using enzyme-linked immunosorbent assay to identify beet leafhopper populations carrying beet curly top virus, *Pl. Dis.* **66:**940–941.
50. Murant, A.F., Roberts, I.M., and Elnagar, S., 1976, Association of virus-like particles with the foregut of the aphid *Cavariella aegopodii* transmitting the semi-persistent viruses anthriscus yellows and parsnip yellow fleck, *J. Gen. Virol.* **31:**47–57.
51. O'Loughlin, G.T., and Chambers, T.C., 1967, The systemic infection of an aphid by a plant virus, *Virology* **33:**262–271.
52. Omura, T., Hibino, H., Usugi, T., Inoue, H., Morinaka, T., Tzurumachi, S., Ong, C.A., Putta, M., Tsuchizaki, T., and Saito, Y., 1984, Detection of rice viruses in plants and individual insect vectors by latex flocculation test. *Pl. Dis.* **68:**374–378.
53. Orlob, G.B., 1968, Relationship between *Tetranychus urticae* Koch and some plant viruses, *Virology* **35:**121–133.
54. Orlob, G.B., and Takahashi, Y., 1971, Location of plant viruses in the two spotted spider mite *Tetranychus urticae* Koch, *Phytopathol.* **72:**21–28.
55. Ossiannilson, F., 1968, Is tobacco mosaic virus not imbibed by aphids and leafhoppers? *Kungl. Lantbruks. Hogskol. Ann.* **24:**369–374.
56. Ozel, M., 1971, Vergleichende elektronenmikroskopische Untersuchungen an Rhabdoviren pflanzlicher und tierischer Herkunft. I. Erste elektronenmikroskopische Ergebnisse mit dem pflanzlichen Modell Sowthistle Yellow Vein Virus (SYVV) and seinem Vektor *Hyperomyzus lactucae* (L), *Zentralbl. Bakteriol. Parasitenk. Infektionskr. Abt. 1.* **217:**160–174.
57. Paliwal, Y.C., 1972, Brome mosaic virus infection in the wheat curl mit *Aceria tulipae,* a non-vector of the virus, *J. Invertebr. Pathol.* **20:**288–302.
58. Paliwal, Y.C., 1980a, Fate of plant viruses in mite vectors and nonvectors, in Harris, K.F., and Maramorosch, K. (eds): Vectors of Plant Pathogens, New York, Academic Press, pp. 357–373.
59. Paliwal, Y.C., 1980b, Relationship of wheat streak mosaic virus and barley stripe mosaic viruses to vector and nonvector eriophyid mites, *Arch. Virol.* **63:**123–132.
60. Paliwal, Y.C., 1982, Detection of barley yellow dwarf virus in aphids by serologically specific electron microscopy, *Can. J. Bot.* **60:**179–185.
61. Paliwal, Y.C., 1987, Immunoelectron microscopy of plant viruses and mycoplasmas, in Harris, K.F. (ed): Current Topics in Vector Research, Volume 3, New York, Springer-Verlag, pp. 217–249.
62. Paliwal, Y.C., and Slykhuis, J.T., 1967, Localization of wheat streak mosaic virus in the alimentary canal of its vector *Aceria tulipae Keifer, Virology* **32:**344–353.
63. Pirone, T.P., 1977, Accessory factors in nonpersistent virus transmission, in Harris K.F., and Maramorosch, K. (eds): Aphids as Virus Vectors, New York, Academic Press, pp. 224–235.
64. Pirone, T.P., and Thornbury, D.W., 1988, Quantity of virus required for aphid transmission of a potyvirus, *Phytopathology* **78:**104–107.
65. Plumb, R.T., 1981, Problems in the use of sensitive serological methods for detecting viruses in vectors, in Plumb, R.T. (ed): *Proc. 3rd Conf. Virus Dis.* Gramin, Europe, Rothamsted, Hertfordshire, England, pp. 123–126.
66. Plumb, R.T., Lennon, E.A., and Gutteridge, R.A., 1986, Forecasting barley yellow dwarf virus by monitoring vector populations and infectivity, in

McLean, G.D., Garrett, R.G., and Ruesink, W.G. (eds): Plant Virus Epidemics, Sydney, Academic Press, pp. 387–398.

67. Rivera, C., Kozuka, Y., and Gamez, R. 1981, Rayado fino virus: Detection in salivary glands and evidence of increase in virus titre in the leafhopper vector *Dalbulus maidis, Turrialba* **31:**78–80.
68. Roberts, I.M., and Brown, D.J.F., 1980, Detection of six nepoviruses in their nematode vectors by immunosorbent electron microscopy, *Ann. Appl. Biol.* **96:**187–192.
69. Roberts, I.M., and Harrison, B.D., 1979, Detection of potato leafroll and potato mop-top viruses by immunosorbent electron microscopy, *Ann. Appl. Biol.* **93:**289–297.
70. Scott, H.A., and Fulton, J.P., 1978, Comparison of the relationship of southern bean mosaic virus and the cowpea strain of tobacco mosaic virus with the bean leaf beetle, *Virology* **84:**207–209.
71. Shikata, E., Maramorosch, K., and Granados, R.R., 1966, Electron microscopy of pea enation mosaic virus in plants and aphid vectors, *Virology* **29:**426–436.
72. Shikata, E., Orenski, S.W., Hirumi, H., Mitsuhashi, J., and Maramorosch, K., 1964, Electron micrographs of wound tumor virus in an animal host and in a plant tumor, *Virology* **23:**441–444.
73. Sinha, R.C., 1965, Recovery of potato yellow dwarf virus from hemolymph and internal organs of an insect vector, *Virology* **27:**118–119.
74. Sinha, R.C., 1966, Sequential infection and distribution of wound tumor virus in the internal organs of a vector after ingestion of virus, *Virology* **26:**673–686.
75. Sinha, R.C., and Black, L.M., 1962, Studies on the smear technique for detecting virus antigens in an insect vector by use of fluorescent antibodies, *Virology* **17:**582–587.
76. Sinha, R.C., and Reddy, D.V.R., 1964, Improved fluorescent smear technique and its application in detecting virus antigens in an insect vector, *Virology* **24:**626–634.
77. Slack, S.A., and Scott, H.A., 1971, Hemolymph as a reservoir for the cowpea strain of southern bean mosaic virus in the bean leaf beetles, *Phytopathology* **61:**538–540.
78. Stein-Margolina, V.A., 1973, Bromegrass mosaic virus in the mite *Aceria tritici* (electron microscopical investigations), *Izvestiya Akad. Nauk, USSR* **2:**189–195.
79. Stein-Margolina, V.A., 1975, Phytopathogenic viruses of cereals in the gall mite *Aceria tritici* Shev, *Tag. Berl. Akad. Landwirtsch.—Wiss DDR Berlin* **134:**181–198.
80. Stobbs, L.W., Cross, G.W., and Manocha, M.S., 1982, Specificity and methods of transmission of cucumber necrosis virus by *Olpidium radicale* zoospores, *Can. J. Plant Pathol.* **4:**134–142.
81. Storey, H.H., 1933, Investigation of the mechanism of the transmission of plant viruses by insect vectors I, *Proc. Roy. Soc. B* **113:**463–485.
82. Takahashi, Y., and Orlob, G.B., 1969, Distribution of wheat streak mosaic virus-like particles in *Aceria tulipae, Virology* **38:**230–240.
83. Tamada, T., and Harrison, B.D., 1981, Quantitative studies on the uptake and retention of potato leafroll virus by aphids in laboratory and field conditions, *Ann. Appl. Biol.* **98:**261–276.

84. Taylor, C.E., and Robertson, W.M., 1969, The location of raspberry ringspot and tomato black ring viruses in the nematode vector, *Longidorus elongatus* (de Man), *Ann. Appl. Biol.* **64:**233–237.
85. Taylor, C.E., and Robertson, W.M., 1970, The location of tobacco rattle virus in the nematode vector, *Trichodorus pachydermus* Seinhorst, *J. Gen. Virol.* **6:**179–182.
86. Taylor, C.E., and Robertson, W.M., 1974, Electron microscopy evidence for the association of tobacco severe etch virus with the maxillae in *Myzus persicae* (Sulz), *Phytopathol. Z.* **80:**257–266.
87. Taylor, C.E., Robertson, W.M., and Roca, F., 1976, Specific association of artichoke Italian latent virus with the odontostyle of its vector *Longidorus attenuatus, Nematol. Medit.* **4:**23–30.
88. Temmink, J.H.M., 1971, An ultrastructural study of *Olpidium brassicae* and its transmission of tobacco necrosis virus, *Meded. Landbouwhogesch. Wageningen* **71:**1–135.
89. Temmink, J.H.M., Campbell, R.N., and Smith, P.R., 1970, Specificity and site of in vitro acquisiton of tobacco necrosis virus by zoospores of *Olpidium brassicae, J. Gen. Virol.* **9:**201–213.
90. Torrance, L., 1987, Use of enzyme amplification in an ELISA to increase sensitivity of detection of barley yellow dwarf virus in oats and in individual vector aphids, *J. Virol. Methods* **15:**131–138.
91. Torrance, L., and Jones, R.A.C., 1982, Increased sensitivity of detection of plant viruses obtained by using a fluorogenic substrate in enzyme-linked immunosorbent assay, *Ann. Appl. Biol.* **101:**501–509.
92. Torrance, L., Plumb, R.T., Lennon, E.A., and Gutteridge, R.A., 1986, A comparison of ELISA with transmission tests to detect barley yellow dwarf virus-carrying aphids, in Jones, R.A.C., and Torrance, L. (eds): Developments and Applications in Virus Testing, Wellesbourne, Warwick, England, Association of Applied Biologists, pp. 165–176.
93. Walters, H.J., 1969, Beetle transmission of plant viruses, *Adv. Virus Res.* **15:**339–363.
94. Whitcomb, R.F., and Black, L.M., 1969, Synthesis and assay of wound-tumor soluble antigen in an insect vector, *Virology* **15:**136–145.

8
Insect-Borne Viruses of Rice

Hiroyuki Hibino

Introduction

Rice, *Oryza sativa* L., one of the most important cereals, is grown widely, especially in Asia. Rice is consumed directly as a food and supplies about 20% of human total calorie consumption in the world, and in Asia, about 36%. Rice is cultivated in diverse environments: in flooded fields, uplands, deep-water areas, and tidal wetlands; in lowlands and to highlands; in the tropics and in the temperate region, to 53°N in latitude. Areas planted for rice in the world is 148 million ha, including 113 million ha in Asia, 7 million ha in South America, 5 million ha in Africa, and 2 million ha in North and Central America (World Statistics, 1985). Rice is also grown in Europe, the USSR, and the Oceania regions of the central and south Pacific.

A total of 13 rice viruses are known worldwide (Table 8.1). Rice virus diseases have occurred, one after another, in all major rice-growing countries. The first outbreak of rice dwarf virus (RDV) was recorded in 1897 in Japan (99) and that of rice stripe virus (RStV) in 1903 (119). A rice black-streaked dwarf virus (RBSDV) outbreak occurred for the first time in Japan in 1941 (120). Other virus diseases were recognized as important threats to rice production during the mid 1950s to 1980. The first outbreak of rice hoja blanca virus (RHBV) occurred in 1956 in Latin America (51). A major outbreak of tungro, a composite disease caused by rice tungro bacilliform virus (RTBV) and rice tungro spherical virus (RTSV) (71, 157), was recorded for the first time in 1957 in the Philippines, during 1965–1966 in Thailand, and in 1969 in Bangladesh, India, Indonesia, and Malaysia (123, 161). A rice transitory yellowing virus (RTYV) outbreak was recorded from 1960 to 1962 in Taiwan (26, 196) and in 1964 in Southern China (34). Rice yellow mottle virus (RYMV) was first observed in 1966 in Kenya (10), in 1976 in Sierra Leone, in 1977 on the Ivory Coast, and in 1978 in Nigeria (12, 174). An outbreak of rice grassy stunt virus (RGSV) occurred from 1970 to 1973 in India, Indonesia, and the Philip-

Hiroyuki Hibino, National Agriculture Research Center, Kannondai, Tsukuba 305, Japan
 Advances in Disease Vector Research, Volume 6.

TABLE 8.1. Geographical distribution of insect-borne viruses of rice.

Virus	Vector	Distribution
Asia		
Dwarf virus (RDV)	Leafhoppers	China, Japan, Korea, Nepal
Gall dwarf virus (RGDV)	Leafhoppers	China, Malaysia, Thailand
Black-streaked dwarf virus (RBSDV)	Planthoppers	China, Japan, Korea
Ragged stunt virus (RRSV)	Planthoppers	South and Southeast Asia, China, Japan, Taiwan
Bunchy stunt virus (RBSV)	Leafhoppers	China
Transitory yellowing virus (RTYV)	Leafhoppers	China, Japan, Taiwan, Thailand
Tungro bacilliform virus (RTBV)	Leafhoppers	South and Southeast Asia, China?
Tungro spherical virus (RTSV)	Leafhoppers	South and Southeast Asia, China, Japan
Stripe virus (RStV)	Planthoppers	China, Japan, Korea, Taiwan, USSR
Grassy stunt virus (RGSV)	Planthoppers	South and Southeast Asia, China, Japan, Taiwan
Europe		
Giallume virus (RGV)	Aphids	Italy, Spain?
Africa		
Yellow mottle virus (RYMV)	Beetles	Africa
America		
Hoja blanca virus (RHBV)	Planthoppers	South and Central America

pines (161). Rice ragged stunt virus (RRSV) outbreak occurred in 1977 in Southeast Asia (72, 126), and a rice gall dwarf virus (RGDV) outbreak in 1981 in China (35).

In many rice-growing countries, the damage caused by these rice virus diseases was more severe after the late 1960s. Increases in virus incidence generally occur after high yielding cultivars are introduced and rice cultivation is intensified. Cultivation was intensified by the introduction of double cropping rice, a wider cultivation of rice cultivars with similar genetic backgrounds, application of fertilizer, and the development of irrigation systems. Such improvements apparently favored the insect vectors of these viruses. Now, virus diseases are the major threat to stable rice production in irrigated areas in the tropics and subtropics, as well as in the temperate region.

All the economically important rice viruses are insect-borne (Table 8.2). Rice dwarf virus is the first plant virus for which a relation to leafhopper vector was described (99). The virus disease was first considered to be caused by leafhoppers (201). In 1910, Ando (4) found that rice dwarf was caused by an unknown causal agent carried by the leafhopper. This was the first discovery of virus transmission by insect vectors. Fukushi (42, 43) found transovarial passage of rice dwarf virus and assumed multiplication of the virus in the leafhopper.

TABLE 8.2. Vector species, vector relation, and transmissibility via eggs of insect-borne rice viruses.

Virus	Vector species	Vector relation	Egg transmission
Leafhopper-borne			
RDV	*Nephotettix cincticeps, N. nigropictus, N. virescens, Recilia dorsalis*	Propagative	Yes
RGDV	*N. nigropictus, R. dorsalis, N. virescens, N. malayanus, N. cincticeps*	Propagative	Yes
RBSV	*N. virescens, N. cincticeps*	Propagative	No
RTYV	*N. cincticeps, N. nigropictus, N. virescens*	Propagative	No
RTBV	*N. virescens, N. nigropictus, R. dorsalis, N. malayanus, N. parvus, N. cincticeps*	Noncirculative	No
RTSV	*N. virescens, N. nigropictus, R. dorsalis, N. cincticeps, N. malayanus, N. parvus*	Noncirculative	No
Planthopper-borne			
RBSDV	*Laodelphax striatellus, Unkanodes sapporonus, U. albifascia, Terthron albovittatus*	Propagative	No
RRSV	*Nilaparvata lugens, N. bakeri*	Propagative	No
RStV	*L. striatellus, U. sapporonus, U. albifascia, Terthron albovittatus*	Propagative	Yes
RHBV	*Sogatodes orizicola, S. cubanus*	Propagative	Yes
RGSV	*N. lugens, N. bakeri, N. muiri*	Propagative	No
Aphid-borne			
RGV	*Rhopalosiphum padi, Sitobion avenae, Metopolophium dithodium*	Circulative	No
Beetles-borne			
RYMV	*Apophylis* spp., *Chaetocnema* spp., *Sesselia pussilla, Trichispa sericea*	Noncirculative	No

Geographical Distribution

Out of 13 insect-borne rice viruses, 10 of them occur in Asia (Table 8.1), but only rice hoja blanca virus (RHBV)(136) is known to occur on the American continent. Rice yellow mottle virus (12) spreads widely in rice-growing areas of Africa. Rice giallume virus (RGV)(158) occurs in northwest Italy.

The geographical distribution of rice viruses is related to the distribution of their vectors and the availability of virus reservoirs during the off season. In the tropics, rice stubble generally produces new growth, which provides food for vector insects and serves as a source of virus infection for the next season. In the subtropics and in warm areas in temperate regions, rice stubble may wither during winter. Some stubble, however, produces new growth in spring and serves as a virus source.

In temperate regions where virus-infected rice stubble dies out, the viruses overwinter in vector insects or in alternative hosts. Rice black-streaked dwarf virus persists on winter wheat and barley on which the vector planthopper also overwinters. Rice dwarf virus and RStV persist in their vector insects, and congenitally infected insects overwinter and serve as a source for virus infection the following season. Congenitally infective insects do not seem to be a stable source for the next season because the viruses have a deleterious effect on the insects. After major outbreak of RDV and RStV, the percentage of virus-infective vectors in field populations gradually decrease in the years following (108). A biannual weed, water foxtail, seems to serve as an alternative host of RDV. In temperate regions, RRSV and RGSV do not overwinter either in rice or the vector planthopper. They are introduced through long-distance migratory flights by the virus-carrying vector planthopper *Nilaparvata lugens,* as the rice season begins. Probably, long-distance movements of the planthopper also occurs in the tropics and subtropics (173). *Laodelphax striatellus,* a vector of RStV and RBSDV, and probably *Sogatodes orizicola,* a vector of RHBV, are capable of long-distance migratory flight (111).

The subtropical RTYV, RDV, and rice bunchy virus (RBSV) probably have the potential to cause epidemics in the tropics, whereas RYMV is a potential danger outside Africa in areas where its vector chrysomelid beetles are common.

Vector Species

Of the 15 rice viruses occurring in the world, 13 are insect-borne (Table 8.2). Rice necrosis mosaic and rice stripe necrosis viruses are probably transmitted by a fungus, *Polymixa graminis* (40, 87).

Six rice viruses are leafhopper-borne, all of them in Asia. Each of the six viruses can be transmitted experimentally by more than one leafhopper species, many of which have a little or no biological relations with rice. The principal vectors of most leafhopper-borne viruses are either *Nephotettix virescens* or *Nephotettix cincticeps.* The former is the predominant species in rice fields in the tropics, the latter in the subtropics and temperate regions. *Nephotettix virescens* is nearly monophagous to rice, but *N. cincticeps* colonizes on weeds and other cereals, especially during the winter. *Nephotettix nigropictus* and *Recilia dorsalis* are common in rice fields but there are generally fewer of them than *N. virescens* or *N. cincticeps*. *Nephotettix* spp. also transmit the yellow dwarf mycoplasma agent and *R. dorsalis* transmits the orange leaf mycoplasma agent (123, 161).

Five rice viruses are planthopper-borne (Table 8.2). Each virus has two to four planthopper species as a vector. Except for *N. lugens, L. striatellus*, and *S. orizicola*, all other vectors have little or no biological relations with rice, and *L. striatellus* and *S. orizicola* also colonize on weeds and other cereals. However, *N. lugens* is nearly monophagous to rice and it moves from one place to another through migratory flight and colonizes newly planted rice fields (110, 173).

Rice giallume virus is transmitted by aphids, and RYMV by several species of chrysomelid beetles (11, 158).

Vector–Virus Interactions

Persistent Transmission

Out of 13 insect-borne rice viruses, 11 persist in vector insects (Table 8.2). All the persistent viruses except RGV multiply in their vectors. However, RGV is considered to be circulative, not propagative in its aphid vectors, although no confirmative studies have been performed. Vector insects that acquire propagative viruses become infective one to two weeks after feeding on virus source plants. In the fields, the propagative viruses generally occur sporadically and spread slowly. Because of the presence of natural enemies, the average life of vector insects in the field is generally short, and most insects that acquire viruses die before they become infective. The propagative viruses, however, have the advantage of long-distance dispersal by vector insects. The viruses persist in the vector insects throughout the dormancy period, or off-season when the insects feed on nonhost plants. In vector insects, the propagative viruses can be seen in a number of organs, including salivary glands, nerves, muscles, intestine, Malphigian tubes, mycetome, and fat body (20, 48, 49, 73, 144, 148, 178, 181, 185).

Four rice viruses can be transmitted via the eggs of their vectors (Table 8.2). Transovarial passage of virus particles occurs in leafhoppers for RDV and RGDV (42, 43, 88, 143, 188) and in planthoppers for RHBV and RStV (52, 188, 220). The percentage of congenitally infective progeny of infective females is generally high. A close relationship between trans-ovarial passage of RDV and symbiotic microorganisms in the vector leafhopper has been suggested (144). At the yolk-forming stage of egg development in the leafhopper, the symbiotes invade the ovariole. Rice dwarf virus particles have been found in close association with symbiotes in the mycetocytes of the ovariole. The mycetocytes containing RDV were enclosed in the oocyte. Also, RDV-like particles were observed in the oocyte, together with bacteroid symbiotes at an earlier stage of oocyte development. A specific affinity between virus particles and symbiotic microorganisms may be required for the transovarial passage of the virus by the vector.

Control of rice viruses that pass through vector eggs is generally difficult. Once a virus disease becomes epidemic, the epidemic tends to continue for several consecutive years (108, 226). Vector populations that include congenitally virus-infective insects do not lose the virus when subsequent generations feed on nonhost plants. In the case of viruses that do not pass through eggs, virus-infective populations lose the virus after a generation advance on nonhost plants.

Semipersistent Transmission

Three viruses, RTBV, RTSV, and RYMV, are semipersistent in the vector insects. *Nephotettix virescens* acquires RTBV and RTSV in a rather short access feeding period and transmits the viruses immediately after feeding (72, 74). The leafhopper retains the virus for three to four days. However, RYMV can be transmitted by many beetles, which retain the virus for five to eight days. Interactions between the virus and vector beetles seem less specific because RYMV is mechanically transmissible (11, 12). Generally, semipersistent viruses occur in patches and spread rapidly through fields. Since semipersistent viruses do not persist long in their vectors, the percentage of infective vectors in a field reflects the virus incidence in the field.

Leafhoppers infected with RTBV and RTSV lose the viruses upon molting (14, 65). The infectivity of leafhoppers can be neutralized by allowing the leafhoppers to feed on antiserum against RTBV or RTSV (68). Probably, RTBV and RTSV are specifically adsorbed on the surface of the cuticular walls of leafhopper mouths or stylets and they are ejected when the leafhoppers feed on plants. The specific site for the adsorption is not known.

Rice tungro bacilliform virus depends on the presence of RTSV for transmission by the leafhopper (14, 65, 71). Leafhoppers acquire RTBV when they are exposed first to RTSV sources and then to RTBV sources or when they are exposed to source plants with both RTBV and RTSV. Recent studies suggest that RTSV infectivity and the helper activity for RTBV transmission (acquisition) by the leafhopper are differentiable (14, 65, 68). Leafhoppers retain RTSV for two to three days, whereas they retain helper activity for seven days (14, 65). When leafhoppers that have fed on RTSV-infected plants were fed anti-RTSV serum, the leafhoppers lost RTSV infectivity but retained the ability to acquire RTBV. Thus, RTSV itself is unlikely to be the bearer of helper activity. A helper component is probably produced in plants infected with RTSV. The hypothetical helper component may be essential for RTBV to be adsorbed on the wall of the leafhopper mouth.

Vector–virus interaction of RTBV, RTSV, and some other plant viruses that are semipersistent in their vectors are highly specific. Although, the presence of helper components has not been well documented, all these viruses may require specific helper components for transmission by their vector insects. The helper components may be proteins, which are coded on virus genome, as are those of potato virus Y and cauliflower mosaic virus (166). It is not known whether RTSV has its own helper component in addition to the one for RTBV transmission. Recently, Hunt et al. (84) reported the possible involvement of a helper component for leafhopper transmission of maize chlorotic dwarf virus, which is similar to RTSV in particle structure, properties, and relations to the vector (54).

Nucleic Acids and Proteins

Some insect-borne rice viruses have been analyzed for their nucleic acid and protein components (Table 8.3). All of them except RTBV have RNA. RDV and RGDV have 12 RNA segments in double-stranded (ds) form (62, 168). Of the 12 segments in RDV and RGDV, the seven larger ones are similar in size. Both viruses have seven proteins, and the molecular weight distributions of their major polypeptides are also similar. Antiserum to SDS-dissociated RDV is serologically related to a RGDV polypeptide that occurs on the surface of the virus particles. Antiserum to intact RDV, however, did not react with RGDV (131), which indicates that it is evolutionally related to both RDV and RGDV.

Both RBSDV and RRSV have 10 ds RNA segments, but the molecular size distribution of their segments differ (103, 156, 169). Particles of RBSDV and other members of Fijivirus have double shells (183), but double-shell particles of RRSV have not been clearly demonstrated (132). Since RRSV is similar to Fijivirus in symptomatology, vector relations, and relations to the cells (133), it was postulated to be a member of Fijivirus (63, 72, 187), although further precise study is required to establish the exact taxonomic position of RRSV (132).

However, RStV, RHBV, and RGSV have been tentatively grouped in the tenui (rice stripe) virus group (55, 76, 136). Nucleic acid extracted from RStV and RGSV was first reported to be in a single-stranded (ss) form (202, 205). Recent investigations, however, have indicated an association of four ss RNA and four ds RNA with RStV (92), which indicates the presence of both "plus" and "minus" polarity RNAs in virus particles, as suggested for maize stripe virus, another member of the tenui virus group (37).

TABLE 8.3. Morphology and components of insect-borne rice viruses.

Virus	Morphology (size)	Nucleic acid Type[a]	Nucleic acid Number	Number of proteins
Phytoreoviruses				
RDV	Icosahedral (70 nm)	ds RNA	12	7
RGDV	Icosahedral (65 nm)	ds RNA	12	7
Fijiviruses				
RBSDV	Isometric (80 nm)	ds RNA	10	–
RRSV	Isometric (65 nm)	ds RNA	10	5
Tenuiviruses (rice stripe virus group viruses)				
RStV	Filament (8 nm × ?)	ss, ds RNA	4 + 4	1
RHBV	Filament (3 nm × ?)	ss RNA	–	1
RGSV	Filament (6 nm × ?)	ss RNA	4	1(2)
Plant rhabudoviruses				
RTYV	Bullet-shaped (180–210 × 94 nm)	RNA	–	4
Luteoviruses				
RGV	Isometric (30 nm)	–	–	–
Unclassified viruses				
RBSV	Spherical? (60 nm?)	–	–	–
RTBV	Small bacilliform (100–300 × 30 nm)	DNA	1	1
RTSV	Isometric (30 nm)	ss RNA	1	2(3)
RYMV	Isometric (30 nm)	–	–	–

[a] ss, single stranded; ds, double stranded.

Rice Black-streaked Dwarf Virus

This virus (183) occurs in Central China, central and southern Japan, and Korea (120, 178). Generally, RBSDV infection occurs sporadically, or in small patches in rice fields, and affects rice, maize, wheat, and barley. It is particularly damaging to rice and maize. Outbreaks of the disease were recorded in 1941 and the year 1965–1967 in Japan, and in 1963 and the year 1965–1967 in China.

Rice black-streaked dwarf virus (RBSDV) is a member of the Fijivirus subgroup of plant Reoviruses. Its particles are isometric, 75 to 80 nm in diameter, and have double shells (183, 185) and contain 10 ds RNA as a genome, with a total molecular size of about 19×10^6 (169). The particles are associated with an RNA-dependent RNA polymerase activity (204, 208). Rice black-streaked dwarf virus is serologically related to maize rough dwarf virus, the viruses of cereal tillering disease, and pangola stunt virus (127, 134).

Rice plants infected with RBSDV show pronounced stunting, darkening of leaves, twisting of leaf tips, splitting of the leaf margin, and galls along the veins on the underside of leaf blades and the outer surface of sheaths and columns (120, 188). The symptoms are milder in some cultivars (140, 141). The galls appear usually on older leaves as waxy white, irregularly elongated protuberance, which later turn brownish or black. The galls results from hyperplasia and hypertrophy of phloem tissues (98). Inclusion bodies about 6.5 μm in diameter occur in the gall cell cytoplasm (98). The RBSDV particles are scattered or arranged in crystalline arrays near the viroplasmic inclusions (183, 185), but particles in inclusions have no outer coat. The virus is also seen in tubules in the cytoplasm. In RBSDV-infected *L. striatellus,* virus particles occur within or around viroplasmic inclusions and in tubules in the cytoplasm of the cells (178, 183, 185).

Rice black-streaked dwarf virus is not mechanically transmissible. It is transmitted by the smaller brown planthopper *L. striatellus, Unkanodes sapporonus,* and *Unkanodes albifascia* (78, 120, 188, 189, 190). *Laodelphax striatellus* is the major vector of RBSDV and the other vectors have a little or no biological relations with rice. The virus multiples in the planthoppers (115) but it is not transmitted via the eggs. Its incubation period in the planthoppers is 7 to 35 days. The proportion of infective planthoppers in populations exposed to RBSDV ranges from 30 to 89% in *L. striatellus,* 34% in *U. sapporonus,* and 32% in *U. albifascia.* The presence of RBSDV in single *L. striatellus* can be determined serologically (211), whereas its infectivity can be assayed by the microinjection method (114).

After the rice is harvested, the infective *L. striatellus* moves to weeds, barley, and wheat, where it transmits RBSDV (90). After overwintering on these plants, the first-generation nymphs of *L. striatellus* acquire RBSDV from barley and wheat plants and then move to the newly planted rice plants. There is a higher incidence of RBSDV in early planted rice in Japan and in the second rice crop in China.

Rice Bunchy Stunt Virus

This virus occurs in southern China (213, 217). So far, RBSV has not been well characterized, but virus-like particles, 35 to 60 nm in diameter, occur in the phloem cells of bunchy stunt virus-infected rice plants.

The virus-infected rice plants show stunting, increased tillering, and narrow, short leaves (22, 212, 213). These symptoms differ, depending on the cultivar. In severe cases, infected plants produce nodal branches and bunches of narrow and short leaves and have a bird nest-like appearance.

Rice bunchy stunt virus (RBSV) is transmitted by the green leafhoppers *N. virescens* and *N. cincticeps* (22, 212, 213, 215, 217). It is circulative and probably multiples in the leafhoppers, although it is not transmitted via leafhopper eggs. Its incubation period in *N. cincticeps* is 8 to 25 days (average 12 days). The proportion of infective leafhoppers in populations exposed to RBSV ranges from 4 to 39% (average 13%) in *N. cincticeps* and from 3 to 11% (average 8%) in *N. virescens*. Infectivity of RBSV in sap can be assayed by the microinjection method.

The field incidence of RBSV is sporadic. This virus naturally infects rice but not other cereals. In southern China, the bunchy stunt incidence was high in from 1976 to 1979 but very low after 1982. Infected rice stubble overwinters and serves as a source for infection in first-crop rice (217). Overwintered *N. cincticeps* may also serve as another source of infection.

Rice Dwarf Virus

This virus (85) occurs in southern and central China, southern and central Japan, Korea, and Nepal (85, 96, 164, 179). In Japan, dwarf outbreaks occurred infrequently from 1889 to 1930. After 1955, however, the disease reached an epidemic level in southern and central Japan. Incidence was high from 1967 to 1974 and thereafter declined drastically. In China, Korea, and Nepal, RDV generally occurs sporadically in the fields.

Rice dwarf virus (RDV) is a Phytoreovirus. Its particles are icosahedral, and about 70 nm in diameter (47–49, 85, 181). Its outer protein shell consists of 180 structural units arranged regularly. The virus contains 12 ds RNA, with a total molecular size of about 16×10^6 (168), and seven proteins (142). DNA complementary to the 10th largest of the 12 ds RNA has been cloned and its sequence determined (154, 209). The smallest protein is located on the surface of the outer capsid, and antiserum to the intact virion reacts only with this protein (135). The molecular size of the smallest protein of a severe RDV strain (a strain that causes severe crop damage) is larger than that of the ordinary strain (107). Antiserum to dissociated RDV reacted with a protein of RGDV, but the antiserum to intact RDV did not react with the protein (131). Rice dwarf virus particles are associated with RNA polymerase activity (116, 210). The reaction product is RNase sensitive and specifically hybridizes with heat-denatured RDV RNA.

Rice plants infected with RDV show pronounced stunting and fine chlorotic specks on their leaves (43, 85, 188). The specks vary in size and often fuse to form interrupted streaks along the veins. Root growth is markedly arrested and the small roots extend horizontally. Many diminutive tillers may be produced. Infected plants usually survive until harvest but rarely produce panicles; any panicles are produced, they are poor and bear unfilled grains. Starch accumulates in infected leaf tissues.

Rice dwarf virus is not mechanically transmissible. It is transmitted by the green leafhoppers *N. cincticeps, N. virescens, Nephotettix nigropictus,* and *Recilia dorsalis* (44, 143, 188, 201, 216). *Nephotettix cincticeps* is the major vector in Central China, Japan, and Korea. The virus multiplies in the leafhoppers and is transmitted via the eggs (42, 43). Its incubation period is 4 to 58 days (average 12 to 35 days) in *N. cincticeps* and 9 to 42 days (average 10 to 15 days) in *R. dorsalis*. The proportion of infective leafhoppers in populations exposed to RDV ranges from 0 to 69% in *N. cincticeps*, 23% in *N. nigropictus*, from 4 to 35% in *N. virescens* and from 2 to 43% in *R. dorsalis*. The virus that was maintained in vegetatively propagated rice plants without vector transmission could no longer be transmitted by its vector (105).

The proportions of congenitally infective progeny from infective females range from 32 to 100% in *N. cincticeps*, from 0 to 64% in *R. dorsalis*, and 8% in *N. virescens* (42, 43, 143, 188, 216). Rice dwarf virus can be transmitted from a single female via her eggs through six succeeding generations (45). Populations with very high numbers of congenitally infective leafhoppers are obtained by selective breeding (91, 188). The percentage of infective insects in these populations gradually decrease in succeeding generations when maintained on healthy seedlings. Congenitally infective nymphs start to transmit RDV 0 to 20 days after hatching in *N. cincticeps* and 3 to 14 days in *R. dorsalis*. The presence of RDV in single leafhoppers can be determined serologically (121, 149, 180).

Rice dwarf virus has deleterious effects on *N. cincticeps*, with infective leafhoppers having a short life span and oviposition period, higher nymphal mortality, and less fecundity (143).

The infectivity of RDV can be assayed by the microinjection method (46, 104) and by fluorescent antibody focus counting on cell monolayers taken from *N. cincticeps*, *N. nigropictus*, and *R. dorsalis* (106).

Parenchyma cells contain round or oval inclusion bodies 3 to 10 μm in diameter in the cytoplasm (43). The bodies undergo metachromasy when stained with acridine orange, which indicates the presence of ds RNA (198). Rice dwarf virus particles are embedded in inclusion bodies with an electron-dense matrix or they are clustered in membrane-bound bodies in the cytoplasm (48, 49, 181). In infective leafhoppers, RDV particles occur within and around inclusions, in phagocytic vesicles, or in cytoplasmic tubules (48, 49, 144, 181). Viral particles occur in mycetocytes in close association with symbiotic organisms (144). Symbiotes with RDV particles are transmitted to the progeny via the eggs.

The virus naturally infects rice and a few weeds. In addition to rice, 14 gramineous plants can be experimentally infected with RDV (43, 188). The *N. cincticeps* nymphs mainly overwinter on weeds (91). From May to July, the *N. cincticeps* adults move to the newly planted rice. These leafhoppers are congenitally infective or acquires RDV from weeds before they move to rice. In Southern China, RDV incidence is higher in the second rice crop, whereas in Japan, the incidence is higher in early planted rice.

Rice Gall Dwarf Virus

This virus was first found in 1980 in Thailand (150, 151) and later, in southern China and Malaysia (35, 150).

Rice gall dwarf virus (RGDV) Phytoreovirus. Its particles are icosahedral, and about 65 nm in diameter (151, 155). The virus contains 12 ds RNA, with a total molecular size of 17×10^6, and seven proteins (62, 148, 150, 153). RNA-dependent RNA polymerase is associated with the RGDV particles (225), and RNA products by the polymerase hybridize with denatured RGDV RNA. One of the viral proteins reacts with antiserum to dissociated RDV but not with antiserum to intact RDV (131).

Virus-infected plants show pronounced stunting, darkening of the leaves, and small galls on the undersurface of leaf blades and the outer surface of sheaths (150, 151). Galls are usually less than 2 mm long and are 0.4 to 0.5 mm wide. Initially light green, the galls later become white; they result from hyperplasia of the phloem tissue (148).

In infected rice plants, RGDV particles occur in the cytoplasm and the vacuoles of the gall and phloem cells (148). Viroplasmic inclusions and tubules with RGDV particles occur in the cytoplasm of these cells. In infected leafhopper cells, RGDV particles are scattered throughout the cytoplasm or contained within tubules or phagocytic vecisles in the cytoplasm (148).

Rice gall dwarf virus is not mechanically transmissible. It is transmitted by *N. nigropictus, N. malayanus, N. virescens, N. cincticeps,* and *R. dorsalis* (88, 138, 150). The virus multiplies in their leafhopper vectors and is transmitted via the eggs (88, 138, 150). The proportion of infective leafhoppers in populations exposed to RGDV ranges from 2 to 95% in *N. nigropictus,* from 1 to 43% in *N. cincticeps,* 9% in *N. malayanus,* from 0.1 to 14% in *N. virescens,* and from 11 to 33% in *R. dorsalis* (88, 138, 150). The percentage of congenitally infective progeny from infective females ranges from 0 to 100% in *N. nigropictus*. The presence of RGDV in single *N. nigropictus* can be determined serologically (149). Infectivity of RGDV can be assayed by the microinjection method (155) and by fluorescent antibody focus counting on cell monolayers from *N. cincticeps* (152).

Generally, RGDV-infected plants occur sporadically in the fields. During 1981–1982, however, RGDV reached an epidemic level in China (35). In addition to rice, nine gramineous plants can be infected with RGDV (167, 218). The virus naturally infect rice and a weed, *Alopecurus acqualis* (219). In China, infected rice stubble and *A. acqualis* overwinter and are sources of the virus in spring. The percentage of infective leafhoppers in overwintering populations is high for *R. dorsalis* but low for other leafhoppers (219).

Rice Giallume Virus

This virus occurs in northwest Italy (158), and a virus that causes a similar disease occurs in Spain. Rice giallume virus (RGV) is a Luteovirus, and probably a strain of beet western yellow virus that is distribute widely in temperate regions (16), but "giallume" disease is not known in other rice-growing areas. Its isometric particles are 20 to 25 nm in diameter. In RGV-infected rice, the virus particles are restricted to the phloem cells (3, 13), in which the virus particles appear first in the nucleus and then in the cytoplasm (39). These infected cells also contain double-membraned vesicles with fibrils, degenerated mitochondria, and opaque material deposits along the walls of their phloem cells (39). Virus-like particles have also been found in the gut of *Rhopalosiphum padi* following exposure to "giallum"-infected plants (13). The RGV is serologically related to RPV and MAV isolates of barley yellow dwarf virus (172).

Rice plants infected with RGV show yellow to orange discoloration, stunting, and reduced tillering (13, 158). Discolored leaves later become necrotic, and infected plants may die prematurely.

Rice giallume virus is not mechanically transmissible. It is transmitted by the aphids *Rhopalosiphum padi, Sitobion avenae,* and *Metopolophium dithodium* (158, 159), although *R. padi* is the major vector. The virus is circulative but probably does not multiply in the aphids. The proportion of infective *R. padi* in populations exposed to RGV is as high as 78%.

The virus naturally infect rice and a few weeds. In addition to rice, nine gramineous plants can be experimentally infected with RGV (2, 3). Giallume appear in patches in the fields. The patchy appearance is correlated with the presence of the weed *Leersia oryzoides* (160). In spring, new sprouts of *L. oryzoides* appear before rice is sown. The *R. padi* that has colonized on *L. oryzoides,* acquires RGV from infected *L. oryzoides,* and moves to rice. The aphids moves back to *L. oryzoides* in summer when rice plants are less attractive. Diseased plants generally forms patches around RGV-infected *L. oryzoides*. Later, the disease may spread to cover entire fields.

Rice Grassy Stunt Virus

This virus occurs in south and southeast Asia, China, Japan, and Taiwan (66). Grassy stunt virus of rice was first recognized in 1963 in the Philippines (171). Its incidence was high from 1970 to 1977 in Indonesia, from 1973 to 1977 in the Philippines, and during 1973–1974 in India.

Rice grassy stunt virus (RGSV) is a member of the tenui (rice stripe) virus group. Its particles are filamentous, 6 to 8 nm in width (66, 76, 94), and often circular, with a modal contour length of 950 to 1350 nm. The virus contains four ss RNA (205), and one major and one minor protein

(76, 205). Also, RGSV particles are associated with RNA polymerase activity (207). Serologically, the virus is distantly related to RStV (76).

Grassy stunt strains, which cause severe symptoms, occur in India, Indonesia, Philippines, Taiwan, and Thailand (15, 18, 31, 69, 129, 199). A viral strain that causes severe disease in Taiwan is called rice wilted stunt virus (18).

Virus-infected plants show pronounced stunting, with a proliferation of short, erect, narrow, and pale green or pale yellow leaves (66, 171). These leaves often have numerous small, irregular dark-brown spots. Infected plants produce few panicles. Any panicles that are present bear dark-brown, unfilled grains.

In RGSV-infected rice plants, mesophyll cells contain masses of fibrils in the nuclei and cytoplasm (165). Fibrils are often seen in membrane-bound bodies in the cytoplasm. Tubules associated with particles, 18 to 25 nm in diameter, can be seen in the sieve tubes (76, 165). Similar particles are found in crystalline arrays in the fat body and tracheae of *N. lugens* following exposure to RGSV (186).

Rice grassy stunt virus is principally transmitted by the brown planthopper *N. lugens* (171), which is one of the most important pests in Asia. It also causes feeding damage or "hopper burn" in rice. When there is a high incidence of RGSV, the rice crop is damaged by it, as well as by another brown planthopper-borne virus, RRSV.

The virus is also transmitted by *N. bakeri* and *N. muiri* (93), although *N. bakeri* and *N. muiri* have little biological relations with rice. It multiplies in the planthoppers but is not transmitted via the eggs. Its retention period in the planthopper is 5 to 25 days (average 11 days). After feeding on RGSV-infected rice plants, from 5 to 60% of *N. lugens* and 69% of *N. muiri* become infective. About 50% of the *N. lugens* vectors that are carrying RGSV transmit the virus when tested for infectivity (94, 149). The presence of RGSV in single planthoppers can be determined serologically (38, 76, 94, 149), while its infectivity can be assayed by the microinjection method (76, 94).

Rice is the only known crop naturally infected with RGSV, although 16 *Oryza* spp. and 5 weeds have been experimentally infected with RGSV (7, 124). The planthopper is nearly monophagous of rice. The virus and the planthopper are generally endemic in areas where rice is grown throughout the year. Infected rice plants are the source of RGSV and *N. lugens* planthoppers disperse RGSV through migratory flight. In the early summer, long-distance migratory flight of the planthopper occurs from the endemic areas to areas where summer rice is grown (110, 173). Some of the immigrant planthoppers carry RGSV (80, 94).

Rice Hoja Blanca Virus

This virus (137) occurs in South and Central America; it also was seen from 1957 to 1959 in the United States (51). Hoja blanca of rice was first recognized in 1935 in Colombia and reached an epidemic level from 1956 to 1967 in Latin America. During 1965–1967, however, the direct damage caused by its vector, *Sogatodes orizicola,* was more severe than that caused by one hoja blanca virus (RHBV) (95, 226). From 1968 to 1980, hoja blanca disease ceased to be an economic problem. In 1981, the disease reappeared in some Latin American countries, and reached an epidemic level from 1981 to 1985. In 1957, 1959, and 1962, *S. orizicola* were observed in the southern rice-growing areas of the United States.

Rice hoja blanca virus is a member of the tenui (rice stripe) virus group. Its particles are filamentous, 3 nm wide, with a spiral or helical configuration (136, 137), and contain RNA and one major and one minor protein. Virus-infected plants produce a large amount of a virus-specific protein (36), which is not serologically related to the viral proteins. The virus is serologically related to *Echinochloa* hoja blanca virus but not to other members of the rice stripe virus group.

Rice plants infected with RHBV show stunting, chlorotic or yellow striping and mottling of leaves, and brownish or rotten roots (51). Plants infected early in their growth stage die prematurely. Panicles of infected plants are smaller, discolored, and deformed, often exsert incompletely, and bear few or no grains. Rice cells infected with RHBV contain a large mass of fine filaments, 8 to 10 nm in width, in the nuclei and cytoplasm (184). Similar masses also occur in the intestine of infective planthoppers.

Rice hoja blanca virus is not mechanically transmissible. It is transmitted by *S. orizicola* (1, 128). Another planthopper, *Sogatodes cubanus,* experimentally transmits RHBV with difficulty and does not survive long on rice (32, 52). The RHBV multiples in the planthoppers and is transmitted via the eggs (52, 192). As many as 60 to 100% of the progeny of infective females are congenitally infected with RHBV. The percentage of infective planthoppers in colonies can be increased up to 90 to 100% by selective breeding (52, 61). Because it is difficult to obtain a RHBV-free colony, the relations of RHBV to its vectors are not well understood, but study using virus-free *S. orizicola* indicated a long incubation period in the planthopper (30 to 36 days) (52).

The infective *S. orizicola* vector has a short life span and low fertility, fecundity, and egg hatchability (95, 192). The cyclic nature of hoja blanca epidemics can be explained by the deleterious effects of RHBV on *S. orizicola* (95).

Rice Ragged Stunt Virus

This virus (133) was first recognized in 1976 in Indonesia (72). It reached epidemic levels in 1977 in Indonesia and the Philippines (72, 126) and from 1979 to 1982 in Thailand (30). Generally, rice ragged stunt virus (RRSV) occurs in south and southeast Asia, China, Japan, and Taiwan (133). After 1983, the RRSV incidence declined to a low level in these countries.

The virus particles are isometric, 63 to 65 nm in diameter (63, 72, 73,

102, 103, 132), and contain 10 ds RNA, with a total molecular size of 23×10^6 (103, 156). The virus contains five major proteins (56), three are core proteins and two are component of spikes on the core. The RNA polymerase activity is associated with RRSV particles (208).

Rice plants infected with RRSV show stunting, twisting of leaf tips, ragged leaves, and vein swelling or galls on the outer surface of sheaths and the underside of leaf blades, at the base (72, 73, 126). These galls result from hyperplasia of the phloem tissues (73, 126), and amorphous inclusions can be seen in the gall cells (73). Panicle emergence is delayed and often incomplete. The panicles are short and bear unfilled grains.

Infected plants retain their green color after heading, and generate nodal branches. The symptoms in some cultivars are milder (163).

Virus particles are restricted to the phloem and gall cells in rice plants (63, 73, 187). In the rice cells, the particles are scattered in the cytoplasm or embedded in viroplasmic inclusions in the cytoplasm. In infected planthoppers, RRSV particles occur within or around cytoplasmic inclusions in the cells (63, 73); they also occur in tubules and phagocytic vesicles in the cytoplasm of fat body cells.

Rice ragged stunt virus is not mechanically transmissible. It is transmitted by the brown planthopper *N. lugens* and *N. bakeri* (72, 126, 139), although *N. bakeri* has few biological relations with rice. The virus multiplies in planthoppers (162) and is not transmitted via the eggs. Its incubation period in the planthopper is 4 to 33 days (average 9 days). The proportion of infective planthoppers in populations exposed to RRSV ranges from 12 to 48% in *N. lugens* and is 5% in *N. bakeri*. The presence of RRSV in single *N. lugens* can be determined serologically (38, 70, 149). About 50% of *N. lugens* carrying RRSV are infective (70). The infectivity of RRSV can be assayed by the microinjection method (70, 177).

In addition to rice, all *Oryza* spp. and 17 other gramineous plants can be experimentally infected with RRSV (6, 63, 101, 125), but natural infection of weeds and other cereals is rare. However, *N. lugens* is nearly monophagous of rice. The virus is generally endemic in areas where rice is grown continuously throughout the year, and its vector (*N. lugens*) is capable of dispersing RRSV as well as RGSV through long-distance migratory flights.

Rice Stripe Virus

This virus (203) occurs in China, Japan, Korea, Taiwan, and the USSR. Its incidence was high from 1960 to 1972 and 1977 to 1985 in Japan (108, 112) and during 1964–1965 and 1973–1974 in Korea (164).

Rice stripe virus (RStV) particles are flexuous filaments, about 8 nm in width, with branches (117, 118). Filaments have a helical configuration and are often circular (76, 117, 118). Virus particles are associated with four ss RNA and four ds RNA (92). This indicates the presence of four ss

RNA with both plus and minus forms in RStV, as is also seen in maize stripe virus (37). The RStV contain one major protein (117, 202), and its RNA polymerase activity is associated with the RStV particles (206). Rice stripe virus is closely related, serologically, to maize stripe virus (55) and distantly related to RGSV (76). Infected plants produce a large amount of a virus-specific protein, which is not serologically related to the viral protein (113).

Rice plants infected with RStV show chlorotic stripe, mottling and necrotic streaking of leaves, drooping of unfolded and twisted leaves, general chlorosis, and lack of vigor (188). Mottling may also occur on sheaths. When infection occurs at the early growth stage, the plants may die prematurely or be stunted, and tiller numbers of infected plants are usually reduced. Infected plants produce poor panicles. The symptoms are milder in some rice cultivars (176, 194).

The inclusion bodies of RStV-infected rice cells (77, 100) are generally homgeneous and round, oval, elongated, needle-like, ring, or eight-figured in shape. Under the electron microscope, infected cells can be seen to contain large masses of fine granules in the cytoplasm and nuclei and electron-dense crystals in the cytoplasm and vacuole (222). Fine granules are likely to correspond to the virus, the crystals to a noncapsid protein produced in infected plants.

Rice stripe virus is mechanically transmissible on rice with difficulty (146). It is transmitted by the smaller brown planthopper *L. aodelphax striatellus, Terthron albovittata, Unkanodes albifascia,* and *Unkanodes sapporonus* (79, 119, 188, 189, 190, 191, 203, 220); *L. striatellus,* however, is the major vector. The virus multiples in the planthopper (147) and is transmitted via the eggs (119, 188, 220). Its incubation period in the planthoppers is 5 to 21 days. The proportion of infective planthoppers in populations exposed to RStV ranges from 10 to 60% in *L. striatellus,* from 6 to 25% in *U. albifascia,* and is 6% in *U. sapporonus,* and 2% in *T. albovittata. Laodelphax striatellus* colonies with very high or very low transmission efficiencies can be obtained by selective breeding (109, 203). The proportion of congenitally infective progeny from infective females is 80 to 100% in *L. striatellus* and 61% in *U. albifascia* (75, 85, 188, 220). The presence of RStV in single planthoppers can be determined serologically (109, 149, 224). The proportion of planthoppers carrying the virus that are infective is 15 to 77% in *L. striatellus* (109, 149) and 5 to 21% in *U. albifascia* (79). Infectivity of RStV can be assayed by the microinjection method (118, 147, 195, 202).

Laodelphax striatellus overwinter in wheat, barley, and weeds (145). In Japan, RStV incidence is high in early planted rice, and major infection in the early rice occurs in June via first-generation adult vectors (145, 223). Generally, RStV incidence occurs in areas planted with winter wheat and barley, and its incidence in the subsequent crop is related to the density of infective *L. striatellus* in May and June. Serological detection of RStV in the overwintering or the first-generation *L. striatellus* is of use in forecasting RStV incidence.

Rice Transitory Yellowing Virus

This virus (26, 34, 182) occurs in southern and central China and the Okinawa islands of Japan, Taiwan, and Thailand. An RTYV outbreak occurred from 1960 to 1962 in Taiwan (196). In China, epidemics occurred in the years 1964–1965 in Kwandong; 1965–1966, 1969, and 1973 in Fujian; and from 1970 to 1972 in Chekiang. The incidence was also high from 1973 to 1975 in Taiwan (19).

Rice transitory yellowing virus (RTYV) particles are bullet shaped, 180 to 210 nm in length, and 94 nm in width (20, 21, 182). The virus contains four proteins (59), the largest of which is glycosylated and located in the envelope. The RTYV can be detected in plant extracts by the use of concanavalin A, which specifically combines with its glycosylated protein (58). An enzyme that phosphorylates protein is associated with RTYV particles (60).

Rice plants infected with RTYV show leaf yellowing, reduced tillering, and mild stunting (26, 34). Later, infected plants may recover and appear healthy, but symptoms may develop again after this temporary recovery. Because starch accumulates in infected rice leaves (81), iodine–potassium iodate staining is used to demonstrate the presence of RTYV. Phloem parenchyma cells have large inclusion bodies (197) that contain nuclei.

In infected rice cells, RTYV particles appear between two nuclear membranes (20, 182). In infected leafhopper cells, RTYV particles occur in vacuolate structures in the cytoplasm (21).

The virus is transmitted by *N. cincticeps, N. nigropictus,* and *N. virescens* (19, 23, 24, 26, 33, 34, 86); it multiples in the leafhoppers (82) but is not transmitted via the eggs. Incubation period of RTYV in the leafhoppers is 6 to 25 days. The proportion of infective leafhoppers in populations exposed to RTYV ranges from 10 to 75% in *N. nigropictus,* 10 to 71% in *N. cincticeps,* and 0 to 47% in *N. virescens*. Infective *N. cincticeps* has a longer nymphal period and life span, shorter oviposition, and lower fertility (17). The virus is mechanically transmissible from rice to *Nicotiana rustica* but not from rice plant to rice plant (25). The presence of RTYV in single leafhoppers can be determined serologically (200). Its infectivity can be assayed by the microinjection method (83).

Incidence of the virus is low in first-crop rice, but high in second-crop rice, and the infected rice stubble overwinters and serves as a virus source in the spring (19, 23, 33, 34). Overwintered leafhoppers also serve as a virus source.

Rice Tungro Bacilliform Virus and Rice Tungro Spherical Virus

Tungro (123, 170) is the most important virus disease in South and Southeast Asia. It is called *penyakit habang* in Indonesia, *penyakit merah* in Malaysia, and *yellow orange leaf* in Thailand. It also occurs in China (214). Tungro is a composite disease associated with both viruses (71, 157, 175). Rice tungro spherical virus (RTSV) alone is widespread in the Philippines (9) and probably in other south and southeast Asian countries and southern China. Rice tungro spherical virus was once epidemic in southern Japan, where it was called rice *waika* virus (50).

Rice tungro bacilliform virus (RTBV) particles are bacilliform, 100 to 300 nm in length, and 30 to 35 nm in width (71, 157, 175). Rice tungro spherical virus particles are isometric and 30 nm in diameter (53, 71, 157, 175). Whereas RTBV contains one DNA and one protein, RTSV contains one ss RNA and two (three) proteins (Cabauatan and Hibino, unpublished data). Both viruses multiply independently in rice plants.

In rice field affected with tungro, plants showing symptoms are generally infected with RTBV and RTSV. The symptoms include stunting, yellow or yellow–orange discoloration, and reduced tillering (123, 170). The leaves, especially the younger ones, may show striping or mottling and interveinal chlorosis. Discolored leaves may have irregularly shaped dark-brown blotches. Panicle exsertion is delayed and often incomplete, and the panicles are short, sterile, and discolored. Plants infected with RTBV alone develop similar but milder symptoms than those caused by double infection (71), whereas RTSV alone causes no clear symptoms, except very mild stunting. Grain yield reduction is as high as 100% in doubly infected plants, and 40% in RTSV-infected plants (57). Some cultivars show mild or no clear symptoms even when infected with both viruses at the seedling stage (28, 57).

In rice plants, both RTBV and RTSV particles are restricted to the phloem cells (41, 175). The phloem cells may contain both RTBV and RTSV particles, or one or the other. The RTSV-infected cells also have viroplasm-like inclusions and membraneous masses in the cytoplasm (175, 221). The RTSV particles can be readily obtained in dip preparation, whereas the RTBV particles are scarce in such a preparation (71).

The tungro virus composite are transmitted by *N. virescens, N. cincticeps, N. nigropictus, N. malayanus, N. parvus,* and *R. dorsalis* (123, 170). *Nephotettix virescens* is the major vector of tungro. This leafhopper retains tungro infectivity for 2 to 5 days but loses infectivity after molting (122, 123). Immediately after feeding on doubly infected plants, the leafhopper transmits either both RTBV and RTSV together or RTBV alone or RTSV alone (64, 71, 74). It retains RTBV and RTSV for three to four days (14, 65, 74). It readily acquires RTSV on plants that are only infected with RTSV, but it does not acquire RTBV on plants that are only infected with RTBV. The leafhopper acquires RTBV on the latter plants only when it has been exposed to RTSV-infected plants before feeding on RTBV-infected plants (71). Helper activity for RTBV transmission (acquisition) by the leafhopper is retained by the leafhopper for seven days. After it has fed on antiserum to RTSV, the leafhopper loses RTSV infectivity but retains helper activity (68). After molting, the leafhopper loses RTBV and RTSV infectivities as well as helper activity (14, 65).

The transmission efficiency of RTBV and RTSV differs depending on the vector species and the colonies. The proportion of tungro-infective leafhoppers in populations exposed to tungro-infected plants range from 70 to 80% in *N. virescens*, 0 to 27% in *N. nigropictus*, and 0 to 16% in *R. dorsalis* (123, 193).

Transmission patterns of RTBV and RTSV by *N. virescens* differ, depending on the cultivars used (29, 75). On cultivars susceptible to the leafhopper, leafhoppers exposed to both RTBV and RTSV efficiently transmit both viruses. On resistant cultivars, leafhoppers predominantly transmit RTBV alone, and their transmission efficiency is generally lower. However, leafhoppers exposed to RTSV alone transmit RTSV rather efficiently on resistant cultivars.

Since the mid 1960s, tungro has become increasingly important in south and southeast Asia. Major outbreaks occurred in 1969 in Bangladesh; in 1969 and during 1984–1985 in India; from 1969 to 1971, from 1972 to 1975, in 1980, and during 1983–1984 in Indonesia; in 1969 and during 1982–1983 in Malaysia; in 1957, during 1970–1971, and during 1983–1984 in the Philippines; and from 1965 to 1970 and 1979 in Thailand. A rice waika virus (RTSV) outbreak occurred in the year 1972–1973 in Kyushu, Japan (50).

The tungro-associated viruses infect rice and some other gramineous plants (5). Tungro infection has been controlled by cultivar resistance, application of insecticide to reduce leafhopper density, and cultivation practices. Insecticide application is not always efficient, especially in susceptible cultivars. Vector-resistant cultivars have been planted to control tungro infection (67, 75), although many of these resistant cultivars succumb to tungro after a few years of intensive cultivation (27, 67, 89, 130). Cultivars resistant to tungro infection or tolerant to RTBV have been identified (57), but resistance (tolerance) genes have not been transferred to commercial rice.

Rice Yellow Mottle Virus

This virus (10–12) occurs in Burkina Faso, Ghana, the Ivory Coast, Kenya, Liberia, Niger, Nigeria, Mali, Siera Leone, Tanzania, and the Upper Volta (174). It was first reported in Kenya in 1970 and then in West Africa within a decade, although it has not been reported outside the

African continent. It generally occurs in areas where rice is grown continuously.

Rice yellow mottle virus (RYMV) particles are isometric and about 30 nm in diameter (11, 12). In infected rice plants, the particles are scattered, or aggregated in crystalline arrays, in the cytoplasm of the mesophyll and epidermal cells (11, 12). Infected cells contain long flexuous tubules 10 to 15 nm in diameter and aggregates of fibrils in the cytoplasm. Viral concentration is high in rice plants.

Rice plants infected with RYMV show stunting; reduced tillering; wrinkling, mottling, yellowish streaking, and malformation of leaves; incomplete emergence of panicles; and sterile kernels (10, 11, 174). Many upland rice, including African traditional rice, develop mild symptoms when infected with RYMV.

Rice yellow mottle virus is mechanically transmissible (10–12). It is transmitted by such adult chrysomelid beetles as *Apophylis* spp., *Chaetocnema* spp., *Sesselia pussilla,* and *Trichispa sericea. Chaetocnema pulla* and other *Cheatocnema* spp. are less efficient as vectors but occur abundantly in rice fields (11, 12). However, *S. pusilla* is an efficient vector. The RYMV is not circulative in these beetles. The beetles retain RYMV for five to eight days (average 2 to 3 days).

The virus persists in rice and in a wild rice, *Oryza longistaminata,* which is common in swamp areas and rice fields in Africa. The major RYMV source is probably the rice itself, although RYMV probably harbored on *O. longistaminata* before rice was introduced. Cultivars tolerant of RYMV can be a good control measure for the virus (8, 97).

References

1. Acuña Gale, J., and Ramos-Ledon, L., 1957, Informes de interes general en relacion con el arroz, *Admin. Establ. Arroz,* Cuba, Bol. 4.
2. Amici, A., Faoro, F., Osler, R., and Tornaghi, R., 1978, The "giallume" disease of rice in Italy: New natural hosts of the viral agent, a strain of barley yellow dwarf virus, *Riv. Pat. Veg. Ser.* **14:**127–135.
3. Amici, A., Osler, R., and Belli, G., 1974, An isometric virus associated with the "giallume" disease of *Oryza sativa, Phytopathyol. Z.* **79:**285–288.
4. Ando, H., 1910, On rice dwarf disease (in Japanese), *J. Japan Agric. Soc.* **347,** 1–3.
5. Anjaneyulu, A., Daquioag, R.D., Mesina, M.E., and Hibino, H., 1988, Host plants of rice tungro (RTV)—associated viruses, *Int. Rice Res. Newsl.* **13(4):**30–31.
6. Anjaneyulu, A., Salamat, G.E., Mesina, M.E., Hibino, H., Lubigan, R.T., and Moody, K., 1988, Host range of ragged stunt virus (RSV), *Int. Rice Res. Newsl.* **13(4):**32–33.
7. Anjaneyulu, A., Aguiero, V.M., Mesina, M.E., Hibino, H., Lubigan, R.T., and Moody, K., 1988, Host plants of rice grassy stunt virus (GSV), *Int. Rice Res. Newsl.* **13(4):**37.

8. Attere, A.F., and Fatokun, C.A., 1983, Reaction of *Oryza glaberrima* accessions to rice yellow mottle virus. *Plant Dis.* **67:**420–421.
9. Bajet, N.B., Aguiero, V.M., Daquioag, R.D., Jonson, G.B., Cabunagan, R.C., Mesina, M.E., and Hibino, H., 1986, Occurrence and spread of rice tungro spherical virus in the Philippines, *Plant Dis.* **70:**971–973.
10. Bakker, W., 1970, Rice yellow mottle, a mechanically transmissible virus disease of rice in Kenya, *Neth. J. Plant Pathol.* **76:**53–63.
11. Bakker, W., 1974, Characterization and ecological aspects of rice yellow mottle virus in Kenya, Agricultural Research Reports 829, Centre for Agricultural Publishing and Documentation, Wageningen, p. 152.
12. Bakker, W., 1975, Rice yellow mottle virus. Descriptions of Plant Viruses No. 149, Commonwealth Mycological Institute, UK.
13. Belli, G., Amici, A., Corbetta, G., and Osler, R., 1974, The "giallume" disease of rice (*Oryza sativa* L.). *Mikrobiologija* **11:**101–107.
14. Cabauatan, P.Q., and Hibino, H. 1985, Transmission of rice tungro bacilliform and spherical viruses by *Nephotettix virescens* Distant. *Phil. Phytopathol.* **21:**103–109.
15. Cabauatan, P.Q., and Hibino, H., 1983, Unknown disease of rice transmitted by the brown planthopper in the Philippines, *Int. Rice Res. Newsl.* **8(2):** 12–13.
16. Casper, R., 1988, Luteoviruses, in Koenig, R. (ed.): The Plant viruses Vol. 3, Plenum Publishing, New York, pp. 235–258.
17. Chen, C.C., and Chiu, R. J., 1980, Factors affecting transmission of rice transitory yellowing virus by green leafhoppers. *Plant Prot. Bull.* **22:**297–306.
18. Chen, C.C., and Chiu, R.J., 1982, Three symptomatologic types of rice virus diseases related to grassy stunt in Taiwan, *Plant Dis.* **66:**15–18.
19. Chen, C.C., Ko, W.H., Wang, E.S., Yu, S.M., and Hu, D.Q., 1980, Epidemiological studies on the rice transitory yellowing with special reference to its transmission by the rice green leafhoppers (in Chinese). *Bull. Taichung Dist. Agric. Improv. Sta. New Ser.* **4:**1–61.
20. Chen, M.J., and Shikata, E., 1971, Morphology and intracellular localization of rice transitory yellowing virus, *Virology* **46:**786–796.
21. Chen, M.J., and Shikata, E., 1972, Electron microscopy and recovery of rice transitory yellowing virus from its leafhopper vector, *Nephotettix cincticeps, Virology* **47:**483–486.
22. Chen, S.H., Lin, H.Y., Xie, L.H., and Hu, F.P., 1978, Preliminary report of studies on common dwarf disease of rice (in Chinese), *Scientia Agric. Sinica* **3:**79–83.
23. Chen, S.X., Yuan, Y.L., King, D.D., Chen, K.Y., Lin, R.T., and Kao, T.M., 1979, Studies on the development of epiphytotris of the rice yellow stunt disease (in Chinese), *Acta Phytopathol. Sinica* **9:**41–54.
24. Chiu, R.J., Jean, J.H., Chen M.H., and Lo, T.C., 1968, Transmission of transitory yellowing virus of rice by two leafhoppers, *Phytopathology* **58:**740–745.
25. Chiu, R.J., Lee, R.C.J., Chen, M.J., Chen, C.C., Hsu, Y.H., Lu, Y.T., and Kuo, T.T., 1983, Sap transmission, assay and purification of rice transitory yellowing virus (RTYV). (Abstr.). *IV International Congress of Plant Pathology, Aug. 17–24, 1983 at Melbourne,* p. 112.
26. Chiu, R.J., Lo, T.C., Pi, C.L., and Chen, M.H., 1965, Transitory yellowing

of rice and its transmission by the leafhopper *Nephotettix apicalis* (Motsch.), *Bot. Bull. Acad. Sinica* **6:**1–18.
27. Dahal, G., Aguiero, V.M., Cabunagan, R.C., and Hibino, H., 1988, Varietal reaction to tungro with change in leafhopper "virulence," *Int. Rice Res. Newsl.* **13(5):**12–13.
28. Daquioag, R.D., Cabauatan, P.Q., and Hibino, H., 1986, Balimau Putih, cultivar tolerant of tungro-associated viruses. *Int. Rice Res. Newsl.* **11(6):**8.
29. Daquioag, R.D., Tiongco, E.R., and Hibino, H., 1984, Reaction of several rice varieties to rice tungro virus (RTV) complex. *Int. Rice Res. Newsl.* **9(2):**5–6.
30. Disthaporn, S., 1986, Characteristics of rice ragged stunt in Thailand, *Trop. Agric. Res. Ser.* **19:**160–164.
31. Disthaporn, S., Chettanachit, D., and Putta, M., 1983, Unknown virus-like rice disease in Thailand, *Int. Rice Res. Newsl.* **8(6):**12.
32. Everett, T.R., and Lamey, H.A., 1969, Hoja blanca, in Maramorosch, K. (ed): Viruses, Vectors and Vegetation, Interscience Publishers, New York, pp. 361–377.
33. Faan, H.C., and Pui, W.Y., 1980, A preliminary investigation on the primary source and transmission of rice yellow stunt virus in Kwangtung (in Chinese), *J. South China Agric. College* **1:**2–20.
34. Faan, H.Z., Li, Y.G., Fei, W.Y., Zhang, B.D., He, H.S., Ke, C., Ge, Q.W., Sun, G.F., Zheng, D.S., Ye, B.J., and Lu, B.C., 1965, Preliminary studies on rice yellow stunt in Kwangtung (in Chinese), *Zhi Wu Bao Fu (Plant Protection)* **3:**143–145.
35. Faan, S., Liu, C., Zhou, L., Liu, X., and Zhu, D., 1983, Rice gall dwarf—a new disease epidemic in the west of Guangdong Province of South China *Acta Phytopathol. Sinica* **13:**1–6.
36. Falk, B.W., Morales, F.J., Tsai, J.H., and Niessen, A.I., 1987, Serological and biochemical properties of the capsid and major noncapsid proteins of maize stripe, rice hoja blanca, and Echinochloa hoja blanca viruses, *Phytopathology* **77:**196–201.
37. Falk, B.W., and Tsai, J.H., 1984, Identification of single- and double-stranded RNAs associated with maize stripe virus. *Phytopathology* **74:**909–915.
38. Flores, Z.M., Hibino, H., and Perfect, J., 1986, Rice grassy stunt (GSV) and rice ragged stunt (RSV) carriers, *Int. Rice Res. Newsl.* **11(4):**26–27.
39. Faoro, F., Amici, A., and Tornaghi, R., 1978, Cellular ultrastructural alterations as a means of characterizing the virus causing the "giallume" disease of rice, *J. Submicro Cytol.* **10:**105–154.
40. Fauquet, C., and Thouvenel, J.C., 1983, Presence of a new rod-shaped virus in rice plants with stripe necrosis symptoms in Ivory Coast (in French), *C.R. Acad. Sci. Paris Ser. III.* **296:**575–580.
41. Favali, M.A., Pellegrini, S., and Bassi, M., 1975, Ultrastructural alterations induced by rice tungro virus in rice leaves, *Virology* **66:**502–507.
42. Fukushi, T., 1933, Transmission of the virus through the eggs of an insect vector, *Proc. Impr. Acad. (Tokyo)* **9:**457–460.
43. Fukushi, T., 1934, Studies on the dwarf disease of rice plant, *J. Fac. Agric. Hokkaido Imp. Univ.* **37:**41–164.
44. Fukushi, T., 1937, An insect vector of the dwarf disease of the rice plant, *Proc. Imp. Acad. Japan* **13:**328–331.

45. Fukushi, T., 1969, Relationships between propagative rice viruses and their vectors, in Maramorosch, K. (ed): Viruses, Vectors and Vegetation, Interscience Publishers, New York, pp. 279–301.
46. Fukushi, T., and Kimura, I., 1959, On some properties of the rice dwarf virus, *Proc. Japan Acad.* **35:**482–484.
47. Fukushi, T., and Shikata, E., 1963, Fine structure of rice dwarf virus. *Virology* **21:**500–503.
48. Fukushi, T., Shikata, E., and Kimura, I., 1962, Some morphological characters of rice dwarf virus, *Virology* **18:**192–205.
49. Fukushi, T., Shikata, E., Kimura, I., and Nemoto, M., 1960, Electron microscopic studies on the rice dwarf virus, *Proc. Japan Acad.* **36:**352–357.
50. Furuta, T., 1977, Rice waika, a new virus disease, found in Kyushu, Japan, *Rev. Plant Protec. Res.* **10:**70–82.
51. Galvez, G.E., 1969a, Hoja blanca disease of rice, in: The Virus Diseases of the Rice Plant, John Hopkins, Baltimore/IRRI, Philippines, pp. 35–49.
52. Galvez, G.E., 1969b, Transmission of hoja blanca virus of rice, in The Virus Diseases of the Rice Plant, John Hopkins, Baltimore/IRRI, Philippines, pp. 155–163.
53. Galvez, G.E., 1968, Purification and characterization of rice tungro virus by analytical density gradient centrifugation, *Virology* **35:**418–426.
54. Gingery, R.E., Bradfute, O.E., Gordon, D.T., and Nault, L.R., 1987, Maize chlorotic dwarf virus, *CMI/AAB Descriptions of Plant Viruses* No. 194.
55. Gingery, R.E., Nault, L.R., and Yamashita, S., 1983, Relationship between maize stripe virus and rice stripe virus, *J. Gen. Virol.* **64:**1765–1770.
56. Hagiwara, K., Minobe, Y., Nozu, Y., Hibino, H., Kimura, I., and Omura, T., 1986, Component proteins and structure of rice ragged stunt virus, *J. Gen. Virol.* **67:**1711–1715.
57. Hasanuddin, A., Daquioag, R.D., and Hibino, H., 1989, A method for resistance to tungro (RTV). *Int. Rice Res. Newsl.* **13(6):**13–14.
58. Hayashi, T., and Minobe, Y., 1984, Quantitative analysis of rice transitory yellowing virus using concanavalin A labelled with peroxidase (Abstr.) (in Japanese), *Ann. Phytopathol. Soc. Japan* **50:**438.
59. Hayashi, T., and Minobe, Y., 1985, Protein composition of rice transitory yellowing virus, *Microbiol. Immunol.* **29:**169–172.
60. Hayashi, T., and Minobe, Y., 1987, A protein phosphorylation enzyme activity associated with rice transitory yellowing virus. (Abstr.) (In Japanese), *Ann. Phytopathol. Soc. Japan* **53:**431–432.
61. Hendrick, R.D., Everett, T.R., Lamey, H.A., and Showers, W.B., 1965, An improved method of selecting and breeding for active vectors of hoja blanca virus, *J. Econ. Entomol.* **58:**539–542.
62. Hibi, T., Omura, T., and Saito, Y., 1984, Double-stranded RNA of rice gall dwarf virus, *J. Gen. Virol.* **65:**1585–1590.
63. Hibino, H., 1979, Rice ragged stunt, a new virus disease occurring in Tropical Asia, *Rev. Plant Protec. Res.* **12:**98–110.
64. Hibino, H., 1983, Transmission of two rice tungro-associated viruses and rice waika virus from doubly or singly infected source plants by leafhopper vectors, *Plant Dis.* **67:**774–777.
65. Hibino, H., 1983, Relations of rice tungro bacilliform and rice tungro spherical viruses with their vector *Nephotettix virescens*, *Ann. Phytopathol. Soc. Japan* **49:**545–553.

66. Hibino, H., 1986, Rice grassy stunt virus, *CMI/AAB Descriptions of Plant Viruses* No. 320.
67. Hibino, H., 1988, Towards stable resistance to rice virus diseases, in *Proceedings of the Symposium on Crop Protection in the Tropics, 11th International Congress of Plant Protection, 4–9 Oct. 1988, Manila, Philippines*. (In press)
68. Hibino, H., and Cabauatan, P.Q., 1987, Infectivity neutralization of rice tungro-associated viruses acquired by vector leafhoppers, *Phytopathology* **77:**473–476.
69. Hibino, H., Cabauatan, P.Q., Omura, T., and Tsuchizaki, T., 1985, Rice grassy stunt virus strain causing tungro-like symptoms in the Philippines, *Plant Dis.* **69:**538–541.
70. Hibino, H., and Kimura, I., 1982, Detection of rice ragged stunt virus in insect vectors by enzyme-linked immunosorbent assay, *Phytopathology* **72:**656–659.
71. Hibino, H., Roechan, M., and Sudarisman, S., 1978, Association of two types of virus particles with penyakit habang (tungro disease) of rice in Indonesia, *Phytopathology* **68:**1412–1416.
72. Hibino, H., Roechan, M., Sudarisman, S., and Tantera, D.M., 1977, A virus disease of rice (kerdil hampa) transmitted by brown planthopper, *Nilaparvata lugens* Stal, in Indonesia, Contr. Centr. Res. Inst. Agric. Bogor No. **35,** 15 p.
73. Hibino, H., Saleh, N., and Roechan, M., 1979, Reovirus-like particles associated with rice ragged stunt diseased rice and insect vector cells, *Ann. Phytopathol. Soc. Japan* **45:**228–239.
74. Hibino, H., Saleh, N., and Roechan, M., 1979, Transmission of two kinds of rice tungro-associated viruses by insect vectors, *Phytopathology* **69:**1266–1268.
75. Hibino, H., Tiongco, E.R., Cabunagan, R.C., and Flores, Z.M., 1987, Resistance to rice tungro-associated viruses in rice under experimental and natural conditions, *Phytopathology* **77:**871–875.
76. Hibino, H., Usugi, T., Omura, T., Tsuchizaki, T., Shohara, K., and Iwasaki, M., 1985, Rice grassy stunt virus: A planthopper-borne circular filament, *Phytopathology* **75:**894–899.
77. Hirai, T., Suzuki, N., Kimura, I., Nakazawa, M., and Kashiwagi, Y., 1964, Large inclusion bodies associated with virus disease of rice, *Phytopathology* **54:**367–368.
78. Hirao, J., 1968, Transmission of the rice black-streaked dwarf virus by a new planthopper vector, *Delphacodes (?) albifascia* Matsumura, *Japan J. Appl. Entomol. Zool.* **12:**81–85.
79. Hirao, J., 1968, Transmission of rice stripe virus by a Delphacid planthopper, *Delphacodes (?) albifascia* Matsumura, with notes on the development of the vector species (in Japanese). *Japan J. Appl. Entomol. Zool.* **12:**137–147.
80. Hirao, J., Inoue, H., and Oya, S., 1984, Proportion of viruliferous immigrants of the brown planthopper, *Nilaparvata lugens* Stal (Hemiptera: Delphacidae), transmitting rice grassy stunt virus during 1979–1983, *Appl. Entomol. Zool.* **19:**257–259.
81. Hsieh, S.P.Y., 1966, Accumulation of starch in rice leaves infected with transitory yellowing and its application to differentiate transitory yellowing from suffocating disease, *Plant Protec. Bull.* **8:**205–210.

82. Hsieh, S.P.Y., 1969, Multiplication of the rice transitory yellowing virus in its vector, *Nephotettix apicalis* Motsch, *Plant Protec. Bull.* **11:**159–170.
83. Hsieh, S.P.Y., and Roan, S.C., 1967, Mechanical transmission of rice transitory yellowing virus to its leafhopper vector, *Nephotettix cincticeps* Uhler, *Plant Protec. Bull.* **9:**23–30.
84. Hunt, R.E., Nault, L.R., and Gingery, R.E., 1988, Evidence for infectivity of maize chlorotic dwarf virus and for a helper component in its leafhopper transmission, *Phytopathology* **78:**499–504.
85. Iida, T.T., Shinkai, A., and Kimura, I., 1972, Rice dwarf virus, *CMI/AAB Descriptions of Plant Viruses* No. 102.
86. Inoue, H., 1979, Transmission efficiency of rice transitory yellowing virus by the green rice leafhoppers, *Nephotettix* spp. (Hemiptera: Cicadellidae), *Appl. Entomol. Zool.* **14:**123–126.
87. Inouye, T., and Fujii, S., 1977, Rice necrosis mosaic virus, *CMI/AAB Descriptions of Plant Viruses* No. 172.
88. Inoue, H., and Omura, T., 1982, Transmission of rice gall dwarf virus by the green rice leafhopper, *Plant Dis.* **66:**57–59.
89. Inoue, H., and Ruay-Aree, S., 1977, Bionomics of green rice leafhopper and epidemics of yellow orange leaf virus disease in Thailand, *Trop. Agric. Res. Ser.* **10:**117–121.
90. Ishii, M., and Yoshimura, S. 1973, Epidemiological studies on rice black-streaked dwarf virus in Kanto-Tosan district, Japan (in Japanese), *J. Cent. Agr. Exp. Sta.* **17:**61–121.
91. Ishii, M., Yasuo, S., and Yamaguchi, T., 1970, Epidemiological studies on rice dwarf disease in Kanto-Tosan District, Japan (in Japanese), *J. Centr. Agric. Exp. Sta.* **14:**1–115.
92. Ishikawa, K., Omura, T., and Tsuchizaki, T., 1989, Association of double- and single-stranded RNA's with each four components of rice stripe virus, *Ann. Phytopathol. Soc. Japan* **55.** (In press).
93. Iwasaki, M., Nakano, M., and Shinkai, A., 1980, Transmission of rice grassy stunt by *Nilaparvata muiri* China and *N. bakeri* Muir (Abstr.) (in Japanese), *Ann. Phytopathol. Soc. Japan* **46:**411–412.
94. Iwasaki, M., Nakano, M., and Shinkai, A., 1985, Detection of rice grassy stunt virus in planthopper vectors and rice plants by ELISA, *Ann. Phytopathol. Soc. Japan* **51:**450–458.
95. Jennings, P.R., and Pineda, A., 1971, The effect of the hoja blanca virus on its insect vector, *Phytopathology* **61:**142–143.
96. John, V.T., Heu, M.H., Manandhar, D.N., and Pradhan, R.B., 1978, Symptoms resembling those of rice dwarf disease in the Kathmandu Valley, Nepal, *Int. Rice Res. Newsl.* **3(4):**13–14.
97. John, V.T., Thottappilly, G., Ng, Q., Alluri, K., and Gibbons, J.W., 1985, Varietal reaction to rice yellow mottle virus disease, *FAO Plant Protec. Bull.* **33:**109–111.
98. Kashiwagi, Y., 1966, Staining of inclusion bodies in tumour cell of rice infected with black-streaked dwarf virus, *Ann. Phytopathol. Soc. Japan* **32:**168–170.
99. Katsura, S., 1936, The stunt disease of Japanese rice, the first plant virosis shown to be transmitted by an insect vector, *Phytopathology* **26:**887–895.

100. Kawai, I., 1939, On the inclusion bodies associated with the "Shimahagare" disease of rice plant (in Japanese), *Ann. Phytopathol. Soc. Japan* **9:**97–100.
101. Kawano, S., Senboku, T., and Shikata, E., 1981, Host range of rice ragged stunt virus (in Japanese), *Ann. Phytopathol. Soc. Japan* **47:**697–699.
102. Kawano, S., and Shikata, E., 1982, Purification and morphology of rice ragged stunt virus, *J. Fac. Agric. Hokkaido Univ.* **61:**209–218.
103. Kawano, S., Uyeda, I., and Shikata, E., 1984, Particles structure and double-stranded RNA of rice ragged stunt virus, *J. Fac. Agric. Hokkaido Univ.* **61:**408–418.
104. Kimura, I., 1973, Relationship between concentration of rice dwarf virus and the corresponding OD_{260} values of the virus and its RNA, *Ann. Phytopathol. Soc. Japan* **39:**432–434.
105. Kimura, I., 1976, Loss of vector-transmissibility in an isolate of rice dwarf virus (in Japanese), *Ann. Phytopathol. Soc. Japan* **42:**322–324.
106. Kimura, I., 1986, A study of rice dwarf virus in vector cell monolayers by fluorescent antibody focus counting, *J. Gen. Virol.* **67:**2119–2124.
107. Kimura, I., Minobe, Y., and Omura, T., 1987, Changes in a nucleic acid and a protein component of rice dwarf virus particles associated with an increase in symptom severity, *J. Gen. Virol.* **68:**3211–3215.
108. Kiritani, K., 1983, Changes in cropping practices and the incidences of hopper-borne diseases of rice in Japan, in Plumb, R.T. and Thresh, J.M. (eds): Plant Virus Epidemiology, Blackwell Scientific Publications, Oxford, pp. 239–247.
109. Kishimoto, R., 1967, Genetic variation in the ability of a planthopper vector, *Laodelphax striatellus* (Fallen) to acquire the rice stripe virus, *Virology* **32:**144–152.
110. Kishimoto, R., 1976, Synoptic weather conditions indicating long-distance immigration of planthoppers, *Sogatella furcifera* Horvath and *Nilaparvata lugens* Stal, *Ecol. Entomol.* **1:**95–109.
111. Kishimoto, R., 1981, Development, behavior, population dynamics and control of the brown planthopper, *Nilaparvata lugens* Stal, *Rev. Plant Protec. Res.* **14:**26–58.
112. Kishimoto, R., and Yamada, Y., 1986, A planthopper–rice virus epidemiology model: Rice stripe and small brown planthopper, *Laodelphax striatellus* Fallen, in McLean, G.D., Garrett, R.G., and Ruesink, W.G. (eds): Plant Virus Epidemics: Monitoring, Modeling and Predicting Outbreaks, Academic Press, Sydney, pp. 327–344.
113. Kiso, A., and Yamamoto, T., 1973, Infection and symptom development in rice stripe disease, with special reference to disease-specific protein other than virus, *Rev. Plant Protec. Res.* **6:**75–100.
114. Kitagawa, Y., and Shikata, E., 1969, On some properties of rice black-streaked dwarf virus (in Japanese), *Mem. Faculty Agric. Hokkaido Univ.* **6:**439–445.
115. Kitagawa, Y., and Shikata, E., 1974, Multiplication of rice black-streaked dwarf virus in its plant and insect hosts. (in Japanese), *Ann. Phytopathol. Soc. Japan* **40:**329–336.
116. Kodama, T., and Suzuki, N., 1973, RNA polymerase activity in purified rice dwarf virus, *Ann. Phytopathol. Soc. Japan* **39:**251–258.

117. Koganezawa, H., 1977, Purification and properties of rice stripe virus, *Trop. Agric. Res. Ser.* **10:**151–154.
118. Koganezawa, H., Doi, Y., and Yora, K., 1975, Purification of rice stripe virus (in Japanese). *Ann. Phytopathol. Soc. Japan* **41:**148–154.
119. Kuribayashi, K., 1931, Studies on the rice stripe disease (in Japanese), *Nagano Agric. Exp. Sta. Bull.* **2:**45–69.
120. Kuribayashi, K., and Shinkai, A., 1952, A new disease of rice, black-streaked dwarf (Abstr.) (in Japanese), *Ann. Phytopathol. Soc. Japan* **16:**41.
121. Lin, R.F., Chen, G.Y., Gao, D.M., Jin, D.D., Chen, S.X., and Ruan, Y.L., 1980, Studies on the application of the reversed passive carbon agglutination test to the detection of viruliferous individuals of rice dwarf virus (in Chinese), *Acta Microbiol. Sinica* **20:**173–179.
122. Ling, K.C., 1966, Nonpersistence of the tungro virus of rice in its leafhopper vector, *Nephotettix impicticeps, Phytopathology* **56:**1252–1256.
123. Ling, K.C., 1972, Rice virus diseases, The International Rice Research Institute, Los Baños, Philippines, p. 142.
124. Ling, K.C., Aguiero, V.M., and Lee, S.H., 1970, A mass screening method for testing resistance to grassy stunt disease of rice, *Plant Dis. Rept.* **54:**565–569.
125. Ling, K.C., Tiongco, E.R., and Aguiero, V.M., 1978, Host range of rice ragged stunt virus, *Int. Rice Res. Newsl.* **3(2):**8.
126. Ling, K.C., Tiongco, E.R., and Aguiero, V.M., 1978, Rice ragged stunt, a new virus disease, *Plant Dis. Rept.* **62:**701–705.
127. Luisoni, E., Lovisolo, O., Kitagawa, Y., and Shikata, E., 1973, Serological relationship between maize rough dwarf virus and rice black-streaked dwarf virus, *Virology* **52:**281–283.
128. Malaguti, G., Diaz, H.C., and Angeles, N., 1957, La virosis "hoja blanca" del arroz, *Agron. Trop.* **6:**157–163.
129. Mariappan, V., Hibino, H., and Shanmugam, N., 1984, A new virus disease in India, *Int. Rice Res. Newsl.* **9(6):**9–10.
130. Manwan, I., Sama, S., and Rizvi, S.A., 1985, Use of varietal rotation in the management of tungro disease in Indonesia, *Indonesian Agric. Res. Dev. J.* **7:**43–48.
131. Matsuoka, M., Minobe, Y., and Omura, T., 1985, Reaction of antiserum against SDS-dissociated rice dwarf virus and a polypeptide of rice gall dwarf virus, *Phytopathology* **75:**1125–1127.
132. Milne, R.G., 1980, Does rice ragged stunt virus lack the typical double shell of the Reoviridae? *Intervirology* **14:**331–336.
133. Milne, R.G., Boccardo, G., and Ling, K.C., 1982, Rice ragged stunt virus, *CMI/AAB Descriptions of Plant Viruses* No. 248.
134. Milne, R.G., and Luisoni, E., 1977, Serological relationships among maize rough dwarf-like viruses, *Virology* **80:**12–20.
135. Minobe, Y., Nakajima, T., Omura, T., Nozu, Y., and Kimura, I., 1984, Morphology and antigenicity of RDV particles (Abstr.) (in Japanese), *Ann. Phytopathol. Soc. Japan* **50:**135.
136. Morales, F.J., and Niessen, A.I., 1983, Association of spiral filamentous viruslike particles with rice hoja blanca, *Phytopathology* **73:**971–974.
137. Morales, F.J., and Niessen, A.I., 1985, Rice hoja blanca virus, *CMI/AAB Descriptions of Plant Viruses* No. 299.
138. Morinaka, T., Putta, M., Chettanachit, D., Parejarearn, A., Disthaporn, S.,

Omura, T., and Inoue, H., 1982, Transmission of rice gall dwarf virus by Cicadellid leafhoppers *Recilia dorsalis* and *Nephotettix nigropictus* in Thailand, *Plant Dis.* **66:**703–704.
139. Morinaka, T., Putta, M., Chettanachit, D., Parejarearn, A., and Disthaporn, S., 1983, Transmission of rice ragged stunt disease in Thailand, *Japan Agric. Res. Quarterly* **17:**139–144.
140. Morinaka, T., and Sakurai, Y., 1967, Studies on the varietal resistance to black-streaked dwarf of rice plant. 1. Varietal resistance in field and seedling test (In Japanese), *Bull. Chugoku Agric. Exp. Sta. Ser. E.* **1:**25–42.
141. Morinaka, T., and Sakurai, Y., 1968, Studies on the varietal resistance to black-streaked dwarf of rice plant. 2. Evaluation of varietal resistance of rice plant by seedling test (in Japanese), *Bull. Choguku Agric. Exp. Sta. Ser. E.* **2:**1–19.
142. Nakata, M., Fukunaga, K., and Suzuki, N., 1978, Polypeptide components of rice dwarf virus, *Ann. Phytopathol. Soc. Japan* **44:**288–296.
143. Nasu, S., 1963, Studies on some leafhoppers and planthoppers which transmit virus diseases of rice plant in Japan, *Bull. Kyushu Agric. Exp. Sta.* **8:**153–349.
144. Nasu, S., 1965, Electron microscopic studies on transovarial passage of rice dwarf virus, *Japan J. Appl. Entomol. Zool.* **9:**225–237.
145. Okamoto, D., Hirao, J., Teraguchi, M., and Okada, M., 1967, Studies on the smaller brown planthopper, *Laodelphax striatellus* Fallen, as a vector of rice stripe virus. I. On the life cycle of the smaller brown planthopper (in Japanese), *Bull. Chugoku Agric. Exp. Sta. Ser. E.* **1:**89–113.
146. Okuyama, S., and Asuyama, H., 1959, Mechanical transmission of rice stripe virus on rice plants (Abstr.) (in Japanese), *Ann. Phytopathol. Soc. Japan* **24:**35–36.
147. Okuyama, S., Yora, K., and Asuyama, H., 1968, Multiplication of the stripe virus in its insect vector, *Laodelphax striatellus* Fallen (in Japanese), *Ann. Phytopathol. Soc. Japan* **34:**255–262.
148. Omura, T., Hibino, H., Inoue, H., and Tsuchizaki, T., 1985, Particles of rice gall dwarf virus in thin sections of diseased rice plants and insect vector cells, *J. Gen. Virol.* **66:**2581–2587.
149. Omura, T., Hibino, H., Usugi, T., Inoue, H., Morinaka, T., Tsurumachi, S., Ong, C.A., Putta, M., Tsuchizaki, T., and Saito, Y., 1984, Detection of rice viruses in plants and individual insect vectors by latex flocculation test, *Plant Dis.* **68:**374–378.
150. Omura, T., and Inoue, H., 1985, Rice gall dwarf virus, *CMI/AAB Descriptions of Plant Virus* No. 296.
151. Omura, T., Inoue, H., Morinaka, T., Saito, Y., Chettanachit, D., Putta, M., Parejarearn, A., and Disthaporn, S., 1980, Rice gall dwarf, a new virus disease, *Plant Dis.* **64:**795–797.
152. Omura, T., Kimura, I., Tsuchizaki, T., and Saito, Y., 1988, Infection by rice gall dwarf virus of cultured monolayers of leafhopper cells, *J. Gen. Virol.* **69:**429–432.
153. Omura, T., Minobe, Y., Matsuoka, M., Nozu, Y., Tsuchizaki, T., and Saito, Y., 1985, Location of structural proteins in particles of rice gall dwarf virus, *J. Gen. Virol.* **66:**811–815.
154. Omura, T., Minobe, Y., and Tsuchizaki, T., 1988, Nucleotide sequence of segment 10 of the rice dwarf virus genome, *J. Gen. Virol.* **69:**227–231.

155. Omura, T., Morinaka, T., Inoue, H., and Saito, Y., 1982, Purification and some properties of rice gall dwarf virus, a new Phytoreovirus, *Phytopathology* **72:**1246–1249.
156. Omura, T., Minobe, Y., Kimura, I., Hibino, H., Tsuchizaki, T., and Saito, Y., 1983, Improved purification procedure and RNA segments of rice ragged stunt virus, *Ann. Phytopathol. Soc. Japan* **49:**670–675.
157. Omura, T., Saito, Y., Usugi, T., and Hibino, H., 1983, Purification and serology of rice tungro spherical and rice tungro bacilliform viruses, *Ann. Phytopathol. Soc. Japan* **49:**73–76.
158. Osler, R., 1984, Rice giallume, a disease related to barley yellow dwarf in Italy, in *Barley Yellow Dwarf. A Proceedings of the Workshop.* December 6–8, 1983, CIMMYT Mexico, pp. 125–131.
159. Osler, R., Amici, A., and Belli, G., 1974, Transmission of rice "giallume" by an aphid *Rhopalosiphum padi, Riv. Pato. Veg. Ser. IV* **10:**3–15.
160. Osler, R., Amici, A., and Longoni, C.E., 1980, *Leersia oryzoides,* a natural host and winter reservoir of the rice "giallume" strain of barley yellow dwarf virus, *Phytopathol. Z.* **97:**242–251.
161. Ou, S.H., 1985, Rice Diseases, 2nd ed., Commonwealth Mycological Institute, UK. 380 p.
162. Parejarearn, A., and Hibino, H., 1985, Development of rice ragged stunt virus (RSV) in the vector brown planthopper (BPH), *Int. Rice Res. Newsl.* **10(4):**11–12.
163. Parejarearn, A., and Hibino, H., 1987, Symptoms and yield reduction in tolerant varieties infected with ragged stunt virus (RSV), *Int. Rice Res. Newsl.* **12(4):**14–15.
164. Park, R.K. Jung, Y.T., Jin, Y.D., Kim, D.K., and Lee, D.H., 1982, Studies on the occurrence status of rice virus diseases in Yeongnam Area (in Korean), *Res. Rept. ORD* **24(C):**98–106.
165. Pellegrini, S., and Bassi, M., 1978, Ultrastructure alterations in rice plants affected by "grassy stunt" disease, *Phytopathol. Z.* **92:**247–250.
166. Pirone, T.P., 1977, Accessory factors in nonpersistent virus transmission, in Harris, K.F., and Maramorosch, K. (eds): Aphids as Virus Vectors, Academic Press, New York, pp. 221–235.
167. Putta, M., Chettanachit, D., Omura, T., Inoue, M., Morinaka, T., Honda, Y., Saito, Y., and Disthaporn, S., 1982, Host range of rice gall dwarf virus, *Int. Rice Res. Newsl.* **7(6):**13.
168. Reddy, D.V.R., Kimura, I., and Black, L.M., 1974, Co-electrophoresis of dsRNA from wound tumor and rice dwarf viruses, *Virology* **60:**293–296.
169. Reddy, D.V.R., Shikata, E., Boccardo, G., and Black, L.M., 1975, Coelectrophoresis of double-stranded RNA from maize rough dwarf and rice black-streaked dwarf viruses, *Virology* **67:**279–282.
170. Rivera, C.T., and Ou, S.H., 1965, Leafhopper transmission of "tungro" disease of rice, *Plant Dis. Rept.* **49:**127–131.
171. Rivera, C.T., Ou, S.H., and Iida, T.T., 1966, Grassy stunt disease of rice and its transmission by the planthopper *Nilaparvata lugens* (Sta.), *Plant Dis. Rept.* **50:**453–456.
172. Rochow, W.F., and Duffus, J.E., 1981, Luteoviruses and yellow disease, Kurstak, E. (ed): Handbook of Plant Virus, Infections and Comparative Diagnosis, Elsevier/North Holland, Amsterdam, pp. 147–170.
173. Rosenberg, L.J., and Magor, J.I., 1987, Predicting windborne displacements

of the brown planthopper, *Nilaparvata lugens* from synoptic weather data. 1. Long-distance displacements in the north-east monsoon, *J. Animal Ecol.* **56:**39–51.
174. Rossel, H.W., 1986, Rice yellow mottle and African soybean dwarf, newly discovered virus diseases of economic importance in West Africa, *Trop. Agric. Res. Ser.* **19:**146–153.
175. Saito, Y., 1977, Interrelationship among waika disease, tungro and other similar diseases of rice in Asia, *Trop. Agric. Res. Ser.* **10:**129–135.
176. Sakurai, Y., and Ezuka, A., 1964, The seedling test method of varietal resistance of rice plant to rice stripe disease. 2. The resistance of various varieties and strains of rice plant by the method of seeding test (in Japanese), *Bull. Chugoku Agric. Exp. Sta. Ser. A.* **10:**51–70.
177. Senboku, T., Chou, T.G., and Shikata, E., 1979, Some physical properties of rice ragged stunt virus, *Ann. Phytopathol. Soc. Japan* **45:**735–737.
178. Shanghai Institute of Biochemistry, 1974, Studies on the pathogens of the rice black streaked dwarf disease. 1. Viruslike particles in the insect vector, the planthopper *Laodelphax striatellus*, Fallen, *Sci. Sinica* **17:**273–283.
179. Shanghai Institute of Biochemistry and Chekiang Institute of Agricultural Research, 1977, The pathogen of some virus diseases of cereals in China. 1. The pathogen problems of rice dwarf, rice yellow stunt and soil-borne mosaic of barley, *Kexue Tongbao* **22:**93–94.
180. Shanghai Institute of Biochemistry and Chekiang Institute of Agricultural Research, 1978, The pathogens of some virus diseases of cereals in China. IV. Serological determination of the percentage of active individuals transmitting rice dwarf disease in a population of insect vectors (in Chinese), *Acta Biochem. Biophys. Sinica* **10:**355–362.
181. Shikata, E., 1969, Electron microscopic studies of rice viruses, in The Virus Diseases of the Rice Plant. John Hopkins Press, Baltimore/IRRI, Philippines, pp. 223–240.
182. Shikata, E., 1972, Rice transitory yellowing virus, *CMI/AAB Descriptions of Plant Viruses* No. 100.
183. Shikata, E., 1974, Rice black-streaked dwarf virus, *CMI/AAB Descriptions of Plant Viruses* No. 135.
184. Shikata, E., and Galvez, G.E., 1969, Fine flexuous threadlike particles in cells of plants and insect hosts infected with rice hoja blanca virus, *Virology* **39:**635–641.
185. Shikata, E., and Kitagawa, Y., 1977, Rice black-streaked dwarf virus: Its properties, morphology and intracellular localization, *Virology* **77:**826–842.
186. Shikata, E., Senboku, T., and Ishimizu, T., 1980, The causal agent of rice grassy stunt disease, *Proc. Japan Acad. Ser. B* **56:**89–94.
187. Shikata, E., Senboku, T., Kamjaipai, K., Chou, T.G., Tiongco, E.R., and Ling, K.C., 1979, Rice ragged stunt virus—a new member of plant reovirus group, *Ann. Phytopathol. Soc. Japan* **45:**436–443.
188. Shinkai, A., 1962, Studies on insect transmission of rice virus diseases in Japan, *Bull. Natl. Inst. Agric. Sci. (Japan) Ser. C.* **14:**1–112.
189. Shinkai, A., 1966, Transmission of rice black-streaked dwarf, rice stripe, and cereal northern mosaic viruses by *Unkanodes sapporonus* Matsumura (Abstr.) (in Japanese), *Ann. Phytopathol. Soc. Japan* **32:**317.
190. Shinkai, A., 1967, Transmission of rice black-streaked dwarf and rice stripe

viruses by *Ribautodelphax albifascia* (Abstr.) (in Japanese), *Ann. Phytopathol. Soc. Japan* **33:**318.

191. Shinkai, A., 1970, Transmission of rice stripe and northern cereal viruses by *Terthron albovittata* (Mats.) (Abstr.) (in Japanese), *Ann. Phytopathol. Soc. Japan* **36:**375.
192. Showers, W.B., and Everett, T.R., 1967, Transovarial acquisition of hoja blanca virus by the rice Delphacid, *J. Econ. Entomol.* **60:**757–760.
193. Sogawa, K., 1976, Rice tungro virus and its vectors in Tropical Asia, *Rev. Plant Protec. Res.* **9:**21–46.
194. Sonku, Y., 1973, Studies on the varietal resistance to rice stripe disease, the mechanism of infection and multiplication of the causal virus in plant tissues, *Bull. Chugoku Agric. Exp. Sta. Ser. E.* **8:**1–86.
195. Sonku, Y., Hirao, J., and Sakurai, Y., 1967, Studies on rice stripe disease. XIII. Transmission of rice stripe virus by *Delphacodes (?) albifascia* M. (Abstr.) (in Japanese), *Ann. Phytopathol. Soc. Japan* **33:**340.
196. Su, H.J., 1969, Transitory yellowing of rice in Taiwan, in The Virus Diseases of the Rice Plants, John Hopkins, Baltimore/IRRI, Philippines, pp. 13–21.
197. Su, H.J., and Huang, J.H., 1965, Intracellular inclusion bodies in the rice plants affected with transitory yellowing, *Botan. Bull. Acad. Sinica* **6:**170–181.
198. Suzuki, N., Kimura, I., and Kodama, T., 1968, Metachromatic and orthochromatic staining by Azure B of inclusion bodies associated with rice dwarf disease (in Japanese), in, Jubilee Pub. in Commemoration of 60th Birthday of M. Sakamoto, pp. 175–182.
199. Suzuki, Y., Widiarta, N., Raga, N., Nasu, S., and Hibino, H., 1988, Virulent strain of rice grassy stunt (GSV) identified in Indonesia, *Int. Rice Res. Newsl.* **13(1):**24–25.
200. Takahashi, Y., Omura, T., Hayashi, T., Shohara, K., and Tsuchizaki, T., 1988, Detection of rice transitory yellowing virus (RTYV) in infected plant cells and insect vectors by simplified ELISA, *Ann. Phytopathol. Soc. Japan* **54:**217–219.
201. Takata, K., 1895–1896, Results of experiments with dwarf stunt disease of rice plant (in Japanese), *J. Japan Agric. Soc.* **171:**1–4, 1895; **172:**13–32, 1896.
202. Toriyama, S., 1982, Characterization of rice stripe virus: A heavy component carrying infectivity, *J. Gen. Virol.* **61:**187–195.
203. Toriyama, S., 1983, Rice stripe virus, *CMI/AAB Descriptions of Plant Viruses* No. 269.
204. Toriyama, S., 1984, Rice black-streaked dwarf virus virion and RNA polymerase activity (Abstr.) (in Japanese), *Ann. Phytopathol. Soc. Japan* **50:**431.
205. Toriyama, S., 1985, Purification and properties of rice grassy stunt virus (Abstr.) (in Japanese), *Ann. Phytopathol. Soc. Japan* **51:**59.
206. Toriyama, S, 1986, An RNA-dependent RNA polymerase associated with the filamentous nucleoproteins of rice stripe virus, *J. Gen. Virol.* **67:**1247–1255.
207. Toriyama, S., 1987, Ribonucleic acid polymerase activity in filamentous nucleoproteins of rice grassy stunt virus, *J. Gen. Virol.* **68:**925–929.
208. Uyeda, I., Lee, S.Y., Yoshimoto, H., and Shikata, E., 1987, RNA polymerase activity of rice ragged stunt and rice black-streaked dwarf viruses, *Ann. Phytopathol. Soc. Japan* **53:**60–62.

209. Uyeda, I., Matsumura, T., Sano, T., Oshima, K., and Shikata, E., 1987, Nucleotide sequence of rice dwarf virus genome segment 10, *Proc. Japan Acad. Ser. B* **63:**227–230.
210. Uyeda, I., and Shikata, E., 1984, Characterization of RNAs synthesized by the virion-associated transcriptase of rice dwarf virus *in vitro, Virus Res.* **1:**527–532.
211. Woo, Y.B., and Lee, K.W., 1987, Detection of rice black-streaked dwarf virus in rice, maize and insect vectors by enzyme-linked immunosorbent assay (in Korean), *Korean J. Plant Pathol.* **3:**108–113.
212. Xie, L.H., Chen, Z.X., and Lin, Q.Y., 1979, Studies on the dwarf-like disease of rice plants. (in Chinese), *Acta Phytopathol. Sinica* **9:**93–100.
213. Xie, L.H., and Lin, J.Y., 1980, Studies on bunchy stunt disease of rice, a new virus disease of rice plant, *Kexue Tongbao* **25:**785–789.
214. Xie, L.H., and Lin, J.Y., 1982, The occurrence of rice tungro disease (spherical virus) in China (in Chinese), *J. Fujian Agric. Coll.* **3:**15–23.
215. Xie, L.H., and Lin, J.Y., 1983, On the bunchy stunt disease of rice. III. Stability *in vitro* and distribution in host plant of rice bunchy stunt virus (in Chinese), *Acta Phytopathol. Sinica* **13:**15–19.
216. Xie, L.H., Lin, Q.Y., and Guo, J.R., 1981, A new insect vector of rice dwarf virus, *Int. Rice Res. Newsl.* **6(5):**14.
217. Xie, L.H., Lin, Q.Y., and Zhou, Q.L., 1982, On the bunchy stunt disease of rice. II. The distribution, loss, hosts and overwintering of rice bunchy stunt (in Chinese), *Acta Phytopathol. Sinica* **12:**16–20.
218. Xie, S.D., Liu, C.Z., Zhang, S.G., Faan, H.C., and Liu, X.R., 1986, Maize—a new host of rice gall dwarf virus (GDV), *Int. Rice Res. Newsl.* **11(1):**20.
219. Xie, S.D., Zhan, S.G., Liu, X.R., Luan, L.H., Liu, G.Z., Zhou, L.G., and Faan, H.C., 1983, Primary inoculation of rice gall dwarf in Zhanjian (Preliminary report) (in Chinese), *Guangdong Agric. Sci.* **1983(6):**41–42.
220. Yamada, W., and Yamamoto, H., 1955, Studies on the stripe disease of rice plant. I. The virus transmission by an insect, *Delphacodes striatella* Fallen (in Japanese), *Spec. Bull. Okayama Agric. Exp. Sta.* **52:**93–112.
221. Yamashita, S., Doi, Y., and Yora, K., 1977, Some properties and intracellular appearance of rice waika virus, *Ann. Phytopathol. Soc. Japan* **43:**278–290.
222. Yamashita, S., Doi, Y., and Yora, K., 1985, Intracellular appearance of rice stripe virus, *Ann. Phytopathol. Soc. Japan* **51:**637–641.
223. Yasuo, S., Ishii, M., and Yamaguchi, T., 1965, Studies on rice stripe disease. I. Epidemiological and ecological studies on rice stripe in Kanto-Tosan District in central part of Japan (in Japanese), *J. Cent. Agric. Exp. Sta.* **8:**17–108.
224. Yasuo, S., and Yanagita, K., 1963, Serological study on rice stripe and dwarf virus disease. II. Hemagglutination test for rice stripe virus (Abstr.) (In Japanese), *Ann. Phytopathol. Soc. Japan* **28:**84.
225. Yokoyama, M., Nozu, Y., Hashimoto, J., and Omura, T., 1984, *In vitro* transcription by RNA polymerase associated with rice gall dwarf virus, *J. Gen. Virol.* **65:**533–538.
226. Zeigler, R.S., Rubiano, M., and Pineda, A., 1988, A field screening method to evaluate rice breeding lines for resistance to the hoja blanca virus, *Ann. Appl. Biol.* **112:**151–158.

9
Homopteran Transmission of Xylem-Inhabiting Bacteria

Alexander H. Purcell

Introduction

The great majority of bacteria associated with plants are not pathogens, and pathogenicity is the exception, not the rule (53). Of the relatively few bacteria that are pathogens, most do not require mobile vectors such as insects (6, 34, 82). Of the bacterial phytopathogens that have insect vectors, a wide range of relationships have been documented (34). This chapter deals with Homopteran transmission of the relatively few (described so far) types of xylem-limited bacteria (XLB) that are transmitted by insects. Such bacteria are transmitted—not surprisingly—by xylem-feeding insects. The ecological and evolutionary implications of XLB transmission by insects will also be discussed, relevant to the epidemiology of XLB.

Historical Background

As was the case with many vector-borne pathogens, the discovery of the mode of spread (vector transmission) of the plant diseases caused by XLB preceded the discovery of the causal pathogen of these diseases (77). Pierce's disease of grape (PD), now known to be caused by a strain of the bacterium *Xylella fastidiosa* (15, 106), was initially thought to be a viral disease (38). Because the sharpshooter leafhoppers (Cicadellidae: Cicadellinae) found to transmit this pathogen (38) were known to be xylem feeders, Houston et al. (48) investigated the plant tissues in which the pathogen occurred. They found that both vector acquisition and inoculation feeding were successful in transmission experiments only when the insects had access to the xylem beneath the bark. Blocking access to xylem with metal foil prevented acquisition or introduction of the pathogen. Of the suctorial insects tested, all xylem-feeding leafhoppers and spittlebugs proved to be PD vectors, which led to the hypothesis that

---

Alexander H. Purcell, Department of Entomological Sciences, University of California, Berkeley, California 94720, USA.

competence as a PD vector was more a function of vector feeding behavior than phylogeny (25).

Alfalfa dwarf disease (AD) had been described earlier by Weimer (104, 105), who noted the consistent occurrence of "bacterialike bodies" (105) in the xylem vessels of plants with the disease. When it was discovered that the same pathogen caused both PD and AD (38), however, the idea of looking for bacterialike bodies in grapevines with PD appears to have been overlooked. Not until three decades later were bacteria, first described as "Rickettsialike organisms" (31, 45), associated with the disease.

Graft transmission experiments provided evidence that another "viruslike" disease that occurred in the eastern United States, phony disease of peach (PP), was xylem-limited. Only grafted chips or scions that had active xylem—not bud grafts—successfully transmitted the pathogen (51). Like PD, the only other exclusively xylem-borne plant "virus" known at the time, the PP agent was also transmitted by xylem-feeding leafhoppers (98, 99).

The in vitro culture of *X. fastidiosa* from PD grape (15) and PP peach (13, 107) led to the detection of this bacterium in other plant hosts (17, 43, 44, 55, 56, 88–90, 102, 109) and prompted studies of how the bacterium was transmitted by insects (86). Additional plant diseases caused by different strains of the same bacterium, *X. fastidiosa,* were almond leaf scorch (ALS) (18, 64), plum leaf scald (54, 90, 108), and other leaf scalding diseases of deciduous trees (36, 55, 56) as well as wilt disease (PW) of periwinkle (17, 60) (see Table 9.1). Claims that *X. fastidiosa* may cause citrus blight (44) have not be confirmed (24, 29, 30).

TABLE 9.1. Plant diseases caused by xylem-limited bacteria transmitted by insects.

Disease	Pathogen	References
	Xylella fastidiosa	
	Group I.	
Pierce's disease of grape	Cultivable on medium PD3,	15
Alfalfa dwarf	causes symptoms in grape.	15, 31, 38
Almond leaf scorch	Does not cause (infect?) phony peach disease	18, 64
	Group II.	
Phony peach	Grows on PW but not PD3 medium	13, 107
Periwinkle wilt	Does not cause symptoms in grape	17
Ragweed stunt	infects peach	96
Plum leaf scald		90
	Group II.?[a]	
Mulberry, oak, elm		36, 55
Sycamore leaf scorches		56
Sumatra disease of clove	*Pseudomonas* sp. [*solanacearum*?]	22, 50

[a] Mixed reactions reported for one isolate; otherwise appear more closely related to Group II (43).

Pathogens

Table 9.1 summarizes the XLB known to have insect vectors. All the vectors known to date are Homopteran xylem feeders. In addition to the various strains of *X. fastidiosa*, another vector-borne XLB causes Sumatra disease, which is a severe wilt of clove (50). The clove wilt bacterium appears to be closely related to *Pseudomonas solanacearum* (22). The genus *Xylella* is probably closely related to *Xanthamonas*, a member of the Pseudomonaceae (106).

Other categories of XLB, which will not be discussed here, include such XLB as *Clavibacter xyli*, incitant of ratoon stunt disease of sugar cane, that do not appear to have insect vectors (12, 19) and bacteria that enter and move within plant xylem (6) but do not have an obligate affinity for xylem tissue. Some bacteria in the latter group have important insect vectors (34).

Pathogenicity

The mechanisms by which XLB induce symptoms of wilt, chlorosis, necrosis, and stunting has been reviewed recently by Davis (19). Goodwin et al. (32, 33) found no support for the hypothesis (58) that an extracellular toxin of *X. fastidiosa* causes wilting and marginal drying and necrosis of leaf margins in grapevines with PD. In this case, mild but chronic blockage of the xylem system was the most consistent explanation for the physical and chemical changes in the diseased plants. Microscopy of the vascular system from vines with PD leads to much the same conclusion. If only one single or a few serial sections are examined, enough xylem cells free of gums or bacterial aggregates can usually be found to account for adequate water transport, but if serial sections of tracheary elements are examined along their entire length, most of the xylem cells connected along the same axis can be found to be blocked at some point (40).

One of the most surprising discoveries in early research on PD was the very wide symptomless host range of the causal agent. These conclusions were based on the findings that noninfectious sharpshooters acquired the PD "virus" from a large variety of symptomless plants previously fed upon by infectious sharpshooters (27). The PD agent was found to be epxerimentally transmitted to 75 species of plants in 23 families. Only 25% of the species tested gave negative results, a remarkable finding in view of the small number of tests used for some plants. The presence of symptoms in these studies was judged by visually appraising plant size and appearance. More detailed measurements of plant biomass and chemical composition might have revealed more insidious effects of the PD bacterium. The effects of *X. fastidiosa* in ragweed are a mild stunting (96). Typical symptoms of PD, ALS, AD, and PP are slow to appear after insect inoculation, and usually require many months. Experimental transmission in greenhouse plants might not allow enough time for the development of symptoms induced by steadily increasing and systematically distributed bacterial populations within the plant. However, the stresses associated with growing potbound plants for prolonged periods in a heated greenhouse might increase water stress over what is normally realized in field conditions.

Numerous common weeds were found to be hosts for *X. fastidiosa*. Freitag et al. (27) found that 36 plant species were naturally infected, using transmission from experimental hosts by noninfective leafhoppers as a criterion. Naturally infected weeds were detected both by serology (ELISA) and by in vitro isolations of bacteria (43, 88, 89, 102). Pathogenicity tests showed that some of the weeds were pathological hosts (89).

Strain Differences

Isolates of *X. fastidiosa* from the eastern United States differ more widely in their abilities to colonize and cause disease in different kinds of plants. Hopkins (43) has pointed out the present confusing diversity of "strains" of *X. fastidiosa*. Two groupings of strains (52) can be distinguished by their growth on either PW (17) or PD3 (19) media. Only Group I strains could be isolated from plants on both media; Group II strains grew only on PW medium. These correspond approximately with the pathological characters listed in Table 9.1. Studies of intragroup variability are needed (43).

Prolonged in vitro culture produced some *X. fastidiosa* isolates that ranged in their pathogenicity for grape from avirulent to mildly virulent to severely (typical) virulent. Nonvirulent isolates initially multiplied about as quickly within mechanically inoculated grapevines as did virulent isolates, but they did not spread within the inoculated plant, and eventually no bacteria could be recovered from the inoculated plant. These studies suggested that bacterial titers and movement within the plant were directly related to virulence. Exogenous applications of abscisic acid and ethephon increased PD symptom severity and titers of *X. fastidiosa* in *Vitis rotundifolia* (muscadine) but not in *V. vinifera;* however, applications of the plant hormones indoleacetic acid and kinetin prevented PD symptoms and the accumulation of bacterial populations (42). Applications of giberillic acid also reduced symptoms but not XLB population accumulation in peach with PP (28). A hypothetical explanation (42) was that substances that hasten plant maturity and senescence favor bacterial multiplication, whereas substances that retard senescence might prevent or slow populations of *X. fastidiosa*.

Isolates of *X. fastidiosa* vary in their pathogenecity for different plant species. For example, isolates from peach with PP are virulent in periwinkle but not in grape, whereas nearly the opposite is true for PD isolates: mild virulence in periwinkle, no virulence in peach (41). Does the "symptomless" condition of many plant species that have been reported to be hosts based on recovery of bacteria—whether by leafhopper transmission tests or by in vitro isolations—reflect only transient colonization of these plants by *Xyella,* spp. similar to *X. fastidiosa* isolates nonvirulent for grape (41)? Or do they represent chronic low titer infections that never produce symptoms? This is an important question in understanding the epidemiology of the plant diseases involved, as will be discussed later.

The Xylem Environment

Physical Features

The chief function of the xylem tissues of plants is to conduct water and mineral nutrients from the soil to above-ground organs (23). The morphology of xylem vessels can vary widely not only among different plant species but also within the same plant (23, 113). Xylem vessels have characteristically thick walls, which are very large compared to other plant cells. Different-sized vessels transport sap at a rate proportional to the square of their cross-sectional area (113). A vessel that is twice the diameter of another vessel theoretically transports 16 times the volume of sap the smaller vessel transports. In some woody plants, individual vessels can be many meters long, depending upon where they occur in the plant (113). Vessels are interconnected at their axial ends by relatively large openings, and radially by numerous small, membrane-lined pits.

The xylem sap moves predominantly outward towards distal stems and foliar organs. Xylem sap moves along a water potential gradient from roots to the much lower water potentials created by transpiration of water from leaf cell walls (91, 113, 114). The strong negative potentials in transpiring leaves can create great differences in pressures within xylem vessels compared to the outside (atmospheric) pressure—a difference of 10 to 14 atmospheres is common in trees on hot, dry days (66a, 113). At such high potentials, any physical penetration of xylem vessels, such as might be caused by a cut or puncture, will normally allow the entry of air and cause the vessel to cavitate, breaking the thin, cohesive column of xylem sap within the damaged vessel and thus stopping the flow of sap within the vessel (114). This has important consequences for insects or microbes that feed on xylem contents. How do xylem-feeding leafhoppers feed from vessels containing sap at high negative tensions? How do xylem-inhabiting bacteria move from the xylem cells into which they have been introduced if their introduction produces cavitation?

The observations that *Xylella* and other bacteria extensively colonize xylem tissues (12, 31, 40, 45, 46, 49) indicate that they move from the point of inoculation to other xylem cells, but how this occurs is unknown. The lateral cell-to-cell movement of bacteria should require that they penetrate the membrane-lined pits along the sides of vessels (114). The pit membranes are freely permeable to water, but not to particles even smaller than bacteria (113). The regular observation of xylem cells packed with *X. fastidiosa* adjacent to tracheary elements without bacteria is evidence that pit membranes are effective barriers (12), but obviously some lateral movement does occur. One hypothesis is that the bacteria pass through the pit membranes to move laterally between vessels (12), but there is no published ultrastructural evidence that bacteria can actually pass though the pit membrane.

Chemistry of Xylem

Xylem characteristically has very low concentrations of total solutes—from about 0.5 to 20 mg/ml (7, 70)—but concentrations near 1 mg/ml may be more usual (7). Inorganic ions absorbed from the soil solution are mobile in the xylem solution. Nitrogenous compounds are major solutes in xylem sap, chiefly in the form of amino acids and amides, although nitrate ions and ammonia are sometimes detectable (7, 70, 91). Organic acids such as citric and malic acid are thought to be exported products of root metabolism (70). Carbohydrates are typically absent, or present in trace amounts, except for the rising sap of such winter dormant trees as the sugar maple (*Acer saccharum*). Alkaloids manufactured by the plant are also transported in the xylem. The composition and concentrations of various solutes in xylem sap vary with plant species (7), season (67, 70), time of day (70), and location within the plant (47, 70). Plant age and state of health also affects xylem sap composition.

Xylem-limited bacteria have undoubtedly adapted to the dilute concentrations of nutrients available to them in xylem sap. A constant flow of sap, however, may be needed to provide these nutrients (12, 16). The slow growth rate of *X. fastidiosa* in plants or in vitro (16) may reduce the requirements of this XLB for higher concentrations of nutrients. *Xylella fastidiosa* does not occur in juvenile tissues of grapevines (42). This may be due—at least in part—to the chemistry of xylem sap in senescent tissue. The in vitro growth of *X. fastidiosa* in static grape xylem sap is barely detectable (16). In theory, the environment of the insect foregut should be similar to that within actively conducting xylem: a hollow tube with a rapid flow of dilute nutrients. This environment requires that relatively sessile microbes such as *X. fastidiosa* attach securely either to the xylem walls or the insect foregut and extract dilute nutrients from its surroundings. The extracellular matrix material associated with *X. fastidiosa* may aid in the attachment (8, 86). Fimbrae-like structures may also aid in the attachment of this bacterium (8).

The specialized ability to feed predominantly on xylem requires that the feeding insect can locate, penetrate, and suck in sap against strong negative tensions. Each step of this process must be completed without damaging the penetrated cell so as to allow the entry of air or cause an embolism through inducing a plant response to injury that would stop sap transport in the xylem vessel being fed upon (63, 82, 91, 97). The dilute nutrients in xylem sap require that xylem feeders consume very large amounts of sap. The quantities consumed relative to body weight—as much as 100 to 1000 times body weight (47, 66a, 110)—are typically among the largest in the animal kingdom (91, 97). The pump chamber and the muscles that power the pump diaphragm (Figure 9.1) are very large in

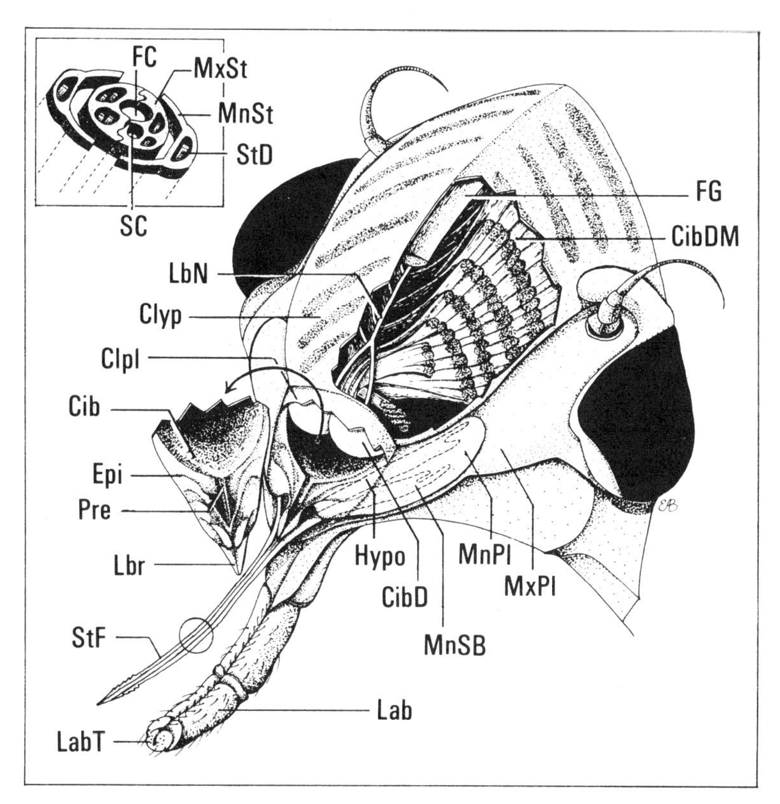

FIGURE 9.1. Anatomy of leafhopper mouthparts and foregut. Cib, cibarium; CibD, diaphragm; CibDM, dilator muscle; Clpl, clypellus; Clyp, clypeus; Epi, epipharynx; FC, food canal; FG, frontal ganglion; Hypo, hypopharynx; Lab, labium; LabT, labial tip; LbN, labral nerve; Lbr, labrum; Mn, mandibular; SB, stylet base; St, stylet; Mx, maxillary; Pl, plate; Pre, precibarium; SC, salivary canal; StD, dendrite; StF, fascicle. Drawing by Elaine Backus (4), University of Missouri; reproduced with permission from E.A. Backus in L.R. Nault and J.G. Rodriguez (eds): The Leafhoppers and Planthoppers, copyright 1985 by Wiley Interscience.

xylem feeders. The resulting feature of a bulbous or "inflated" forehead is a common character to keys to identify the leafhopper subfamily Cicadellinae, which consists entirely of xylem feeders, as far as known (112). Other Homopteran families consist entirely of xylem feeders. Cicadas (Cicadae) and spittlebugs (Cercopidae) also have very large pump muscles, and a correspondingly bulbous forehead. Dilute nutrients undoubtedly have also shaped the evolution in xylem feeders of highly efficient digestive systems. The predominance of amino and organic acids in xylem sap, which are easier to absorb and digest than are proteins and carbohydrates, make this a readily assimilatable nutrient source except for the problem of handling the great quantities of water in which the nutrients are diluted (91, 110). Sharpshooter leafhoppers are highly efficient in using most amino and organic acids: only cysteine, tyrosine, and methionine were digested at less than 90% efficiency; most were digested at nearly 100% efficiency (2b).

Raven (91) calculated that the energy requirements of suctorial feeding on xylem consume a major fraction of the energy content of the ingested sap. It should be kept in mind that the method of xylem collection on which xylem chemistry has been analyzed quantitatively can drastically affect estimates of solute concentrations (70). Very young plant organs can contain many times the concentrations of energy sources, such as amino acids, as do older tissues. The meadow spittlebug *Philaenus spumarius* strongly prefers succulent, actively growing plant tissues, which contains the highest level of amino acids (47). Amino acid concentrations in the xylem of youngest leaves were more than tenfold those in oldest leaves, and nymphal growth and survival was high (>20%/week) on leaves with amino acid concentrations greater than 1 mM. The blue-green sharpshooter (*Graphocephala atropunctata*) strongly prefers young and growing tissues, which influences its host selection probably more than plant species (74). Two species of *Homalodisca* sharpshooters similarly showed strong preferences for succulent terminal tissues among a great variety of plant species (66b).

Vectors

Widely different kinds of microbial parasites of plants—fungi, protozoa, bacteria, and viruses—infect plants via insect vectors of the order of Hemiptera (suborders Heteroptera and Homoptera) (9). A wide variety of plant viruses have sap-sucking vectors. This convergent evolution of diverse microbial parasites toward transmission by sap-sucking insects suggests pervasive selective pressures (82).

An obvious advantage to microbial parasites in using sap-sucking insects as vectors is that Homopteran feeding may breach plants' physical barriers to microbial colonization, but not destroy the continued function or integrity of the host plant tissues (82). Pathogens introduced by vector feeding into dead or nonfunctional vascular cells (xylem or phloem) may be hindered in further colonizing the vascular system. A caterpillar, grasshopper, or other chewing insect removes or obliterates much of the tissue on which it feeds. Even so, some bacteria that can infect plant wounds (e.g., *Erwinia stewartii, Erwinia tracheiphila,* and *Erwinia caratovora*) utilize as vectors chewing insects such as leaf beetles (9, 34). Suctorial insects such as Typhlocybine leafhoppers obliterate the mesophyll or parenchyma cells upon which they feed by ingesting the contents of plant cells, and leave behind an air-filled cell wall (72). In contrast to mesophyll and parenchyma feeders are the many suctorial insects that specialize in feeding upon plant vascular fluids of xylem or phloem (63, 71, 97). These two types of vascular tissues are very different physically and chemically, despite their intimate interconnection in plants' transport systems. Phloem feeders correspondingly differ in many respects from xylem feeders. As will be discussed later, these specialist feeders probably do not feed exclusively on only one type of tissue, but most evidence strongly suggests that specialization predominates (97).

Although cells are damaged by vascular specialists' feeding on phloem, the punctured cells continue to function, as demonstrated by the continued exudation of phloem sap from the severed stylets of feeding aphids (65). "Hopperburn" ensues from the phloem feeding of the leafhopper *Empoasca fabae,* presumably because of physical damage to host plant phloem (9). This suggests that other suctorial phloem feeders that do not induce similar plant reactions do not so severely disturb the function of phloem. For xylem tissues, the best evidence so far that suctorial feeding by xylem specialists does not disrupt the continued function of punctured cells is that feeding is prolonged over long periods, with prodigious amounts of excrement produced (66a).

Obviously, xylem-feeding insects have solved the problem of penetrating functioning xylem without reducing the flow of sap below a nutritionally acceptable level. Without experimental data we can only speculate that the penetration of plant tissues by a sharpshooter, for example, does not introduce air into the wound. Numerous investigators have noted the logic that xylem feeders should tap xylem sap nondestructively (66a, 82, 91, 97).

Like many Homoptera, sharpshooters secrete a gelling salivary component (63, 71) during feeding penetration. This saliva quickly solidifies and then is penetrated by the insect's stylets, to form a salivary sheath (61, 62, 63). Presumably, the salivary sheath keeps fluids from cells outside the target cells from entering the stream of sap being sucked in by the feeding insect (63). This process may also allow penetration of a xylem cell by the stylets without permitting air to enter the penetrated cell. Aphids fill the void left by withdrawn stylets with sheath saliva (61). It is not known if sharpshooter leafhoppers do the same upon withdrawal from feeding. Electronic monitoring of feeding of the blue-green sharpshooter revealed that the sustained ingestion signals (waveforms) did not closely correlate with the amount of fluid excreted (10). The explanation may be that some feeding connections were inefficient in preventing cavitation, so that sustained feeding (pumping) provided little if any ingestion of fluid. The size of the vessel penetrated and the intracellular xylem tension would also determine the volumetric rate at which sap could be pumped. Upon histological examination, Crane (10) found that some xylem vessels fed upon by the blue-green sharpshooter were "occluded" but that others were not. The histological methods used would not have shown if the xylem vessels contained air embolisms before fixation. Neither would they have shown if the vessels remained functional after sharpshooter feeding. Whether xylem vessels continue to function normally after sharpshooter feeding remains to be resolved.

Virtually all species of suctorial xylem feeders that have been tested as vectors of *X. fastidiosa* have been proven capable of transmitting the bacterium to plants (25, 77). It is logical that bacteria with such a fastidious requirement for xylem tissue should have xylem-feeding insects as vectors. But it is not clear why other sap-feeding suctorial insects that contact the xylem are not vectors. Other Homopterans have been observed to probe xylem. Ingestion from xylem cells has been documented for some species that have been monitored electronically. For example, Ullman (100) found that the pear psylla (*Cacopsylla pyricola*) ingested fluid from xylem in 19% (on peach) to 24% (on pear) of its feeding probes. The pear psylla spent an average of 43% of its total probing time ingesting from xylem. A number of histological studies have been made to determine where leafhopper feeding tracks terminate in plant tissues. Deltacephaline leafhoppers such as *Nephotettix cincticeps*, inserted their stylets in xylem during 61% of its feeding probes, whereas four other phloem-feeding leafhopper species had over 10% of their probes terminate in xylem (68). *Euscelidius variegatus* probed xylem tissue (2a), as did *Euscelis* sp. (21% of probes), whereas *Orosius argentatus* and two other leafhopper species were observed to regularly build a salivary sheath terminating in xylem cells (20). Yet tests of such leafhoppers as *E. variegatus* and other species not noted for prolonged xylem ingestion have never shown that these species transmitted *Xylella* to plants (79, 93). Clearly, regular penetration of xylem tissues, even with attempted ingestion, is not adequate for bacterial transmission.

For bacteria introduced into a xylem vessel to survive after stylet withdrawal, the vessel must remain functional or the bacteria must move to other vessels that will continue to function after withdrawal. This may explain why Homopterans not specialized for xylem feeding are not competent vectors. One possibility is that they damage xylem transport in probed vessels. For XLB to multiply, a continuous flux of xylem sap from which they extract nutrients is probably required (16). Embolisms created during insect feeding would interrupt this flux. However, XLB may be able to infect and move from damaged vessels. Evidence for this is that mechanical introduction of bacterial suspensions with a needle puncture is very efficient in infecting grape with *X. fastidiosa* (41), although the numbers of bacteria introduced via needle inoculation may be much greater than the numbers introduced by suctorial insects, as will be discussed later. Another reason that non-xylem feeders do not transmit XLB may be that they do not pump adequate amounts of xylem sap, either because their cibarial pumps (Figure 9.1) cannot overcome the high negative pressures of the xylem stream or because they do not regularly ingest from the penetrated xylem tissues.

The massive cibarial pumps of xylem feeders are the site of growth of *X. fastidiosa* (Figure 9.2) (8, 86, 111). Other portions of the foregut,

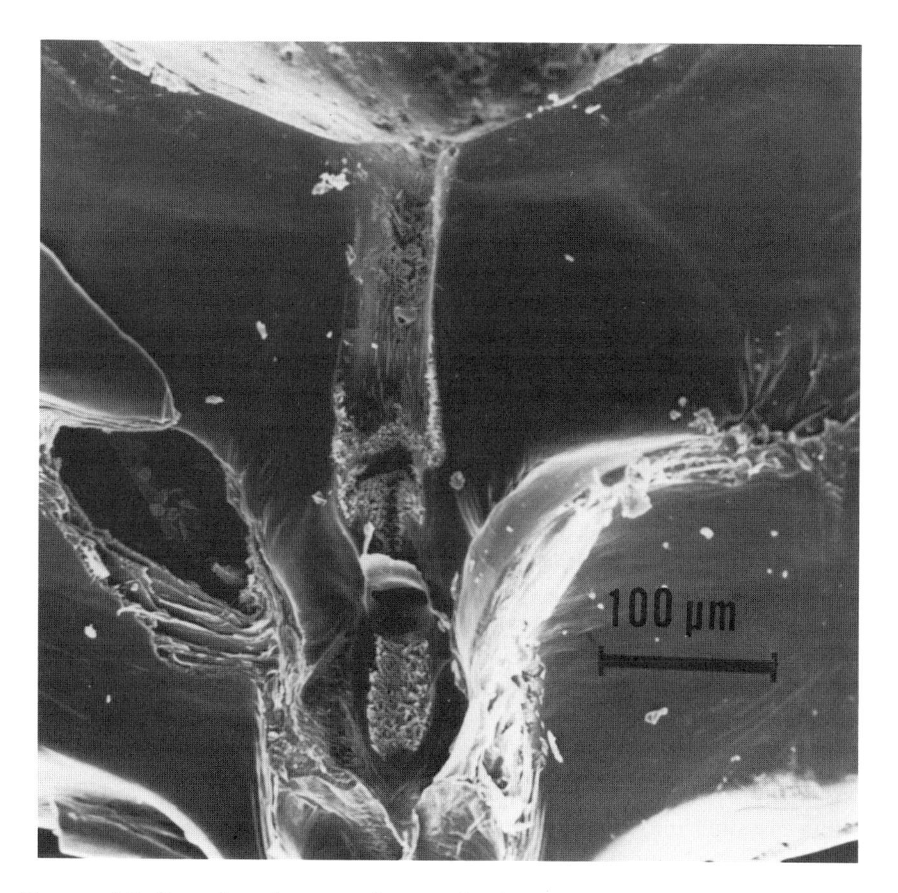

FIGURE 9.2. Scanning electron micrograph of the precibarium of *Homaladisca coagulata* colonized by *Xylella fastidiosa* (plum leaf strain). Epipharynx below precibarial valve; floor of cibarial pump visible at top; × 200 (bar, 100 microns). Photograph courtesy of R.H. Brlansky, University of Florida.

however, may be more directly involved in the transmission of XLB. Two features of vector transmission of *X. fastidiosa* point out the importance of the foregut (stylets through the esophagus). The first is that acquisition and inoculation of bacteria can be accomplished very quickly: there is no evidence that a latent period is required (77, 85, 93). Secondly, transmission is not transstadial, that is, infectious (capable of transmitting) insects do not transmit after molting (85). These two features of vector transmission strongly suggest that the foregut is the site from which the bacteria are introduced by the feeding vector into the plant. Further evidence that circulative movement of bacteria through the hemocoel is not necessary for transmission was provided by experiments in which the injection of either *Xylella* (86) or other bacteria that could be acquired and transmitted in membrane feeding systems (84) failed to result in bacterial transmission to plants or artificial feeding solutions.

What area of the foregut is most important to transmission? Microscopy of the foregut of the sharpshooter PD vector *G. atropunctata* revealed that *X. fastidiosa* gradually colonized the insect's pump chamber after the leafhoppers fed on grapevines with symptoms of PD. The visually impressive mats of bacteria lining the pump entrance and other regions of the cibarial chamber (8, 86) would by themselves suggest that this area might be the source of inoculum for vector transmission. Extensive bacterial colonization of the foregut also would explain the persistence of infectivity in vectors, virtually over their adult lifetimes (93). The transmission efficiency of individual sharpshooters, however, was not related to the degree of colonization (86). Some insects had no bacteria visible in the pump chamber, yet transmitted, whereas others with heavy mats of bacteria failed to transmit.

Xylella bacteria have also been observed in the precibaria of sharpshooters (Figure 9.3) (8) anterior to the precibarial valve (Figure 9.1). A functional analysis of the morphology of the foregut suggests that the precibarium may be the site from which bacteria are egested into plants during feeding by infectious xylem feeders. McLean and Kinsey (62) have discussed in detail the evidence for mechanisms of aphid ingestion and egestion. Their arguments are directly relevant to the homologous features of xylem-feeding Homoptera. The central cibarial pump is flanked by two functional valves: the precibarial valve (3, 4) and the esophageal valve (94, 103). During fluid uptake, the precibarial valve opens to admit ingested fluid into the expanding pump chamber. The precibarial valve closes when the pump chamber contracts and fluid is thus expelled into the midgut via the esophagus. The esophageal valve (62) closes to prevent fluid from moving from the midgut to the pump. The esophagus of sharpshooter leafhoppers is thin, membranous, and lightly muscled (Purcell, unpublished). It probably collapses when subjected to pump suction (62). Thus, flow from the midgut is preempted by the check valve action of the esophagus or esophageal valve.

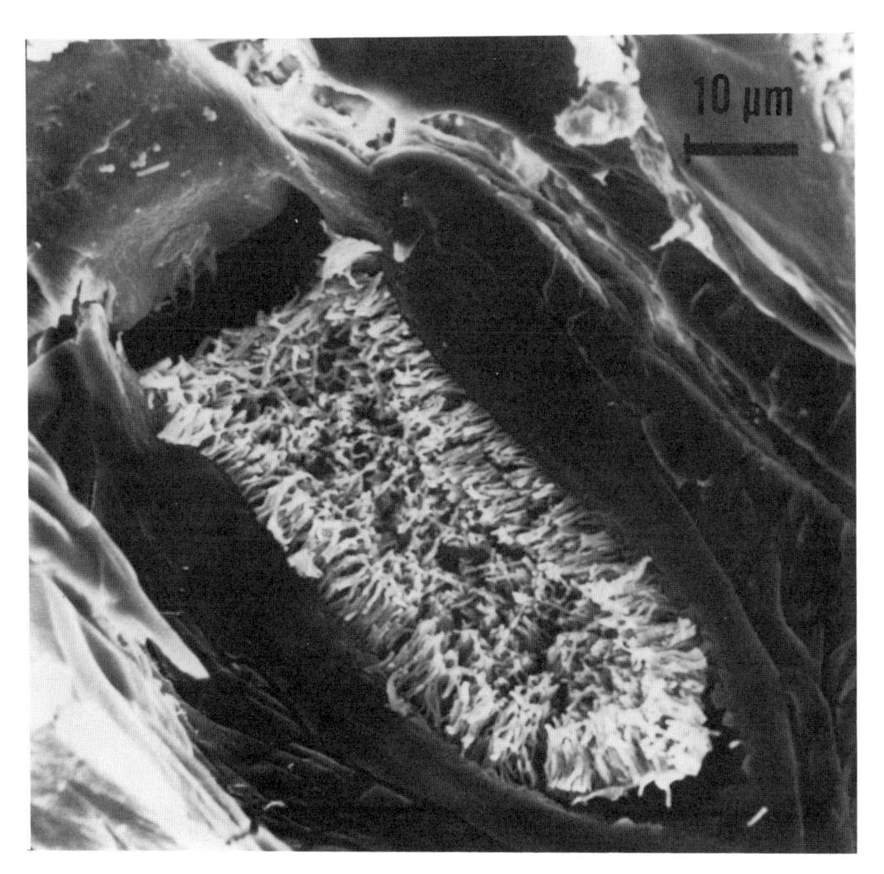

FIGURE 9.3. Enlargement of anterior (distal) precibarium from Figure 9.2, colonized by a dense mat of *X. fastidiosa;* × 1200 (bar, 10 microns). Photograph courtesy of R.H. Brlansky, University of Florida.

The esophageal valve should not prevent ingested fluid from being pumped into the midgut (62). During exploratory "sample probes," sap can be drawn up to the pump entrance, evaluated by the chemosensory sensilla that line the precibarium, and then egested so as to allow another sample to be taken before sustained ingestion proceeds (5, 62). Logically, egestion—or *extravasation* suggested as a term to describe this process (62)—after chemosensory sampling of fluid within the precibarium (4, 5) is most likely when *Xylella* would be transmitted. Sharpshooters have thicker, more heavily sclerotized precibaria than do phloem-feeding or mesophyll-feeding leafhoppers, presumably to brace the precibarium against the powerful force of the cibarial dilators (4). The relatively long channel in the precibarium leading from the cibarium to the precibarial valve would channel fluid flow more forcefully than would the relatively short homologous channel found in mesophyll feeders. One cause of reverse flow of xylem sap from insect mouthparts to xylem may be the negative tensions within the xylem. The high tensions of xylem sap should pull on the fluid stream within the food canal of xylem-feeding insects. Miscoordination in opening the precibarial valve and dilating the cibarium would tend to suck fluid from the feeding insect's food canal back into the plant. How much fluid returned to the xylem stream would be determined by the amount of hesitation or delay in dilating the cibarium after the precibarial valve is open and by how long the esophagus and esophageal valve remained open until closed by the lower pressure in the cibarium. Experimental proof is needed to support these conjectures.

There has not been much research over the past 10 years on mechanisms of transmission of XLB and practically none on the strains of *X. fastidiosa* shown to belong to Group II in Table 9.1. These strains include those that induce PP in peach. Are the transmission characteristics of these strains the same as for strains causing PD in grape? Transmission efficiency of PD strains to grapevines is high (85, 93), whereas those of PP strains to peach are reported as much lower (99). Phony disease of peach transmission also is reported to have a latent period of at least eight days (99). If confirmed, this would be quite different from that of PD strains. Are there strain differences in transmissibility? Do members of other families of xylem-feeders transmit XLB in the same fashion as the sharpshooters? Some xylem-feeding leafhopper species other than those of the subfamily Cicadellinae were much less efficient PD vectors than were sharpshooters. Species of *Pagaronia* and *Oncometopia* were much less efficient than Cicadellinine leafhoppers in PD transmission to grape (93). It should be interesting to determine vector transmission characteristics (e.g., latency, persistence, effect of molting, etc.) of the clove wilt bacterium (*Pseudomonas* sp.) by Machaerotid spittlebugs and other xylem feeders. *Xylella fastidiosa* has served as the predominant model for XLB transmission, a situation that may not be justified for other, as yet unstudied XLB.

Another question relative to transmission specificity is what makes some XLB transmissible by xylem feeders and others not transmissible. *Clavibacter xyli* (14), for example, meets all requirements of a true XLB but was not transmitted by sharpshooters (Purcell, unpublished data). Attachment and detachment of XLB to and from the walls of the foregut during the feeding process must be critical to the persistence and multiplication of bacteria within the foregut. Average flow velocities of ingested sap calculated from excretion rates and foregut dimensions can be as high as 9 cm/sec (86); and instantaneous velocities during more rapid feeding (see, for example, ref. 66a) should be at least several times faster. During microscopic examinations of sharpshooter pump chambers, microbes other than *Xylella*-like bacteria are rarely seen (8, 86). Apparently *Xylella* can anchor themselves to the cuticle lining the foregut, even in such high flow velocity locations as the entrance to the pump chamber (86) or within the precibarium (8); *X. fastidiosa* also was recovered from sharpshooter bodies posterior to the head (86). Serological tests such as ELISA (69) can detect *X. fastidiosa* in insects (88, 111), but such tests do not necessarily reveal the ability of the insect to transmit the bacteria. Infectivity may have been lost with molting, but bacterial infection of the midgut may have been retained.

Studies of vector transmission (21) of the bacterial pathogen of Sumatra disease would allow a comparision of the transmission of this *Pseudomonas*-like bacterium (22) to the transmission of *Xylella*.

Epidemiology

It should be evident from the foregoing discussion that vector specificity for *X. fastidiosa* extends to a large number of insect species, many of which are abundant. Because of this and the highly efficient transmission of the bacterium, it is hard to understand why *Xylella*-caused plant diseases are not more widespread. For example, large, contiguous areas of central and eastern North America do not have PD, AD, or PP, despite an abundance of suctorial xylem feeders on many hosts of these bacterial pathogens. The boundary of the geographical limits of PD in the eastern United States can nearly be superimposed on winter isotherms or other measures of minimum winter temperatures, which supports the hypothesis that winter temperatures may be important in limiting PD (39, 77, 78, 83). Experimental evidence of the effect of cold on PD (75) and the history of failed attempts to transmit the PP agent in a region in which PP was a rare disease (99), further support this hypothesis. However, there still is no satisfactory experimental evidence to support or refute the hypothesis that winter climate governs the geographical distribution of diseases caused by *Xylella*.

Further support for the effects of climate on the epidemiology of PD came from field experiments in which only leafhopper inoculations of *X. fastidiosa* (PD strain) that were made early in the growing season persisted as chronic infections the following growing season (80). Apparently, late-season PD infections of grape did not survive winter dormancy. The findings that some infected vines completely recover helped to reconcile attempts to model (80) the consequences of vector densities and natural infectivity (73) with observed rates and patterns of disease spread under California conditions. The seasonal effects on the survivability of PD infections also accounts (80) for the discrepancies between vector distribution within a given vineyard and the subsequent spread of PD (73). Differences among grape cultivars in the incidence of PD and the degree to which insecticidal control of vectors is effective (76) also depend upon the seasonal response of the cultivar in eliminating infections during winter dormancy (80).

The distributions of PD, AD, and ALS within California are not congruent. Pierce's disease of grape occurs throughout the state, although the incidence varies tremendously from locality to locality, but AD and ALS do not overlap (79). In view of the common and widespread occurrence of symptomless PD hosts, along with widespread populations of vector species, the consistent rarity of PD in some regions of California must wait to be explained by factors not yet appreciated.

In Florida, sharpshooters collected in or near vineyards were naturally infective with PD strains of *X. fastidiosa* during summer months after the time of year (summer) during which the leafhoppers were first able to acquire the pathogen from diseased grapevines (1). If this is when most chronic infections are established, the overwintering survival of midsummer inoculations of grapevines in Florida must be higher than that observed in California. In coastal California, a high percentage of the principal vector species, *G. atropunctata* (*Hordnia circellata*), that enter vineyards in early spring are naturally infective (73), which presumably explains the pattern of primary spread (from outside vineyards to within vineyards) observed in such localities (73, 80).

On a global basis, Pierce's disease appear to be limited to the Americas (37, 39). The absence of PD from mild winter regions of Europe seems anomalous (81). All the great plagues of the grapevine that are of American origin—powdery mildew, downy mildew, and phylloxera—found their way to Europe and the rest of the world's grape-growing regions, along with imported grapevines. But PD has not been definitively recorded from Europe, despite the importation of tolerant grape cultivars that could be expected to harbor the causal bacterium as well as the (continuing) importation of live plants that are symptomless hosts of *X. fastidiosa* and the abundance of xylem feeders in and around vineyards. The survivability of *X. fastidiosa* in "symptomless hosts" should be of concern in devising quarantine procedures to prevent the spread of PD to continents where it has not occurred. An understanding of why PD has not appeared in Europe might also suggest new strategies for its control.

Undiscovered Bacteria in Plant Xylem

Of the many and varied bacteria associated with plants, pathogens have received the most attention (30, 35, 53). Even nitrogen-fixing bacteria that are symbiotic in root nodules of legumes (*Rhizobium* spp.) were initially interpreted as disease symptoms (26). The reasons for this emphasis on harmful bacteria–plant relationships are practical and economic. New discoveries, however, continue to change our perspective of the impact of bacteria on plants. Some epiphytic bacteria form nuclei for ice crystals that injure plant tissue as "frost damage" (59). Other studies have demonstrated that bacterial epiphytes (101) or rhizosphere (root zone) bacteria (92, 95) can serve in the biological control of plant disease or improve plant nutrition. These bacteria not long ago were considered to be "neutral" microbial associates of plants. Without much trepidation, one can predict that numerous other bacteria–plant associations will be discovered. Some of these may prove to be mild—and thus easily overlooked—pathogens. Other nonpathogenic or even beneficial relationships may be found.

Genetically engineered endophytic bacteria may provide a means of producing and delivering novel gene products to plants without transforming the plant genome (11). For this purpose, the genetically engineered bacteria would have to colonize their host plants systemically, and harmlessly, a role probably more suited to the inhabitants of xylem than of other vascular tissues. Disease, directly or indirectly, has led to all discoveries so far of XLB. Conversely, we can postulate that the lack of disease in symptomless infections of XLB has and will continue to obscure the occurrence of XLB. In the future, we may seek to discover and explain important effects or uses of nonpathogenic XLB. For this to happen, it is likely that model systems based on *Xylella* to explain vector transmission may be inadequate. What are the origins and roles of bacteria frequently isolated from the gut of the sharpshooter leafhoppers (57, 84)? The immense diversity of xylem-feeding leafhoppers in the American tropics alone (112) is staggering to contemplate, even though still little explored. Surely there must be XLB and other microbes awaiting discovery amidst this diversity!

Summary

1. The fact that bacterial phytopathogens have numerous insect vectors with common functional features implies adaptation to an insect transmission mechanism, not a particular vector species.
2. It is still not clear why only suctorial insects specialized for xylem feeding are XLB vectors. An improved understanding of the vector transmission of XLB should come from studies using a spectrum of vectors and host plants and of species and strains of XLB, including XLB that are not vector transmissible.
3. *Xylella fastidiosa* strain differences need to be investigated, not only from the perspective of procaryote taxonomy, physiology, and phytopathology, but also with respect to vector transmission and epidemiological consequences.
4. Epidemiological data suggest that the seasonality of infection is important to disease spread, but more evidence is needed to confirm or expand upon this suggestion. How does seasonality of chronic infection vary with regional and year-to-year climatic variation?
5. The fate and spread of XLB in symptomless plant hosts are little understood, but probably critical components of the epidemiology of

vector-borne XLB. These are also important in the formulation of quarantine strategies.

6. Biotechnology should provide new incentives and methods for exploring xylem-inhabiting prokaryotes and other endophytic microbes for beneficial uses, not simply as pathogens. We know relatively little of the composition of the "normal" microflora of xylem and the effects of these microbes in plants. Some of these may be insect transmitted; others may not.

Acknowledgments. I thank Michael Davis, Paul Goodwin, Stanley Kostka, and Barry Hill for helpful suggestions and comments and Elaine Backus and Ron Brlansky for providing the figures.

References

1. Adlerz, W.C., and Hopkins, D.L., 1979, Natural infectivity of two sharpshooter vectors of Pierce's disease of grape in Florida, *J. Econ. Entomol.* **72:**916–919.
2a. Alivizatos, A.S., 1982, Feeding behavior of the spiroplasma vectors Dalbulus maidis and Euscelidius variegatus in vivo and in vitro, *Ann. Inst. Phytopathol. Benaki,* **13:**128–144.
2b. Anderson, P.C., Brodbeck, B.V., and Mizell, R.F., 1989, Metabolism of amino acids, organic acids and sugars extracted from the xylem fluid of four host plants by adult *Homaladisca coagulata* (Say) (Homoptera: Cicadellidae), *Entomol. Exp. Appl.* In press.
3. Backus, E.A., 1983, The sensory systems and feeding behavior of leafhoppers. II. A comparison of the sensillar morphologies of several species (Homoptera: Cicadellidae), *J. Morphol.* **176:**3–14.
4. Backus, E.A., 1985, Anatomical and sensory mechanisms of planthopper and leafhopper feeding, in Nault, L.R., and Rodriguez, J.G. (eds): The Leafhoppers and Planthoppers, Wiley-Interscience, New York, pp. 163–194.
5. Backus, E.A., and McLean, D.L., 1985, Behaviorial evidence that the precibarial sensilla of leafhoppers are chemosensory and function in host discrimination, *Entomol. Exp. Appl.* **37:**219–228.
6. Billing, E., 1982, Entry and establishment of pathogenic bacteria in plants, in Rhodes-Roberts, M.E., and Skinner, F.A. (eds): Bacteria and Plants, Academic Press, New York, pp. 51–70.
7. Bollard, E.G., 1960, Transport in the xylem, *Ann. Rev. Plant Physiol.* **11:**141–161.
8. Brlansky, R.H., Timmer, L.W., French, W.J., and McCoy, R.E., 1983, Colonization of the sharpshooter vectors, *Oncometopia nigricans* and *Homalodisca coagulata,* by xylem-limited bacteria, *Phytophypathology* **73:**530–535.
9. Carter, W., 1973, Insects in Relation to Plant Disease, Wiley, New York, 759 p.
10. Crane, P.S., 1970, The feeding behavior of the blue-green sharpshooter *Hordnia circellata* (Baker) (Homoptera: Cicadellidae), Ph.D. dissertation, University of California, Davis, 132 p.

11. Crawford, M., 1988, ARS prodded into open, *Science* **239:**719.
12. Davis, M.J., 1989, Host colonization and pathogenesis in plant diseases caused by fastidious xylem-inhabiting bacteria, in Tjamos, E.C. and Beckman, C.H. (eds): Vascular Wilt Diseases in Plants, Springer-Verlag, Berlin, pp. 33–50.
13. Davis, M.J., French, W.J. and Schaad, N.W., 1981, Axenic culture of the bacteria associated with phony disease of peach and plum leaf scald, *Curr. Microbiol.* **6:**309–314.
14. Davis, M.J., Gillaspie, A.G. Jr., Vidaver, A.K., and Harris, R.W., 1984, *Clavibacter:* a new genus containing some phytopathogenic coryneform bacteria, including Clavibacter xyli subsp. xyli sp. nov., subsp. nov. and Clavibacter xyli subsp. cynodontis subsp. nov., pathogens that cause ratoon stunting disease of sugarcane and Bermudagrass stunting disease, *Int. J. Syst. Bacteriol.* **34:**107–117.
15. Davis, M.J., Purcell, A.H., and Thompson, S.V., 1978, Pierce's disease of grapevines: Isolation of the causal bacterium, *Science* **199:**75–77.
16. Davis, M.J., Purcell, A.H., and Thompson, S.V., 1980, Isolation media for Pierce's disease bacterium, *Phytopathology* **70:**427–429.
17. Davis, M.J., Raju, B.C., Brlansky, R.H., Lee, R.F., Timmer, I.W., Norris, R.C., and McCoy, R.E., 1983, Periwinkle wilt bacterium: Axenic culture, pathogenicity, and relationships to other gram-negative, xylem-inhabiting bacteria, *Phytopathology* **73:**1510–1515.
18. Davis, M.J., Thomson, S.V., and Purcell, A.H., 1979, Etiological role of the xylem-limited bacterium causing Pierce's disease in almond leaf scorch, *Phytopathology* **70:**472–475.
19. Davis, M.J., Whitcomb, R.F., and Gillaspie, A.G., Jr., 1981, Fastidious bacteria of plant vascular tissue and invertebrates (including so-called rickettsia-like bacteria), in Starr, M.P., et al. (eds): The Prokaryotes: A Handbook on Habits, Isolation, and Identification of Bacteria, Springer-Verlag, Heidelberg, pp. 2172–2188.
20. Day, M.F., Irzykiewicz, H., and McKinnon, A., 1952, Observations of the feeding of the virus vector *Orosius argentatus* (Evans) and comparisons with certain other Jassids, *Austral. J. Sci. Res.* **5:**128–142.
21. Eden-Green, S.J., Balfas, R., and Jamalius, 1986, Transmission of xylem-limited bacteria causing Sumatra disease of cloves in Indonesia by tube-building cercopids, *Hindola* spp. (Homoptera: Machaerotidae), in Wilson, M.R., and Nault, L.R. (eds): *Proc. 2nd Int. Workshop on Leafhoppers and Planthoppers of Economic Importance,* Provo, Utah, July 28–Aug. 1, 1986, Commonwealth Institute of Entomology, London, pp. 101–107.
22. Eden-Green, S.J., Supriad, Hasnam, N., and Hunt, P., 1985, Serological relationship between the xylem-limited bacterium causing Sumatra disease of cloves in Indonesia and *Pseudomonas solanacearum* in Civerolo, L., Collmer, A., Davis, R.E., and Gillaspie, A.G. (eds): Plant Pathogenic Bacteria, Curr. Plant Sci. Biotech. Agr., Martinus Nijhoff, Dordrecht, Netherlands, pp. 357–363.
23. Esau, K., 1977, Anatomy of Seed Plants, 2nd ed., Wiley, New York, 550 p.
24. Feldman, A.W., Hanks, R.W., Good, G.E., and Brown, G.E., 1977, Occurrence of a bacterium in YTD-affected as well as healthy citrus trees, *Plant Dis. Reptr.* **61:**546–550.

25. Frazier, N.W., 1965, Xylem viruses and their insect vectors, in Hewitt, W.B. (ed): Proc. Int. Conf. on Virus and Vector on Perennial Hosts, with Special Reference to Vitis, University of California, Davis, Davis, California, pp. 91–99.
26. Fred, E.B., Baldwin, I.L., and McCoy, E.M., 1932, Root nodule bacteria and leguminous plants, University of Wisconsin Studies in Science, No. 5, Madison, WI, 343 p.
27. Freitag, J.H., 1951, Host range of Pierce's disease virus of grapes as determined by insect transmission, *Phytopathology* **41:**920–934.
28. French, W.J., and Stassi, D.L., 1978, Response of phony-infected peach trees to gibberillic acid, *Hortscience* **13:**158–159.
29. Gardner, J.M., Chandler, J.L., and Feldman, A.W., 1985, Growth responses and vascular plugging of citrus inoculated with rhizobacteria and xylem-resident bacteria, Plant and Soil **86:**333–345.
30. Gardner, J.M., Feldman, A.W., and Zablotowicz, R.M., 1982, Identity and behavior of xylem-residing bacteria in rough lemon roots of Florida citrus trees, *Appl. Environ. Microbiol.* **43:**1335–1342.
31. Goheen, A.C., Nyland, G., and Lowe, S.K., 1973, Association of a rickettsialike organism with Pierce's disease of grapevines and alfalfa dwarf and heat therapy of the disease in grapevines, *Phytopathology* **63:**341–345.
32. Goodwin, P.H., De Vay, J.E., and Meredith, C.P., 1988, Roles of water stress and phytotoxins in the development of Pierce's disease of the grapevine, *Physiol. Mol. Plant Pathol.* **32:**1–15.
33. Goodwin, P.H., De Vay, J.E., and Meredith, C.P., 1988, Physiological response of *Vitis vinifera* cv "Chardonnay" to infection by the Pierce's disease bacterium, *Physiol. Mol. Plant Pathol.* **32:**17–32.
34. Harrision, M.D., Brewer, J.W., and Merrill, L.D., 1980, Insect involvement in the transmission of bacterial pathogens, in Harris, K.F., and Maramorosch, K. (eds): Vectors of Plant Pathogens, Academic Press, New York, pp. 201–292.
35. Hayward, A.C., 1974, Latent infections by bacteria, *Annu. Rev. Phytopathol.* **12:**87–97.
36. Hearon, S.S., Sherald, J.L., and Kostka, S.J., 1980, Association of xylem-limited bacteria with elm, sycamore, and oak leaf scorch, *Can. J. Bot.* **18:**1986–1993.
37. Hewitt, W.B., 1958, The probable home of Pierce's disease virus, *Plant Dis. Reptr.* **42:**211–215.
38. Hewitt, W.B., Houston, B.R., Frazier, N.W., and Freitag, J.H., 1956, Leafhopper transmission of the virus causing Pierce's disease of grape and dwarf of alfalfa, *Phytopathology* **36:**117–128.
39. Hopkins, D.L., 1976, Pierce's disease of grapevines, *Am. Wine Soc. J.* **8:**26–27.
40. Hopkins, D.L., 1979, Disease caused by leafhopper-borne, Rickettsialike bacteria, *Annu. Rev. Plant Pathol.* **17:**277–294.
41. Hopkins, D.L., 1984, Variability of virulence in grapevine among isolates of Pierce's disease bacterium, *Phytopathology* **74:**1395–1398.
42. Hopkins, D.L., 1985, Effects of plant growth regulators on development of Pierce's disease symptoms in grapevine, *Plant Dis.* **69:**944–946.
43. Hopkins, D.L., 1988, Natural hosts of *Xylella fastidiosa* in Florida, *Plant Dis.* **72:**429–431.

44. Hopkins, D.L., 1988, Production of diagnostic symptoms of blight in citrus inoculated with *Xylella fastidiosa, Plant Dis.***72:** 432-435.
45. Hopkins, D.L., and Mollenhauer, H.H., 1973, Rickettsia-like bacterium associated with Pierce's disease of grapes, *Science* **179:**298–300.
46. Hopkins, D.L., Mollenhauer, H.H., and Mortensen, J.A., 1974, Tolerance to Pierce's disease and the associated Rickettsia-like bacterium in muscadine grape, *J. Am. Soc. Hort. Sci.* **99:**436–439.
47. Horsfield, D., 1977, Relationships between feeding of *Philaenus spumarius* (L.) and the amino acid composition of xylem sap, *Ecol. Entomol.* **2:**259–266.
48. Houston, B.R., Esau, K., and Hewitt, W.B., 1947, The mode of vector feeding and the tissues involved in the transmission of Pierce's disease virus in grape and alfalfa, *Phytopathology* **37:**247–253.
49. Huang, P.-Y., Milholland, R.D., and Daykin, M.E., 1986, Structural and morphological changes associated with the Pierce's disease bacterium in bunch and muscadine grape tissue, *Phytopathology* **76:**1232–1238.
50. Hunt, P., Bennett, C.P.A., Syamsu, H., and Nurwenda, E., 1987, Sumatra disease in cloves induced by a xylem-limited bacterium following mechanical inoculation, *Plant Pathol.* **36:**154–163.
51. Hutchins, L.M., Cochran, L.C., Turner, W.F., and Weinberger, J.H., 1953, Transmission of phony disease virus from tops of certain affected peach and plum trees, *Phytopathology* **43:**691–696.
52. Kamper, S.M., French, W.J., and deKloet, S.R., 1985, Genetic relationships of some fastidious xylem-limited bacteria, *Int. J. System. Bacteriol.* **35:**185–188.
53. Kennedy, R.W., and Lacy, G.H., 1982, Phytopathogenic prokaryotes: An overview, in Mount, M.S., and Lacy, G.H. (eds): Phytopathogenic Prokaryotes, Volume I, Academic Press, New York, pp. 3–17.
54. Kitajima, E.W., Bakarcic, M., and Fernandez-Valiela, M.V., 1975, Association of rickettsialike bacteria with plum leaf scald disease, *Phytopathology* **65:**476–479.
55. Kostka, S.J., Tattar, T.A., and Sherald, J.L., 1986, Elm leaf scorch: Abnormal physiology in American elms infected with fastidious, xylem-inhabiting bacteria, *Can. J. For. Res.* **16:**1088–1091.
56. Kostka, S.J., Tattar, T.A., Sherald, J.L., and Hurtt, S.S., 1986, Mulberry leaf scorch, a new disease caused by a fastitious, xylem-inhabiting bacterium, *Plant Dis.* **70:**690–693.
57. Latorre-Guzman, B.A., Kado, C.I., and Kunkee, R.E., 1977, *Lactobacillus hordniae,* a new species from the leafhopper (*Hordnia circellata*), *Intl. J. Syst. Bacteriol.* **27:**362.
58. Lee, R.F., Raju, B.C., and Nyland, G., 1978, Phytotoxin(s) produced by the Pierce's disease organism in culture, *Phytopathol. News* **12:**218 (Abstr.).
59. Lindow, S.E., 1982, Epiphytic ice nucleation-active bacteria, in Mount, M.S., and Lacy, G.H. (eds): Phytopathogenic Prokaryotes, Volume I, Academic Press, New York, pp. 335–362.
60. McCoy, R.E., Thomas, D.L., Tsai, J.H., and French, W.J., 1978, Periwinkle wilt, a new disease associated with xylem deliminted Rickettsialike bacteria transmitted by a sharpshooter, *Plant Dis. Reptr.* **62:**1022–1026.
61. McLean, D.L., and Kinsey, M.G., 1967, Probing behavior of the pea aphid, *Acyrthosiphon pisum.* I. Definitive correlation of electronically recorded

waveforms with aphid probing activities, *Ann. Entomol. Soc. Am.* **60:**400–406.

62. McLean, D.L., and Kinsey, M., 1984, The precibarial valve and its role in the feeding behavior of the pea aphid, *Acyrthosiphon pisum, Bull. Entomol. Soc. Am.* **30:**26–31.
63. Miles, P.W., 1968, Insect secretions in plants, *Annu. Rev. Entomol.* **6:**137–164.
64. Mircetich, S.M., Lowe, S.K., Moller, W.J., and Nyland, G., 1976, Etiology of almond leaf scorch and transmission of the causal agent, *Phytopathology* **66:**17–24.
65. Mittler, T.E., 1957, Studies on the feeding and nutrition of *Tuberolachnus salignus* (Gmelin) (Homoptera, Aphididae). I. The uptake of phloem sap, *J. Exp. Biol.* **34:**334–341.

66a. Mittler, T.E., 1967, Water tension in plants—an entomological approach, *Ann. Entomol. Soc. Am.* **57:**139–141.

66b. Mizell, R.F., and French, W.J., 1986, Leafhopper vectors of phony peach disease: Feeding site preference and survival on infected and uninfected peach, and seasonal response to selected host plants, *J. Entomol. Res.* **22:**11–22.

67. Moreno, J., and Garcia-Martinez, J.L., 1983, Seasonal variations of nitrogenous compounds in the xylem sap of citrus, *Physiologia Plantarum* **59:**669–675.
68. Naito, A., 1977, Feeding habits of leafhoppers, *J. Agric. Res. Qtly.* **11:**115–119.
69. Nomé, S.F., Raju, B.C., Goheen, A.C., Nyland, G., and Docampo, D., 1980, Enzyme-linked immunosorbent assay for Pierce's disease bacteria in plant tissues, *Phytopathology* **70:**746–749.
70. Pate, J.R., 1976, Nutrients and metabolites of fluids recovered from xylem and phloem: Significance in relation to long distance transport in plants, in Wardlow, I.F., and Passiours, J.B. (eds): Transport and Transfer Processes in Plants, Academic Press, New York, pp. 253–281.
71. Pollard, D.G., 1973, Plant penetration by feeding aphids (Hemiptera: Aphidoidea): A review, *Bull. Entomol. Res.* **62:**631–714.
72. Pollard, D.G., 1968, Stylet penetration and feeding damage of *Eupteryx melissae* Curtis (Hemiptera: Cicadellidae) on sage, *Bull. Entomol. Res.* **58:**55–71.
73. Purcell, A.H., 1975, Role of the blue-green sharpshooter, *Hordnia circellata* in the epidemiology of Pierce's disease of grapevines, *Environ. Entomol.* **4:**745–752.
74. Purcell, A.H., 1976, Seasonal changes in host plant preferences of the blue-green sharpshooter *Hordnia circellata, Pan Pac. Entomol.* **52:** 33–37.
75. Purcell, A.H., 1977, Cold therapy of Pierce's disease of grapevines, *Plant Dis. Reptr.* **61:**514–518.
76. Purcell, A.H., 1979, Control of the blue-green sharpshooter *Graphocephala atropunctata* and effects on the spread of Pierce's disease of grapevines, *J. Econ. Entomol.* **72:**887–892.
77. Purcell, A.H., 1979, Leafhopper vectors of xylem-borne plant pathogens, in Maramorosch, K., and Harris, K.F. (eds): Leafhopper Vectors and Plant Disease Agents, Academic Press, New York, pp. 603–625.

78. Purcell, A.H., 1980, Environmental therapy for Pierce's disease of grapevines, *Plant Dis.* **64:**388–390.
79. Purcell, A.H., 1980, Almond leaf scorch: Leafhopper and spittlebug vectors, *J. Econ. Entomol.* **73:**834–838.
80. Purcell, A.H., 1981, Vector preference and inoculation efficiency as components of varietal resistance to Pierce's disease in European grapes, *Phytopathology* **71:**429–435.
81. Purcell, A.H., 1982, Advances in the understanding of Pierce's disease and its insect vectors, in Grape and Wine Centennial Symp. Proc., University of California, Davis, June, 1980, pp. 46–50.
82. Purcell, A.H., 1982, Evolution of the insect vector relationship, in Mount, M.S., and Lacy, G.H. (eds): Phytopathogenic Prokaryotes, Academic Press, New York, pp. 121–156.
83. Purcell, A.H., 1985, The ecology of bacterial and mycoplasma plant diseases spread by leafhoppers and planthoppers, in Nault, L.R., and Rodriguez, J.G. (eds): The Leafhoppers and Planthoppers, Wiley, New York, pp. 351–380.
84. Purcell, A.H., and Finlay, A.H., 1979, Acquisition and transmission of bacteria through artificial membranes by leafhopper vectors of Pierce's disease, *Entomol. Exp. Appl.* **25:**188–195.
85. Purcell, A.H., and Finlay, A.H., 1979, Evidence for noncirculative transmission of Pierce's disease bacterium by sharpshooter leafhoppers, *Phytopathology* **69:**393–395.
86. Purcell, A.H., Finlay, A.H., and McLean, D.L., 1979, Pierce's disease bacterium: Mechanism of transmission by leafhopper vectors, *Science* **206:**839–841.
87. Purcell, A.H., and Frazier, N.W., 1985, Habitats and dispersal of the leafhopper vectors of Pierce's disease in the San Joaquin Valley, *Hilgardia* **53:**1–32.
88. Raju, B.C., Goheen, A.C., and Frazier, N.W., 1983, Occurrence of Pierce's disease bacteria in plants and vectors in California, *Phytopathology* **73:**1309–1313.
89. Raju, B.C., Nomé, S.F., Docampo, D.M., Goheen, A.C., Nyland, G., and Lowe, S.K., 1980, Alternative hosts of Pierce's disease of grapevines that occur adjacent to grape growing areas in California, *Am. J. Enol. Vitic.* **31:**144–148.
90. Raju, B.C., Wells, J.M., Nyland, G., Brlansky, R.H., and Lowe, S.K., 1982, Plum leaf scald: Isolation, culture, and pathogenicity of the causal agent, *Phytopathology* **72:**1460–1466.
91. Raven, J.A., 1984, Phytophages of xylem and phloem: A comparison of animal and plant sap-feeders, Adv. Ecol. Res. **13:**135–234.
92. Schroth, M.N., and Hancock, J.G., 1982, Disease suppressive soil and root colonizing bacteria, *Science* 216:1376–1381.
93. Severin, H.H.P., 1949, Transmission of the virus of Pierce's disease by leafhoppers, *Hilgardia* **19:**190–202.
94. Snodgrass, R.E., 1935, Principles of Insect Morphology, McGraw-Hill, New York, 667 p.
95. Suslow, T., 1982, Role of root-colonizing bacteria in plant growth, in Mount, M.S., and Lacy, G.H. (eds): Phytopathogenic Prokaryotes, Volume I, Academic Press, New York, pp. 187–223.
96. Timmer, L.W., Brlansky, R.H., Lee, R.F., and Raju, B.C., 1983, A

fastidious, xylem-limited bacterium infecting ragweed, *Phytopathology* **73:**975–979.
97. Tonkyn, D.W., and Whitcomb, R.F., 1987, Feeding strategies and the guild concept among vascular feeding insects and microorganisms, *Curr. Topics Vector Res.* **4:**179–199.
98. Turner, W.F., 1949, Insect vectors of phony peach disease, Science **109:** 87–88.
99. Turner, W.F., and Pollard, H.N., 1959, Insect transmission of phony peach disease, *U.S. Dept. Agric. Tech. Bull. 1193.*
100. Ullman, D., 1988, Feeding behavior of the winter-form pear psylla, *Psylla pyricola* (Homoptera: Psyllidae), on reproductive and transitory plant hosts, *Environ. Entomol.* **17:**675–678.
101. Vidaver, A.K., 1982, Biological control of plant pathogens with prokaryotes, in Mount, M.S., and Lacy, G.H. (eds): Phytopathogenic Prokaryotes, Volume II, Academic Press, New York, pp. 387–397.
102. Weaver, D.J., Raju, B.C., and Wells, J.M., 1980, Occurrence in Johnson-grass of Rickettsia-like bacteria related to the phony peach disease organism, *Plant Dis.* **64:**485–487.
103. Weber, H., 1930, Biologie der Hemipteren, Springer-Verlag, Berlin 543 p.
104. Weimar, J.L., 1933, Effect of environmental and cultural factors on the dwarf disease of alfalfa, *J. Agr. Res.* **47:**351–368.
105. Weimar, J.L., 1937, Effect of the dwarf disease on the alfalfa plant, *J. Agr. Res.* **55:**87–104.
106. Wells, J.M., Raju, B.C., Hung, H-Y., Weisberg, W.G., Mandelco-Paul, L., and Brenner, D.J., 1987, *Xylella fastidiosa* new genus new species Gram-negative xylem-limited fastidious plant bacteria related to *Xanthamonas* spp., *Int. J. Syst. Bacteriol.* **37:**136–143.
107. Wells, J.M., Raju, B.C., and Nyland, G., 1983, Isolation, culture, and pathogenicity of the bacterium causing phony disease of peach, *Phytopathology* **73:**859–862.
108. Wells, J.M., Raju, B.C., Thompson, J.M., and Lowe, S.K., 1981, Etiology of phony peach and plum leaf scald diseases, *Phytopathology* **71:**1156–1161.
109. Wells, J.M., Weaver D.J., and Raju B.C., 1980, Distribution of Rickettsia-like bacteria in peach and their occurrence in plum, cherry, and some perennial weeds, *Phytopathology* **70:**817–820.
110. Wiegert, R.G., 1964, The ingestion of xylem sap by meadow spittlebugs, *Philaenus spumarius* (L.), *Am. Mid. Natl.* **71:**422–428.
111. Yonce, C.E., and Chang, C.J., 1987, Detection of xylem-limited bacteria from sharpshooter leafhoppers and their feeding hosts in peach environs monitored by culture isolations and ELISA techniques, *Environ. Entomol.* **16:**68–71.
112. Young, D.A., 1977, Taxonomic study of the Cicadellinae (Homoptera: Cicadellidae). Part 2. New World Cicadellini and the genus *Cicadella,* N. Carolina Agr. Exp. Stn. Tech. *Bull.* 239, Raleigh, N. Carolina, 1135 p.
113. Zimmermann, M.H., 1983, Xylem Structure and the Ascent of Sap, Springer-Verlag, New York, 143 p.
114. Zimmermann, M.H., and McDonough, J., 1978, Dysfunction in the flow of food, in Horsfall, J.G., and Cowling E.B. (eds): Plant Disease An Advanced Treatise, Volume III, Academic Press, New York, pp. 117–139.

10
Plant Pathogenic Spiroplasmas and Their Leafhopper Vectors

Deborah A. Golino and George N. Oldfield

Introduction

In the last two decades, it has become clear that the genus *Spiroplasma* is one of great size and diversity. Since the first species was cultured in 1970 (46, 130), these helical, motile, wall-less procaryotes have been discovered in or on a multitude of hosts (27, 61, 151).

For many years, this taxon was known to biologists only by the disease symptoms produced in infected plants. The first known observation of citrus stubborn disease, now known to be caused by *Spiroplasma citri* (129), was made about 1915 (40). Later, Alstatt (4) reported a new disease of corn, later to be named corn stunt, which has since been demonstrated to be caused by *Spiroplasma kunkelii* (148). The helicity of two spiroplasmas, the Drosophila sex-ratio organism (120, 150) and the 277F tick spiroplasma (118), had been described by the late 1960s when the landmark discoveries of Doi et al. (35) and Ishiie et al. (67) triggered a reconsideration of the etiology of many plant diseases hitherto assumed to be caused by viruses. A second critical step was the axenic cultivation of the *Spiroplasma* species that causes citrus stubborn (46, 130), which was named *S. citri* (129). In the years that have followed, numerous new species and/or groups of spiroplasmas have been described (28, 139, 140, 151).

It is our intention in this chapter to focus on the plant pathogenic spiroplasmas and their interaction with leafhopper vectors. Since an understanding of pathogens, vectors, and their interactions requires an appreciation of each group of organisms, a brief overview of the diversity of spiroplasmas is presented. Leafhopper species that are known to transmit spiroplasmas are introduced. A section on transmission terminology is provided to aid in a clear examination of the vector status of each species. The historical development of information concerning *S.*

Deborah A. Golino, USDA-ARS, Department of Plant Pathology, University of California, Davis, California 95616, USA.
George N. Oldfield, Department of Plant Pathology, University of California, Riverside, California 92521, USA.

citri and *S. kunkelii* has been extensively reviewed by others (19, 32, 142). Therefore, we summarize the current information available regarding each species, the diseases they cause, their vectors, and some of the other plant pathogens that have been ecologically associated with spiroplasmas. This background provides the basis for the subsequent discussion of spiroplasma–leafhopper relationships.

Spiroplasmas and Their Habitats

An understanding of spiroplasmas and their taxonomic status is easily confused by a study of the historical references. For many years, in the absence of cultivable disease agents, plant diseases caused by spiroplasmas were believed to have a viral etiology (39, 40, 77). A later report, in which microscopic observations of the organisms were presented, confused spiroplasmas with spirochetes (120). Once cultured, the lack of a cell wall suggested to some workers that spiroplasmas were similar to L-phase variants (L-forms) of bacteria, an idea that did not last long (89). More recently, spiroplasmas were included with the nonhelical plant pathogenic mollicutes in the general group referred to as the "mycoplasma-like organisms" or MLOs. Serological and genetic studies (26, 71, 91, 126) have now established that the relationship between the spiroplasmas and the MLOs is more distant than was once believed. Although an understanding of the exact relationship of these two types of leafhopper-associated, wall-less procaryotes awaits further characterization of the MLOs (which have yet to be axenically cultured), the MLOs are expected to share with the spiroplasmas membership in the class Mollicutes. However, the spiroplasmas are no longer referred to as MLOs; they comprise instead a unique family—Spiroplasmataceae. This group has been reviewed extensively (10, 11, 140, 145, 146, 151). Spiroplasmas and other taxa in the class Mollicutes, are now accepted as having emerged from a common, Gram-positive ancestor related to certain clostridia (44, 127, 143, 152).

The common ecological character that is shared by the spiroplasmas as a group is their association with arthropods. It is somewhat ironic that these organisms first came to the attention of biologists in the role of plant pathogens; few species of this taxon, thought by some to be one of the largest in the world (61), are known to infect plants. Indeed, a logical argument can be made for the hypothesis that the plant pathogenic spiroplasmas are of relatively recent origin (143).

Unique morphological, physiological and biochemical properties have been used to establish groups and sub-groups within the genus *Spiroplasma* (151). A number of distinct serotypes can be distinguished using techniques developed specifically for the Spiroplasmas. These include the deformation, metabolic inhibition, and growth inhibition tests (134, 144, 150). Subject to frequent additions, the current total number of major groups in the Spiroplasmataceae is 23 (139). All the plant pathogenic spiroplasmas fall into Group I (13), increased to eight subgroups in 1987 by the addition of a Syrian pathogen, *Spiroplasma phoeniceum* (131). The other five subgroups of Group I have not been shown to be plant pathogens and may be insect pathogens only. Species epithets have been accepted, according to guidelines proposed by the ICSB Subcommittee on the Taxonomy of Mollicutes (63, 64, 65), for the three subgroups of Group I that are known to be plant pathogens.

Leafhoppers That Transmit Spiroplasmas

The reported leafhopper vectors of plant pathogens represent eight subfamilies of Cicadellidae (103, 103a); all those species reported as natural or experimental vectors of spiroplasmas belong to the subfamily Deltocephalineae, the Cicadellid subtaxon that includes the greatest number of vector genera and vector species (33, 103). A brief account of natural vectors, and those experimental vectors for which plant host and/or geographical ranges coexist with their vectored spiroplasmas, which suggests their possible role as natural vectors, follows.

Baldulus tripsici

Along with the vector species of the genus *Dalbulus,* this species is an efficient experimental vector of *S. kunkelii* (93). Found only in the eastern United States, it is a specialist, reported only from *Tripsacum dactyloides* (74, 74a).

Circulifer (-Neoalitaurus) tenellus

Widely distributed in western North America, *C. tenellus* has been reported from numerous locations in North and South Africa and countries of the Near East by Young and Frazier (154). It occurs sporadically east of the Rocky Mountains and in Florida and Puerto Rico. Its geographical distribution coincides largely with that of *S. citri* and many of its reproductive hosts (especially several genera of Old World Brassicaceae) have been experimentally inoculated with *S. citri* and are recognized hosts of both the leafhopper and the spiroplasma in California (107). Primarily an inhabitant of semiarid areas, *C. tenellus* is a multivoltine species in most or all of its geographical range, producing four generations a year in Arizona (128). Citrus does not seem to be a host of *C. tenellus* since our attempts to rear *C. tenellus* on citrus have failed repeatedly. However, we have observed adults congregating near the tops of small citrus plants in young orchards in the San Joaquin Valley of California during autumn migrations so apparently some association with citrus does occur under field conditions.

An intensely vagile species, *C. tenellus* is a generalist feeder with a survival strategy that requires that it move from one to another host throughout the year.

Dalbulus Species

Seven of the eight species of *Dalbulus* tested are efficient experimental vectors of *S. kunkelii* (56, 76, 81, 93, 100). *Dalbulus maidis,* the most widely distributed member of the genus (100), develops high populations on maize throughout the neotropics (81) and occurs as far south as Argentina. In the late growing season, it is found in maize fields of the southeastern United States (119) and in California (18, 45). Annual and perennial teosintes are hosts (6, 97). The perennial *Tripsacum dactyloides* is also a host, but development is slower than on maize. The virtual absence of the perennial host *Tripsacum* north of the southeastern United States may contribute to the restriction of its numbers farther north (138). It occurs widely in Mexico below 2000 m (6) and is most prevalent below 750 m. Nielson (102) reported 10 to 12 generations a year under greenhouse conditions in Arizona. *Dalbulus maidis* and *Dalbulus elimatus,* the two *Dalbulus* species that are well adapted to maize (81), are presently the only species shown to be associated with *S. kunkelii* (56). A maize specialist that has been collected from two perennial *Zea* species by Nault and DeLong (97), *D. elimatus* disperses from maize at the end of the season and feeds on wild and cultivated plants (6). The more restricted *D. maidis* is largely dependent on irrigated maize during the growing season and volunteer maize during the dry season (50). North of Mexico, *D. elimatus* occurs in the southwestern United States (6). Nault et al. (97) reported that *D. maidis* was always more abundant than *D. elimatus* at elevations above 1600 m in the Mexican state of Jalisco. These two maize specialists develop faster from egg to adult on maize and on *Tripsacum* than several other *Dalbulus* species that are *Tripsacum* specialists, but will reproduce on maize in the laboratory and will experimentally transmit *S. kunkelii* (94). Of these, *Dalbulus tripsacoides* is found on several species of *Tripsacum* in the eastern and western coastal areas of central Mexico; *Dalbulus guevari* and *Dalbulus longulus* occur on maize and on *Tripsacum* in westcentral Mexico, and along the eastern coast, respectively (94). Of the two other *Dalbulus* species reported to be experimental vectors of *S. kunkelii, Dalbulus quinquenotatus* is found on seven species of *Tripsacum* and is broadly distributed in Mexico; *Dalbulus guzmani* is known from just two *Tripsacum* species.

For an excellent account of the host range, phylogeny, biogeography, ecology, and reproductive biology of *Dalbulus* species, the reader is referred to Nault (94).

Euscelidius variegatus

Recorded (102) from diverse areas in Europe, Asia, North Africa, and the western United States, *E. variegatus,* which was reportedly introduced into North America, appears to be restricted to the Pacific states and Utah

(153). Nymphs and adults are found in grassy alfalfa fields and on sage, rosemary, and marjoram (34). The present authors and others have reared it routinely on barley in the greenhouse.

Macrosteles fascifrons

This well-known vector of the aster yellows MLO in North America (75) is widespread in the northern United States, Canada, and Alaska (62). Some of the reports of transmission of aster yellows by this species refer to the "short-winged" form, which recently was described as a separate species, *Macrosteles severini* (62). Reports by Kunkel in the eastern United States, where Hamilton (62) recognizes the widespread existence of *M. fascifrons,* indicate that this leafhopper has an extremely wide host range and that it reproduces in the laboratory (45, 75) on both dicotyledonous plants (including aster, lettuce, sowthistle, plantain, dandelion, calendula, and African daisy) and grasses (wheat, oat, rye, and barley). Its migration from south to north in the spring is well documented (141). As a common species in Illinois, it may play a more important role than *C. tenellus* in disseminating *S. citri* from brittle root-diseased horseradish in Illinois (41).

Neoaliturus haematoceps

This species ranges widely from Spain and Morocco, eastward through the Mediterranean countries to southern USSR (92, 102). Its geographical range widely overlaps that of *C. tenellus,* but it prefers areas of relatively more succulent, lush vegetation (115).

Scaphytopius Species

In the early 1970s in California, *Scaphytopius nitridus* and *Scaphytopius acutus delongi* came to the attention of entomologists attempting to establish the means by which stubborn disease was spread (68) (W. Ewart, personal communication). *Scaphytopius nitridus* occupies a niche on citrus in southern California and Arizona similar to that occupied by *S. acutus delongi* on San Joaquin Valley citrus. In southern California, Kaloostian and Pierce (68) reported two generations a year. Four generations were produced in the laboratory from March to October, when egg laying ceased, to begin again in February (104). An excellent experimental vector, there is only minimal evidence of its capacity as a natural vector (110, 111). Nielson and Morgan (104) state that it ranges from El Salvador north to Arizona, California, and Texas in the United States. Celery is an excellent laboratory rearing host. Nielson and Morgan found that it reproduces on *Cassia didymobotrya* in Arizona. *Scaphytopius acutus delongi,* suspected by some to be a western American species separate from *Scaphytopius acutus,* which would then be considered to be restricted to eastern North America (137), has not been found on citrus outside the San Joaquin Valley in California. The present authors have observed *S. acutus delongi* adults and nymphs on citrus midwinter, spring, and summer. Little else is known about its association with citrus. Although its role as a natural vector has not been demonstrated, its presence in citrus groves where *S. citri*-infected trees occur suggests that it may serve to spread that agent from citrus to citrus. The less-studied species *Scaphytopius californiensis* was collected by the present authors from a weed-infested alfalfa field in San Bernardino County, California and reared in the laboratory on red clover (108). Its absence from collections of leafhoppers from hosts of *S. citri* in citrus environs in California suggests that it plays at most only a minor role in the maintenance of *S. citri* in nature.

Other Species

Three inefficient experimental vectors of *S. kunkelii, Exitianus exitiosus, Graminella nigrifrons,* and *Stirellus bicolor,* are commonly encountered on maize in the southwestern United States and are members of genera that evolved in tropical grasslands. *Exitianus exitiosus* is a generalist feeder capable of feeding on *Zea* and many other grasses; it overwinters in temperate regions of the United States as an adult (R.F. Whitcomb in prcss).

Graminella nigrifrons, one of the most common and ubiquitous leafhoppers in the eastern and central United States (74, 74a), has many grass hosts and overwinters as an adult in the eastern United States (135). It reproduces on corn and uses Johnson grass as an overwintering host (Whitcomb, in press). Three generations per year were reported by Boyd and Pitre (14).

Stirellus bicolor occurs mainly on the broomsedge *Andropogon virginicus* but can feed on a wide variety of grasses. In the southeastern United States, it spreads from perennial grasses to corn fields. In temperate regions, it overwinters as an adult (Whitcomb, in press).

Transmission Terminology

The transmission process is a complex series of events, many of which are poorly understood, and for which terminology is often insufficient to differentiate between observations that are similar but not identical. Because the use of common phrases varies, a definition of terms is a useful component of articles on the leafhopper transmission of mollicutes (52, 122, 123). Before discussing the transmission of each plant pathogenic spiroplasma species, we will, therefore, discuss the terminologies of vector status and pathogen detection, two areas which we feel are often confused. Perhaps publication of these views will stimulate greater discussion of this problem and eventually a consensus among researchers in the field will be reached. The following discussion is phrased in terms of spiroplasmas, but could well refer to MLOs or to viral pathogens.

If a field-collected leafhopper placed on a healthy indicator plant feeds and, after an incubation period has passed, and in the absence of other spiroplasma vectors, the plant develops disease caused by a known plant pathogenic spiroplasma, that leafhopper can be said to be a *natural* vector of the specific spiroplasma. The criterion we feel is important is that, in a nonmanipulated environment, the vector passed through all the stages of transmission from acquisition and latency through inoculation, as detailed by Purcell (123).

If a laboratory-reared leafhopper caged on a spiroplasma-infected plant is forced to feed and, after appropriate periods of latency and inoculation, it transmits the pathogen to a healthy plant, that leafhopper species should be described as an *experimental* vector of the spiroplasma. Until it has been demonstrated that those same events occur in nature, the status of that leafhopper species as a natural vector is undetermined.

There are additional *artificial* ways in which leafhoppers can be manipulated to transmit spiroplasmas. Leafhoppers will feed through parafilm membranes and directly ingest appropriately prepared spiroplasmas grown in culture (83, 84, 125). If the isolate has not lost pathogenicity as a result of cultural passage (19, 38, 80), transmission may occur. Even more directly, cultured spiroplasmas may be injected into the hemocoel of candidate leafhoppers. During the latent period, spiroplasmas invade a succession of the leafhopper's tissues and, eventually, reach the salivary secretions of the insect at which time they may be transmitted to plants (84, 86). In this chapter, leafhoppers that have only been demonstrated to transmit spiroplasmas after this type of manipulation will be referred to as *artificial* vectors. Some leafhoppers that are not transmission-competent after plant feeding will transmit spiroplasmas after injection (see Table 10.2). Therefore, not all leafhoppers that transmit after injection are natural vectors. In particular, the injection process bypasses a portion of the pathway to the salivary glands of the vector (i.e., from the stylet channels through the foregut, midgut, and, perhaps most significantly, the gut wall of the injected insect), which is thought to be an important stage in natural transmission. In general, confusion can be avoided by stating the specific technique used to produce inoculative insects and by avoiding the unqualified use of the word *vector* for insects that will transmit in this fashion.

The importance of distinguishing among *natural, experimental,* and *artificial* vectors becomes clearer as we discuss the series of events that result in spiroplasma transmission by leafhoppers. Specificity enters into these events in some degree throughout each step of the process. Plant pathogenic spiroplasmas grow and multiply in the phloem of susceptible plants (57, 58, 66, 78). Leafhoppers, either in natural circumstances or under experimental conditions, insert their stylets directly into phloem sieve tube elements and feed on the spiroplasma-containing phloem sap. Not all leafhoppers will feed on all plants. The type of feeding that occurs, and the amount of phloem sap ingested, is in part determined by the host plant species. Although the research we can draw from in this area is by no means complete, three simplified outcomes of ingestion can be hypothesized. (a) Gut infections with spiroplasmas may be short term and extracellular without increasing titer. (b) Spiroplasmas may multiply in the leafhopper, invading any of a series of tissues including the gut epithelium, hemolymph, salivary glands, and/or brain but are not transmitted to healthy plants during feeding. (c) All steps for transmission may be completed, including acquisition, gut passage, multiplication in several tissues with eventual invasion of the salivary glands (latency), and transmission via salivary secretions to a suitable host plant.

Although the details of all these hypothesized processes need further study, an awareness of the multiplicity of outcomes of spiroplasma ingestion by a leafhopper can be particularly useful in interpreting data accumulated during field or laboratory studies. The presence of spiroplasmas in the body of a leafhopper, as detected by enzyme-linked immunosorbent assay (ELISA), gene probes, microscopic observation, or direct culturing, does not necessarily indicate that the species of leafhopper is a natural vector of the detected spiroplasma. In all three of the possible outcomes of acquisition outlined in the above paragraph, for at least a short time after ingestion, spiroplasmas could be detected with some of these techniques. However, only in case (c) above would actual transmission occur. This will be discussed in additional detail in the spiroplasma–leafhopper section of this chapter.

Spiroplasma citri

The biology and ecology of *S. citri* has been extensively reviewed (10, 12, 19, 145, 146). In this section, the established literature concerning the diseases caused by *S. citri* and its transmission by leafhoppers and other associated plant pathogens will be summarized without any attempt to provide the historical perspective available elsewhere.

Diseases

Spiroplasma citri, which was proven in the early 1970s to be associated with citrus stubborn disease (46, 130), has since been demonstrated to have an extremely broad plant host range. Horseradish brittle root disease has also been clearly associated with infection by this spiroplasma

(41, 124). In addition, disease has been described in a plethora of other plant hosts both naturally infected and experimentally inoculated with *S. citri* (3, 19, 21, 105, 109, 113, 114). It is probable that *S. citri* causes other agriculturally significant plant diseases that have not yet been recognized. The authors suspect that the relatively slow rate of spread of *S. citri* limits its economic importance mostly to perennial crops and crops such as horseradish that are vegetatively propagated. Although we have observed infection of such annual crops as turnip, radish, and broccoli (21, 69, 73), the long latent period of *S. citri* in plant hosts and the low percentage of infected plants seem to prevent significant crop loss. Since the effect upon most species of plants that have been experimentally infected is severe (21), the possibility does exist that under unusual epidemiological conditions, significant losses due to *S. citri* infection of annual crop plants could occur.

Citrus stubborn disease, the most economically important plant disease caused by *S. citri,* has been observed largely in the southwestern United States and in Mediterranean countries. Affected trees (59) can be stunted as a result of severe internode shortening. Chlorosis, similar in appearance to symptoms brought about in citrus by zinc deficiency, may be observed. A proliferation of shoots from axillary buds is also common, and off-season flushes of vegetative growth and flowers may indicate infection. Fruit quality and yields are both reduced on infected trees. Symptoms in the field are not reliable for diagnosis even by workers experienced with the disease (59), since all citrus stubborn disease symptoms can be caused by other pathogens or by environmental stresses.

The disease is most obvious when infected budwood is used as a scion; in these cases, normal orchard establishment does not occur. Young trees that are naturally infected may also develop severe systemic infections. In more mature trees, it is thought that primary infections are rarely established. Symptoms in older trees are often seen in single limbs in an irregular pattern. Effects of the disease upon yield may be significant, even though symptoms may be very mild or absent much of the year.

Varieties of citrus that may be naturally infected include sweet orange, grapefruit, mandarin orange, tangelo, pummelo, and shaddock (22, 24, 25). In addition, the disease can be graft transmitted to lemon, lime, Troyer citrange, and trifoliate rootstocks (20).

Because symptomatology is unreliable and neither serology nor culturing (D.J. Gumpf, personal communication) is consistent as a diagnostic test in citrus, graft inoculation of sensitive indicator varieties in greenhouse tests is still a useful technique in determining the disease status of individual trees (149). Studies have demonstrated that the titer of spiroplasmas varies seasonally in citrus (19, 60). The authors have found that leafhoppers acquire spiroplasmas at very low rates from citrus. It is possible that citrus is largely a "dead-end" host of *S. citri* and rarely

serves as the initial source of inoculum in the field. The distribution of infected trees in orchards supports the idea that little tree-to-tree infection occurs (60).

The second plant disease that occurs naturally and has been established to be caused by *S. citri* is brittle root disease of horseradish (42, 124). In horseradish (*Armoracia rustica* Gaertn., Mey., and Scherb.), *S. citri* causes stunting, yellowing, and tissue necrosis, which eventually leads to the death of the infected plant. It has been demonstrated that strains of spiroplasmas that are serologically indistinguishable from citrus-infecting strains of *S. citri* can be cultured from brittle root-diseased horseradish; in addition, citrus-infecting strains of *S. citri* will cause brittle root symptoms in horseradish (136).

Vectors

The transmission of a spiroplasma by a leafhopper under natural conditions involves many types of interaction between plant, pathogen, and insect. The first experimental transmissions of *S. citri* that fulfilled Koch's postulates were accomplished by injecting the leafhopper *Euscelis plebejus* (Fallen) with the cultured organism (30). Note that this particular leafhopper species has never been shown to transmit the organism after natural feeding on infected plants (see Table 10.2). In this chapter, we discuss the need for a careful examination of the experimental data available on transmission of spiroplasmas and attempt to distinguish between various modes of transmission and detection. The case of *E. plebejus,* the first leafhopper demonstrated to transmit *S. citri,* is an excellent example of some of the distinctions we would like to emphasize. In this section, we will discuss each species of leafhoppers that have been demonstrated to acquire *S. citri* from plants either under field conditions or during experimental feeding. All species capable of artificial transmission will be discussed together.

Circulifer Tenellus

At this time, there is strong evidence that *C. tenellus* (Baker), commonly known as the beet leafhopper, is the most important natural vector of *S. citri* in the western United States (107). It was the only species of leafhopper repeatedly found naturally inoculative for *S. citri* during many years of field collecting and plant indexing (110). Collections of this leafhopper made throughout the year successfully transmitted *S. citri* to indicator plants (110) and *C. tenellus* collected from Washington State and Utah have been shown to harbor the spiroplasma (107, 108). *Spiroplasma citri* strains from Illinois are transmitted by *C. tenellus,* a leafhopper species that is occasionally collected in that state (41).

Owing to the widely overlapping host ranges of *C. tenellus* and *S. citri,* and their similar geographical distributions, *C. tenellus* probably is very active in the dissemination of *S. citri* in most of the areas where it occurs in North America. Its role in the Mediterranean area may be less important, since it is rarely encountered there and other *Cirulifer* spp. are abundant.

Macrosteles Species

Macrosteles fascifrons (Stål) experimentally transmits isolates of *S. citri* obtained from brittle root-diseased horseradish in Illinois (106). No evidence is available concerning the natural transmission of *S. citri* by this species. A western America relative [reported as *Macrosteles fascifrons* but renamed *Macrosteles severini* Hamilton (62)] failed to transmit California isolates of *S. citri* (110).

A recent paper by Hamilton (62) served to emphasize the importance of depositing voucher specimens representing taxa upon which reports of transmission capabilities are based. Hamilton refers to *M. fascifrons* (Stål) as "a Juncus-feeding transboreal species, in Asia restricted to eastern Siberia extensively confused with the Aster Leafhopper (*M. quadrilineatus* Forbes." The identity of the *Macrosteles* used to transmit strains of *S. citri* (106) has not yet been clarified in the light of this taxonomic revision.

Neoaliturus Haematoceps

In the Old World, *N. haematoceps* has been shown to be a natural as well as experimental vector of *S. citri* (17, 43). Naturally inoculative insects have been collected and shown to transmit to periwinkle (*Catharanthus roseus*). Natural transmission has been reported by leafhoppers collected in Corsica, France (17), and *S. citri* has been cultured from leafhoppers collected in Morocco and Syria (12).

Scaphytopius Species

Scaphytopius acutus delongi (Young), *S. californiensis* Hepner, and *S. nitridus* (Delong) have each been demonstrated to transmit *S. citri* under similar experimental conditions (69, 70, 107, 111, 113). These conditions consist of an acquisition feeding period of at least one or two days on a diseased host plant, a latent period on the plant species the leafhopper was originally reared on, and a transmission feeding period upon an indicator plant. Disease symptoms, confirmed by spiroplasma cultivation, were used to determine the success of transmission. Detailed information on the vector relations between *S. nitridus* and *S. citri* is available (111). In addition, much of the information available on the experimental host range of *S. citri* (21) was obtained using this vector.

Of the three *Scaphytopius* species, no information is available that would support the natural inoculativity of *Scaphytopius californiensis* (107) and none of the 572 field-collected *S. acutus delongi* fed on young periwinkle plants transmitted *S. citri* (110). For *S. nitridus,* some limited data exist, which suggest that this leafhopper may be a natural, if minor, vector of *S. citri.* In one instance, California field-collected *S. nitridus* were reported to transmit to indicator plants (110). Also in Arizona, the progeny of field-collected *S. nitridus* contained spiroplasmas when cultures were initiated from them (3); this might indicate that the parent leafhoppers transmitted *S. citri* to the host plant on which eggs were laid, which allowed the emerging nymphs to acquire *S. citri* as they fed.

Artificial Vectors

It is evident (see Table 10.2) that any leafhopper species that can serve as a natural or experimental vector of *S. citri* has also been proven, when tested, to serve as an artificial vector. Also, three species of leafhoppers will transmit *S. citri* after injection but have not been proven capable of acquiring the spiroplasma from infected plants or by membrane feeding.

Euscelidius variegatus and *E. plebejus* have been used as artificial vectors following injection with cultured *S. citri.* (86, 87, 133). Both species were chosen as experimental vectors for their ease in laboratory rearing, their wide plant host ranges, and their demonstrated ability to transmit MLO diseases (86, 102). Although neither can transmit *S. citri* following plant feeding (see Table 10.2), their ability to transmit *S. citri* after injection has made them useful laboratory vectors, particularly in early host range studies of this spiroplasma (30, 85, 86, 87, 133).

In addition, *Colladonus montanus* (124a) apparently can serve as an artificial vector, although it is unable to transmit after feeding on infected plants.

Leafhoppers That Harbor *S. Citri*

Leafhopper species are known from which *S. citri* has been cultured, although they did not transmit the organism to indicator plants when experimentally exposed through plant feeding, injection, or membrane feeding. Rana et al. (125) reported that *S. citri* could be cultured from membrane fed *Acertagallia* spp., *Actinopterus angulatus, Erythroneura variabilis, Hordnia circellata* (Baker), and *Spissistilus festinus* (Say); *Empoasca* spp. were negative for *S. citri*. Oldfield et al. (110) showed that *Aceratagallia curvata, Graminella sonora, Euscelidius variegatus, Macrosteles fascifrons,* and *Ollarianus strictus* could acquire *S. citri* from infected plants and harbor it in their bodies for up to three weeks, as assayed by culturing. In the same experiment, *Empoasca* spp. apparently did not acquire *S. citri*. This suggests a specificity for the ability to harbor *S. citri.* Attempted spiroplasma cultivation from field collections of the same leafhopper species over many years has resulted in only rare isolations from *O. strictus;* no *S. citri* cultures were obtained from the other species (110).

Other Associated Pathogens

A number of plant pathogens have been demonstrated to be transmitted by the same leafhoppers that transmit *S. citri.* During studies involving strains of *S. citri* obtained from the field using leafhoppers, the possiblity of dual infection of source plants or vectors should be considered.

Curly Top Virus

Circulifer tenellus is known to transmit two other plant pathogens. The first, sugar beet curly top virus (CTV), causes economically important diseases that damage a broad range of host plants including sugar beet, tomato, potato, beans, and cucurbits (8). Curly top virus is a phloem-limited geminivirus, transmitted in a persistent fashion by *C. tenellus.* In the western United States, extensive spray programs are directed at *C. tenellus* in an effort to control curly top disease. We have often wondered to what extent the damage caused by *S. citri* in California is ameliorated by these spray programs.

Beet Leafhopper Transmitted Virescence Agent

Another *C. tenellus*-borne plant pathogen known to co-infect plants with *S. citri* is a mycoplasma-like organism (MLO), the beet leafhopper transmitted virescence agent (BLTVA). This agent from California has been observed repeatedly in field associations with *S. citri* (Oldfield and Golino, unpublished observations), and one line has been characterized (53, 54).

Other Pathogens

An MLO seen in field association with *S. citri* in Illinois is also known to be transmitted by *C. tenellus* (36). A third observation of an MLO–*S. citri* association has been made in the Mediterranean (9). At this time, the relationship between these virescence-type MLOs is unknown; clarification awaits a better understanding of the taxonomy of the MLOs as a group (71, 91).

Macrosteles fascifrons is best known as a vector for its ability to transmit the aster yellows (AY) MLO (75). Observations of AY in association with *S. citri* under field conditions in central California (Oldfield, unpublished observations), and the reports of transmission of strains of *S. citri* by *M. fascifrons* (106), raise the possibility of co-transmission of the two agents by *M. fascifrons* under some conditions.

Several other plant pathogens have been observed in ecological association with *S. citri*. *Spiroplasma phoeniceum,* the recently named, third plant pathogenic spiroplasma species (131), discussed later in this chapter, was discovered in a citrus field plot in the indicator plant periwinkle. In this same plot, symptoms of virescence type MLO disease, similar to BLTVA and AY, were observed. Observations of dual infection of MLOs and *S. citri* have also been made in Morocco (9). When the vectors of *S. phoeniceum* and these MLOs are found, their ability to transmit *S. citri* should certainly be examined.

Spiroplasma kunkelii

The literature on *S. kunkelii* has been recently reviewed (142). As with *S. citri,* a summary of current findings on the spiroplasma and its impact upon its plant and insect hosts is provided.

Diseases

Corn stunt spiroplasma can cause severe, economically significant disease in maize throughout the Americans (4, 5, 16). Outbreaks have been reported from the United States in Arizona, California, Florida, Mississippi, and Texas. Other countries in which corn stunt disease has been observed include Colombia (90), Costa Rica (51), Jamaica (37), Mexico (31), and Peru (23, 96).

Spiroplasma kunkelii has a natural host range that is limited to maize (*Zea Mays* L.) and teosinte [*Zea mays mexicana* (Schrad.) Reeves and Manglesd.], the close relative of the more widely domesticated maize. Infected plants are stunted, with chlorotic spots and stripes banding the leaves, in parallel with the leaf veins. Internode shortening such as is observed in *S. citri* infections of citrus is also seen in *S. kunkelii*-infected host plants. Another common symptom is the proliferation of shoots from axillary buds. A symptom that has been observed to vary with cultivar (93) is the reddening of older leaves on infected plants. The yields of *S. kunkelii*-infected corn are dramatically reduced. Although the proliferation of secondary shoots sometimes results in the development of multiple ears, the size and quality of the ears that are produced are so poor as to make them unusable.

Some additional plant species have been shown to serve as hosts of *S. kunkelii*. Other members of the genus *Zea* have been experimentally inoculated. Using an artificial vector, *Euscelidius variegatus,* injected with cultured *S. kunkelii,* Markham et al. (88) were able to infect not only maize but also the dicots broad bean (*Vicia faba* L.) and periwinkle [*Catharanthus roseus* (L.) G. Don.]. When the same artificial vector was used (2, 83), mustard (*Brassica hirta* Moench), pea (Pisum sativum L.), radish (Raphanus sativum L.), and spinach (*Spinacia oleracea L.*) were also found to be hosts. Of additional interest was the ability of *E. variegatus* to transmit *S. kunkelii* from mustard to mustard experimentally.

Much of the literature on corn stunt published in the past was based on a combination of symptom observation and vector studies. The work of Nault and his coworkers (95, 96) has clearly demonstrated that a complex of leafhopper-borne maize pathogens exists in the Americas. Common symptoms and leafhopper vectors exist for *S. kunkelii* and maize bushy stunt MLO (see discussion later in this section), which confuses diagnosis even under closely controlled experimental conditions (93). Techniques now exist to differentiate among the corn stunt spiroplasma, the maize bushy stunt MLO (MBSM), maize rayado fino virus (MRFV), and maize chlorotic dwarf virus (MCDV) (142). Thus, caution must be exercised in interpreting articles on the range and severity of "corn stunt" that were published before *S. kunkelii* was cultured and the complex was understood. Many of the early reports of corn stunt may have involved multiple infections. Because no voucher specimens are available, even the pioneering work of Kunkel (76, 77) can only be assumed to have been done with single infections of *S. kunkelii*. Future studies can be conducted in such a way as to exclude the possibility of misdiagnosis of members of this complex.

Vectors

Dalbulus Species

Only two species, *D. maidis* and *D. elimatus,* are thought to be natural vectors of corn stunt disease. *Dalbulus maidis* was implicated as a vector by Kunkel (76, 77) early in the history of research on corn stunt and is believed to be the primary vector of the pathogen throughout the Americas (96, 98, 142). *Dalbulus maidis* feeds and reproduces on corn, where it can reach populations of such high density that it is a directly damaging pest (18). Although the two species have been observed with similar rates of infection (56), *D. elimatus* has a far more limited geographical range—it does not exist in much of the tropical region where *S. kunkelii* is important.

In reviewing the literature on transmission of *S. kunkelii,* we were unable to find published accounts of field-collected leafhoppers transmitting *S. kunkelii* to indicator plants. The data, which implicates *D. maidis* and *D. elimatus* as vectors, relies on their association with maize, the cultivation of *S. kunkelii* from field-collected leafhoppers, and ELISA tests which complement separate laboratory studies (1, 2, 16, 23, 31, 56, 83, 93, 121). In other words, by the definitions we established earlier in this chapter, there is no data to demonstrate that either *D. maidis* or *D. elimatus* are natural vectors of *S. kunkelii* (Table 10.1).

Additional Experimental Vectors

In addition to *D. maidis* and *D. elimatus,* 13 species of leafhoppers have transmitted *S. kunkelii* under various experimental conditions (see Table 10.2). They are *Cicadulina mbila, Baldulus tripsici, Dalbulus gelbus, Dalbulus guevarai, Dalbulus guzmani, Dalbulus longulus, Dalbulus quinquenotatus, Dalbulus tripsacoides, Euscelidius variegatus, Exitianus exitiosus, Graminella nigrifrons, Macrosteles sexnotatus,* and *Stirellus bicolor.*

Other Associated Pathogens

As with *S. citri, S. kunkelii* has been observed in association with other plant pathogens. Corn stunt, as described in the past, has been demonstrated to consist of a complex of at least four distinct pathogens (96). In addition to *S. kunkelii,* this complex includes a non-cultivable MLO and two viruses. All four pathogens are transmitted by deltacephaline leafhoppers, although they have different specificities, and produce symptoms on corn that are easily confused in a field situation.

Maize Bushy Stunt Mycoplasma-Like Organism

Maize bushy stunt (MBS), an MLO associated with the typical nonhelical cells bound by a unit membrane (15, 96), and sensitive to tetracycline antibiotics (7), has been shown to cause a disease in maize that was once attributed to the ''Mesa Central'' strain of corn stunt (82). Nault (93) has compared disease symptoms, pathogen host ranges, and vector specificities and found that the two pathogens can be distinguished by these biological parameters.

Leaf symptoms of MBS–MLO include leaf reddening, chlorosis, twisting, tearing, and shortening. Multiple tillers develop at both the base of the plant and the leaf axils. Severe stunting dramatically decreases ear size and increases the number of ears per plant. Except under the most exacting growing conditions, these symptoms could easily be confused with those produced by *S. kunkelii*. In contrast to the somewhat broader host range of *S. kunkelii* (as described earlier), the host range of MBS–MLO is limited to maize (*Zea Mays* L.) and Mexican teosinte (*Zea Mays* L. *mexacana*).

The MBS–MLO complex has been experimentally transmitted by *D. maidis, D. elimatus, D. gelbus, D. guevarai, D. longulus, D. quinquenotatus, D. tripsacoides* (81), *Graminella nigrifrons* (56a), and *G. sonorus* (56b). In contrast to *S. kunkelii,* MBS–MLO is not transmissible by *Stirella bicolor* or *E. exitiosus* (93).

Maize Rayado Fino Virus

Like mauze bushy stunt MLO, the disease caused by maize rayado fino virus (RFV) was first believed to be a new strain of corn stunt (5). Gámez

(47) demonstrated that this disease was caused by a virus, which was persistently transmitted by *D. maidis*. This virus is a small, isometric, single-stranded RNA virus of about 31 nm; it is not mechanically transmitted. It can be distinguished from other members of the corn stunt complex by the stipple striping it produces on maize (49, 96). Maize rayado fino virus has been observed infecting corn in the Americas from Texas and Florida, and as far south as Uruguay (51). A review (50) of the relationship between RFV and its vector is available in an earlier volume of this series.

Maize Chlorotic Dwarf Virus

Maize chlorotic dwarf virus (MCDV) causes corn stunt-like symptoms in the southern United States (55). Unlike other members of the corn stunt complex, MCDV is not transmitted by *Dalbulus* species but rather by *G. nigrifrons*. It is a semipersistently transmitted, isometric, single-stranded RNA virus with a host range limited to grasses (99, 101, 117). The distribution of MCDV is more limited than might otherwise be expected from its plant host range and distribution of its vector (96, 142).

Spiroplasma phoeniceum

For many years, only two plant pathogenic spiroplasmas, *S. citri* and *S. kunkelii,* were recognized. In 1983 and 1984, during field experiments in citrus orchards on the Syrian coast, a new plant pathogenic spiroplasma was discovered (132). First observed in periwinkle plants, *Catharanthus roseus,* planted in orchards to determine the rate of transmission of the citrus spiroplasma, it causes symptoms of yellowing and stunting not unlike those produced by *S. citri*. However, the serological properties of the spiroplasmas infecting these plants differed from those of *S. citri,* an observation that led to the description of a new species.

This new spiroplasma has been named *Spiroplasma phoeniceum* (131), since criteria necessary to elevate this spiroplasma to species status have been satisfied. *Spiroplasma phoeniceum* falls in subgroup I of the genus, as do the other plant pathogens (151). The natural vector of *S. phoeniceum* is unknown at this writing. Artificial transmission by injection of spiroplasma broth cultures into *M. fascifrons* has resulted in inoculation of both aster (*Callistephus chinensis*) and periwinkle (*C. roseus*) (R.F. Whitcomb, A.L. Hicks, C. Saillard, and J.M. Bové, personal communication).

Spiroplasma–Leafhopper Relationships

The available information on the transmission of plant pathogenic spiroplasmas by leafhoppers can be divided on the basis of field versus experimental data (Table 10.1 and Table 10.2). Relatively little is known about the natural inoculativity of leafhopper species for any *Spiroplasma* species. This is because far greater resources are required for field collection and identification of individual specimens than for the manipulation of a taxonomically determined insect colony in a laboratory. In addition, when extensive field survey work has been done, relatively low inoculativity may lead to inconclusive results. Experiments that differentiate fully between level of inoculativity and importance as a field vector are difficult to imagine; they would require impossibly large data sets to be meaningful. Even the extensive collaborative efforts to establish the identity of field vectors of *S. citri* in California, which spanned many years and involved large investments of personnel and resources, resulted in only partial data on the complex agroecosystem involved in stubborn disease spread.

In examining both Table 10.1 and Table 10.2, the lack of a relationship between the ability to culture a spiroplasma from a leafhopper and the ability of that leafhopper to transmit the spiroplasma either in the field or under experimental conditions becomes evident. If the hypothetical results of spiroplasma ingestion outlined in the section of transmission

TABLE 10.1. Plant pathogenic spiroplasmas in field-collected leafhoppers.

	Detection method			
Leafhopper	ELISA	Culture	Transmission	References
Spiroplasma citri				
Aceratagallia obscura Oman		–		3
Aceratagallia spp.		–	–	110
Aceratagallia sp.		+		38
Circulifer tenellus (Baker)		+	+	79, 112
Colladonus montanus (Van Duzee)			–	110
Empoasca spp.		–	–	110
Erythroneura spp.			–	110
Euscelis alsius (Ribaut)	+[a]			12
Exitianus capicola (Stål)	+			12
Graminella sonora (Ball)		–	–	110
Neoaliturus haematoceps (Mulsant & Rey)	+		+	12, 17
Ollarianus strictus Ball		+	–	110
Psammotettix striatus L.	+[a]			12
Recilia angustisectus (Linnavuori)	+			12
Scaphytopius acutus delongi Young		–	–	110
Scaphytopius nitridus (De Long)		+	+	110
Toya propinqua (Fieber)	+[a]			12
Spiroplasma kunkelii				
Dalbulus elimatus (Ball)	+			56
Dalbulus maidis (DeLong & Wolcott)	+			56

[a] The techniques used for this report have since been modified; it is possible that these are ELISA false positives (19).

terminology are compared to available data, however, some patterns can be observed. When a leafhopper species has been proven capable of transmitting a spiroplasma species after acquisition from a plant [i.e., (c) in the earlier section], it was also possible to detect spiroplasmas in that species by culturing and/or ELISA. In addition, when tested, if acquisition from plants were successful, both membrane feeding and injection resulted in transmission. In those cases in which a species apparently harbored spiroplasmas but could not transmit them [either (a) or (b) under transmission terminology, which we cannot yet discriminate], we were surprised to find that injection of leafhoppers always resulted in their ability to transmit to indicator plants. It is always possible that negative data on artificial vectors have simply not been published. However, the limited data available on this subject, as displayed in Table 10.2, suggest that injection of the phloem-feeding leafhopper species with one of the plant pathogenic spiroplasmas may bypass any transmission barriers that would be encountered with natural feeding or membrane feeding. Future research on the specificity of the transmission process may clarify this point.

An additional point that should be made regarding leafhopper transmission of spiroplasmas has to do with the importance of a leafhopper species as a proportion of the total leafhopper population versus the role of that species as a vector. There is not necessarily any correlation between the importance of a leafhopper species as a direct damaging pest, or its representation as a proportion of the total leafhopper community in a particular ecology, and its importance as a vector. And, indeed, the fact that one species reproduces on the crop plant in which disease is found, whereas another reproduces only on other species, does not determine which species, if either, is the natural vector of the pathogen.

To cite our own experience with some of the pitfalls discussed above, during the early collaborative efforts to determine the identity of the field vector of *S. citri,* the citrus stubborn disease agent, we found species of leafhoppers from which *S. citri* could be successfully cultured after field collection (110). Of these species, a smaller number could experimentally transmit *S. citri* under laboratory conditions (Table 10.2). Those species that could transmit in the laboratory included *Circulifer tenellus, Scaphytopius acutus delongi, Scaphytopius californiensis,* and *Scaphytopius nitridus*. Early in the project, extra efforts were made to document the transmission biology of *S. nitridus* because it reproduces in citrus. Only after years of testing the inoculativity of field-collected leafhoppers was it decided that *C. tenellus* was most likely the natural vector of stubborn disease, despite the low incidence of the leafhopper in citrus orchards and the inability of the leafhopper to reproduce on citrus.

When the work done by the researchers who discovered *S. phoeniceum* is considered, some insights into the limits of our knowledge about plant pathogenic spiroplasmas emerge. This international working group found

TABLE 10.2. Experimentally established associations of plant pathogenic spiroplasmas with laboratory-reared leafhoppers.

Leafhopper	Assay method						References
	Post-access to plant			Post-injection		Post-membrane	
	Culture	ELISA	Trans.	Culture	Trans.	Trans	
Spiroplasma citri							
Aceratagallia curvata Oman	+		−				110
Circulifer tenellus (Baker)	+		+	+	+	+	80,80a,110,112,125
Colladonus montanus (Van Duzee)					+		124a
Draeculacephela sp.				+			147
Dalbulus elimatus (Ball)				+			147
Euscelidius variegatus (Kirschbaum)	+		−		+		85,86,110
Euscelis plebejus (Fallen)			−	+	+		86,30
Graminella sonora (Ball)	+		−				110
Macrosteles fascifrons (Stål)			+		+		38,106
Macrosteles severnii Hamilton	+		−				107,110
Macrosteles sexnotatus (Fallen)				+			85
Neoaliturus haematoceps (Mulsant & Rey)			+		+		43
Ollarianus strictus Ball	+		−				110
Scaphytopius acutus delongi Young			+[a]				69
Scaphytopius californiensis Hepner			+				107
Scaphytopius nitridus (DeLong)	+		+			+	69,70,125
Texanus spatulatus (Van Duzee)	−		−				110

			Spiroplasma kunkelii				
Cicadulina mbila (Naude)			+				94
Baldulus tripsaci Kramer & Whitcomb			+				81
Dalbulus elimatus (Ball)		+	+				93
Dalbulus gelbus DeLong		+	+				56,81
Dalbulus guzmani DeLong			+				94
Dalbulus guevarai DeLong		+	+				56,81
Dalbulus longulus De Long			+				100
Dalbulus maidis (De Long & Wolcott)		+	+	+	+	+	1,2,56,76
Dalbulus quinquenotatus De Long & Nault		+	+				56,81
Dalbulus tripsacoides De Long & Nault			+				56
Euscelidius variegatus (Kirschbaum)	+		+	+	+	+	83,88
Exitianus exitiosus (Uhler)			+		+		38,93
Graminella nigrifrons (Forbes)			+				93
Macrosteles fascifrons (Stål)			–				77
Macrosteles sexnotatus (Fallen)			+	+	+		83,84
Stirellus bicolor (Van Duzee)			+				93
Perigrinus maidis (Ashmead)			–				77
			Spiroplasma phoenicum				
Macrosteles fascifrons (Stål)			–		+		131

[a] The published account by Kaloostian et al. (69) is of only limited value, since field-collected *Scaphytopius acutus delongi* transmitted *S. citri* after feeding on an experimentally infected greenhouse grown plant. Oldfield and Golino (unpublished) used *S. citri*-free, laboratory-reared insects to transmit *S. citri* from plant to plant.

the new species because they were using a combination of serological techniques and direct culturing of the mollicute. They were perceptive enough to correlate the absence of ELISA reactions in the presence of cultivable spiroplasmas with the possibility that they had discovered a new species, and they had the considerable expertise with the organism necessary to take on the task of definitive identification and characterization. If that group had limited their diagnostic tests to the culture of spiroplasmas from samples associated with plant disease symptoms, *S. phoeniceum* might be unknown today. Given these unusual sets of circumstances, and the relatively short time that spiroplasmas have been known to science, our expectation would be that additional plant pathogenic members of the genus will be discovered in the future. Historical references to any plant disease of spiroplasmal etiology must be analyzed with an awareness not only of the plant pathogens of other taxa (discussed earlier in the chapter) that are ecologically associated with spiroplasmas or their vectors, but also with an awareness of the possibility that more than one plant pathogenic spiroplasma might be present at a given field site.

Concluding Remarks

It is clear that, in spite of the tremendous increase in knowledge about the plant pathogenic spiroplasmas over the last 20 years, the information available about their transmission by leafhoppers under natural conditions is still greatly limited by the technical problems posed by the field studies that would be required to elucidate their biology. In interpreting both experimental data and field data, caution must be taken that the limitations of the techniques at hand are realized. Much remains to be learned about the diversity of these pathogens and their complex ecology in both leafhopper vectors and plant hosts.

Summary

1. Three plant pathogenic spiroplasmas, *Spiroplasma citri, Spiroplasma kunkelii,* and *Spiroplasma phoeniceum* have been described. All are transmitted by leafhoppers.
2. Very little is known about the natural inoculativity of leafhoppers for any of the spiroplasmas that cause plant disease. Geographical and technical difficulties have limited research in this area. Most of our information on leafhopper–spiroplasma relations was obtained under experimental conditions.
3. There is no reason to believe that the diversity of the plant pathogen spiroplasmas is limited to those that have already been described. An

understanding of the nature of these pathogens and the techniques for manipulating them are both quite recent.
4. Studies are needed on the complex ecology of these agents in areas of natural spread.

Acknowledgments. Our thanks to Drs. E.C. Calavan, Norman W. Frazier, Kevin J. Hackett, and R.F. Whitcomb for their reading of the manuscript. The technical assistance of Vicki Butler, Sherri Sandberg, and Alan Yen is appreciated.

References

1. Alivizatos, A.S., and Markham, P.G., 1986, Acquisition and transmission of corn stunt spiroplasma by its leafhopper vector *Dalbulus maidis, Ann. Appl. Biol.* **108:**535–544.
2. Alivizatos, A.S., and Markham, P.G., 1986, Multiplication of corn stunt spiroplasma in *Dalbulus maidis* and transmission *in vitro,* following injection, *Ann. Appl. Biol.* **108:**545–554.
3. Allen, R.M., and Donndelinger, C.R., 1982, Cultivation in vitro of spiroplasmas from six plant hosts and two leafhopper vectors in Arizona, *Plant Dis.* **66:**669–672.
4. Altstatt, G.E., 1945, A new corn disease in the Rio Grande Valley, *Plant Dis. Rept.* **29:**533–534.
5. Ancalmo, O., and Davis, W.C., 1961, Achaparramiento (corn stunt), *Plant Dis. Rept.* **45:**281.
6. Barnes, P., 1954, Biologia, ecologia y distribución de las chicharritas. *Dalbulus elimatus* (Ball) y *Dalbulus maidis* (De. & W.) *Fol. Tech.* II Sec. Agric. Ganad. of Estud. Espec., Mexico, DF, 112 p.
7. Bascope, B., 1977, Agente causal de la llamada "raza mesa central" del achaparramiento del maize. Maestro en Ciencias Thesis, Escuela Nacional de Agricultura, Chapingo, Mexico. 55 pp.
8. Bennett, C.W., 1971, The curly top disease of sugar beet and other plants. Monograph #7. The American Phytopathological Society, St. Paul, MN, 81 p.
9. Bové, J.M., 1981, Mycoplasma infections of plants. *Israel J. Med. Sci.* **17:**572–585.
10. Bové, J.M., 1984, Wall-less prokaryotes of plants. *Ann. Rev. Phytopathol.* **22:**361–396.
11. Bové, J.M., Carle, P., Garnier, M., Laigret, F., Renaudin, J., and Saillard, C., 1989, Molecular and cellular biology of spiroplasmas, In Whitcomb, R.F., and Tully, J.G. (eds):The Mycoplasmas, Volume V, Academic Press, San Diego, pp. 244–364.
12. Bové, J.M., and Saillard, C., 1979, Cell biology of Spiroplasmas, In Whitcomb, R.F., and Tully, J.G. (eds): The Mycoplasmas, Volume III, Academic Press, New York, pp. 83–153.
13. Bové, J.M., Mouches, C., Carle-Junca, P., Degorce-Dumas, J.R., Tully, J.G., and Whitcomb, R.F., 1983, Spiroplasmas in group I: The *Spiroplasma citri* cluster, Yale J. Biol. Med. **56:**573–582.

14. Boyd, F.J., and Pitre, H.N., 1968, Studies on the field biology of *Graminella nigrifrons,* a vector of corn stunt in Mississippi, *Ann Entomol. Soc. Am.* **61:**1423–1427.
15. Bradfute, O.E., and Robertson, D.C., 1977, Electron microscopy as a means for discovery of new maize viruses and virus-like pathogens, In Williams, L.E., Gordon, D.T., and Nault, L. (eds): *Proc. Int. Maize Virus Dis. Colloq. and Workshop,* Ohio State University, Ohio Agric. Res. Dev. Ctr, Wooster, pp. 103–107.
16. Bradfute, O.E., Tsai, J.H., and Gordon, D.T., 1981, Corn stunt spiroplasma and viruses associated with a maize disease epidemic in southern Florida. *Plant Dis.* **65:**837–841.
17. Brun, P., Riolacci, S., Vogel, R., Fos, A. Vignault, J.C., Lallemand, J., and Bové, J.M., 1988, Epidemiology of *Spiroplasma citri* in Corsica, in Timmer, L.W., Gathseg, S.M., and Navarro, L. (eds): *Proc. of the 10th Conf. I.O.C.V.,* Nov. 1986, pp. 300–303.
18. Bushing, R.W., and Burton, V.E., 1974, Leafhopper damage to silage corn in California, *J. Econ. Entomol.* **67:**656–658.
19. Calavan, E.C., and Bové, J.M., 1989, Ecology of *Spiroplasma citri,* in Whitcomb, R.F., and Tully, J.G. (eds): The Mycoplasmas, Volume V, Academic Press, San Diego, pp. 425–485.
20. Calavan, E.C., and Christiansen, D.W., 1966, Effects of stubborn disease on various varieties of citrus trees, *Israel J. Bot.* **15:**121–132.
21. Calavan, E.C., and Oldfield, G.N., 1979, Symptomatology of spiroplasmal plant diseases, in Whitcomb, R.F., and Tully, J.G. eds: The Mycoplasmas, Volume III. Academic Press, New York, pp. 37–64.
22. Carpenter, J.B., 1959, Present status of some investigations on stubborn disease of citrus in the United States, Wallace, J.M. (ed): Citrus Virus Diseases, University of California Press, Division of Agricultural Sciences, Berkeley, pp. 101–107.
23. Castillo, J., 1983, Present knowledge of virus and mollicute diseases of maize in Peru, in Gordon D.T., Knoke, J.K., Nault, L.R., Ritter, R.M. (eds): *Proc. Int. Maize Virus Dis. Colloq. and Workshop,* Ohio State University, Ohio Agric. Res. Dev. Ctr, Wooster, pp. 87–92.
24. Chapot, H., 1959, First studies on the stubborn disease of citrus in some Mediterranean countries, in Wallace, J.M. (ed): Citrus Virus Diseases, University of California Press, Berkeley, pp. 109–117.
25. Chapot, H., Celucchi, V.L., 1964, Maladies, troubles et ravageurs des agrumes au Maroc, Institut National de la Recherche Agronomique, Rabat, p. 359.
26. Chen, T.A., Lei, J.D., and Lin, C.P., 1989, Detection and identification of plant and insect mollicutes, in Whitcomb, R.F., and Tally, J.G. (eds): *The Mycoplasmas,* Volume V, Academic Press, San Diego, pp. 393–424.
27. Clark, T.B., 1984, Diversity of spiroplasma host–parasite relationships, *Israel J. Med. Sci.* **20:**995–997.
28. Clark, T.B., Henegar, R.B., Rosen, L., Hackett, K.J., Whitcomb, R.F., Lowry, J.E., Saillard, C., Bové, J.M., Tully, J.G., and Williamson, D.L., 1987, New spiroplasmas from insects and flowers: isolation, ecology and host associations, *Israel J. Med. Sci.* **23:**687–690.
29. Daniels, M.J., Archer, D.B., and Stemmer, W.P.C., 1982, Interaction of

wall-free prokaryotes with plants, In Rhodes-Roberts, M.E. (ed): Bacteria and Plants, Academic Press, London, pp. 85–99.

30. Daniels, M.J., Markham, P.G., Meddins, B.M., Plaskitt, A.K., Townsend, R., and Bar-Joseph, M., 1973, Axenic culture of a plant pathogenic spiroplasma, *Nature* **244:**523–524.
31. Davis, R.E., 1973, Occurrence of a spiroplasma in corn stunt-infected plants in Mexico, *Plant Dis. Rept.* **57:**333–337.
32. Davis, R.E., and Lee, I.M., 1982, Pathogenecity of spiroplasmas, mycoplasma-like organisms and vascular-limited fastidious walled bacteria, in Mount, M., Lacy, G. (eds): *Phytopathogenic Prokaryotes,* Volume I, Academic Press, New York, pp. 491–513.
33. DeLong, D.M., 1971, The bionomics of leafhoppers, *Ann. Rev. Entomol.* **16:**179–210.
34. DeLong, D.M., Severin, H.H.P., 1947, Characters, distribution and food plants of newly discovered vectors of California aster yellows virus, *Hilgardia* **17:**527–538.
35. Doi, Y., Teranaka, M., Yora, K., and Asuyama, H., 1967, Mycoplasma or PLP group-like microorganisms found in the phloem elements of plants infected with mulberry dwarf, potato witches' broom, aster yellows, or Paulownia witches' broom, *Ann. Phytopathol. Soc. Japan* **33:**259–266.
36. Eastman, C.E., Schultz, G.A., Fletcher, J., Hemmati, K., and Oldfield, G.N., 1984, Virescence of horseradish in Illinois, *Plant Dis.* **68:**968–971.
37. Eden-Green, S.J., 1982, Detection of corn stunt spiroplasma in vivo by ELISA using antisera to extracts from infected corn plants (*Zea mays*), *Plant Pathol.* **31:**289–297.
38. El Bolok, M.M., 1981, Specific and nonspecific transmission of spiroplasmas and mycoplasma-like organisms by leafhoppers (Cicadellidae, Homoptera) with implication for etiology of aster yellows disease, Ph.D. Thesis, Cairo University, Giza, Egypt, pp. 1–125.
39. Fawcett, H.S., 1946, Stubborn disease of citrus, a virosis. *Phytopathology* **36:**675–677.
40. Fawcett, H.S., Perry, J.C., and Johnston, J.C., 1944, The stubborn disease of citrus. *California Citrograph* **29:**146–147.
41. Fletcher, J., 1983, Brittle root of horseradish in Illinois and the distribution of *Spiroplasma citri* in the United States, *Phytopathology* **73:**354–357.
42. Fletcher, J., Schultz, G.A., Davis, R.E., Eastman, C.E., and Goodman, R.M., 1981, Brittle root disease of horseradish: Evidence for an etiological role of *Spiroplasma citri, Phytopathology* **71:**1073–1080.
43. Fos, A., Bové, J.M., Lallemand, J., Saillard, C., Vignault, J.C., Ali, Y., Brun, P., and Vogel, R., 1986, The leafhopper *Neoaliturus haematoceps* is a vector of *Spiroplasma citri* in the Mediterranean area, *Ann. Microbiol.* (*Paris*) **137A:**97–107.
44. Fox, G.E., Stackebrandt, E., Hespell, R.B., Gibson, J., Maniloff, J., Dyer, T.A., Wolfe, R.S., Balch, W.E., Tanner, R.S., Magrum, L.J., Zablen, L.B., Blakemore, R., Gupta, R., Bonen, L., Lewis, B.J., Stahl, D.A., Leuhrsen, L.R., Chen, K.N., and Woese, C.R., 1980, the phylogeny of prokaryotes *Science* **209:**457–463.
45. Frazier, N.W., Freetag, J.H., and Gold, A.H., 1965, Corn naturally

infected by sugarcane mosaic virus in California, *Plant Dis. Rept.* **49:**204–206.

46. Fudl-Allah, A.E.-S.A., Calavan, E.C., and Igwegbe, E.C.K., 1972, Culture of a mycoplasma-like organism associated with stubborn disease of citrus, *Phytopathology* **62:**729–731.
47. Gámez, R., 1969, A new leafhopper borne virus in corn in Central America, *Plant Dis. Rept.* **53:**929–932.
48. Gámez, R., 1973, Transmission of rayado fino virus of maize (*Zea Mays*) by *Dalbulus maidus, Ann. Appl. Biol.* **73:**285–292.
49. Gámez, R., 1977, Leafhopper-transmitted maize rayado Fino virus in Central America In Williams, L.E., Gordon, D.T., and Nault, L.R. (eds): *Proc. Int. Maize Virus Dis. Colloq. and Workshop, Ohio State University, Ohio Agric. Res. Dev. Ctr, Wooster, pp. 15–19.*
50. Gámez, R., and Leon, P., 1983, Maize Rayado Fino virus: Evolution with plant host and insect vector, in Harris, K.F. (ed): Current Topics in Vector Research Volume 1, Praeger, New York, pp. 149–168.
51. Gámez, R., Kitajima, E.W., and Lin, M.T., 1979, The geographical distribution of maize Rayado Fino virus, *Plant Dis. Rept.* **63:**830–833.
52. Gold, R.E., 1979, Leafhopper vectors and western X disease, In Maramorosch, K., Harris, K.F. (eds): Leafhopper Vectors and Plant Disease Agents, Academic Press, New York, pp. 587–602.
53. Golino, D.A., Oldfield, G.N., and Gumpf, D.J., 1988, Induction of flowering through infection by beet leafhopper transmitted virescence agent, *Phytopathology* **78:**285–288.
54. Golino, D.A., Oldfield, G.N., and Gumpf, D.J., 1987, Transmission characteristics of the beet leafhopper transmitted virescence agent (BLTVA), *Phytopathology* **77:**954–957.
55. Gordon, D.T., and Nault, L.R., 1977, Involvement of maize chlorotic dwarf virus and other agents in stunting diseases of *Zea mays* in the United States, *Phytopathology* **67:**27–36.
56. Gordon, D.T., Nault, L.R., Gordon, N.H., and Heady, S.E., 1985, Serological detection of corn stunt spiroplasma and maize Rayado Fino virus in field-collected *Dabulus* spp. from Mexico, *Plant Dis.* **69:**108–111.

56a. Granados, R.R., Granados, J.S., Maramorosch, K., and Reinitz, J., 1968, Corn stunt virus: Transmission by three cicadellid vectors, *J. Econ. Entomol.* **61:**1281–87.

56b. Granados, R.R., Gustin, R.D., Maramorosch, K., and Stoner, W.N. 1968, Transmission of corn stunt virus by the leafhopper *Deltocephalus sonorous* Ball, *Contrib. Boyce Thompson Inst.* **24:**57–59.

57. Granados, R.R., 1969, Chemotherapy of the corn stunt disease, *Phytopathology* **59:**1556 (Abstr.).
58. Granados, R.R., 1969, Electron microscopy of plants and insect vectors infected with the corn stunt disease agent, *Contrib. Boyce Thompson Inst.* **24:**173–188.
59. Gumpf, D.J., and Calavan, E.C., 1981, Stubborn disease of citrus, In Maramorosch, K., and Raychaudhuri, S.P. (eds): Mycoplasma Diseases of Trees and Shrubs, Academic Press, New York, pp. 97–134.
60. Gumpf, D.J., 1988, Stubborn disease caused by *Spiroplasma citri,* In Maramorasch, K., and Raychaudhuri, S.P. (eds): Mycoplasma Diseases of

Crops: Basic and Applied Aspects, Springer-Verlag, New York, pp. 327–342.

61. Hackett, K.J. and Clark, T.B., 1989, The ecology of spiroplasmas, in Whitcomb, R.F., and Tully, J.G. (eds): *The Mycoplasmas,* Volume V, Academic Press, San Diego, pp. 113–200.
62. Hamilton, K.G.A., 1983, Introduced and native leafhoppers common to the Old and new Worlds (Rhyncota: Homoptera: Cicadellidae), *Can. Entomol.* **115:**473–511.
63. ICSB Subcommittee on Taxonomy of *Mollicutes,* 1982, Minutes of the meeting, 27 and 30 August and 5 September 1978, Freiburg and Munich, F.R.G., *Int. J. Syst. Bacteriol.* **32:**261–264.
64. ICSB Subcommittee on Taxonomy of *Mollicutes,* 1984, Minutes of the interim meetings, 2 and 5 September 1980, Custer, South Dakota. *Int. J. Syst. Bacteriol.* **34:**358–360.
65. ICSB Subcommittee on Taxonomy of *Mollicutes,* 1984, *Int. J. Syst. Bacteriol.* **34:** 361-365
66. Igwegbe, E.C.K., and Calavan, E.C. 1970, Occurrence of mycoplasma-like bodies in phloem of stubborn-infected citrus seedlings, *Phytopathology* **60:**1525–1526.
67. Ishiie, T., Doi, Y., Yora, K., and Asuyama, H., 1967, Suppressive effects of antibiotics of tetracycline group on symptom development of mulberry dwarf disease *Ann. Phytopathol. Soc. Japan* **33:**267–275.
68. Kaloostian, G.H., and Pierce H.D., 1972, Notes on *Scaphytopius nitridus* on citrus in California, *J. Econ. Entomol.* **65:**880.
69. Kaloostian, G.H., Oldfield, G.N., Pierce, H.D., and Calavan, E.C., 1979, *Spiroplasma citri* and its transmission to citrus and other plants by leafhoppers, in Maramorosch, K., and Harris, K. (eds): Leafhopper Vectors of Plant Disease Agents, Academic Press, New York, pp. 447–450.
70. Kaloostian, G.H., Oldfield, G.N., Pierce, H.D., Calavan, E.C., Granett, A.L., Rana, G.L., and Gumpf, D.J., 1975, Leafhopper—natural vector of citrus stubborn disease? *Calif. Agric.* **29:**14–15.
71. Kirkpatrick, B., 1989, Strategies for characterizing plant pathogenic mycoplasma-like organisms and their effects on plants in Kosuze, T. and Wester, E.W. (eds): *Plant—Microbe Interactions,* Volume 3, Macmillan, New York, pp. 241–293.
72. Klein, M., Rasooly, R., and Raccah, B., 1988, Transmission of *Spiroplasma citri* the agent of citrus stubborn by a leafhopper of the *Circulifer tenellus* complex in Israel, *Int. Citrus Cong. Middle East.* Tel Aviv. p. 49.
73. Kloepper, J.W., Garrot, D.G., and Oldfield, G.N., 1982, Quantification of plant pathogenic spiroplasmas from infected plants, *Phytopathology* **72:**577–581.
74. Kramer, J.P., 1967, A taxonomic study of *Graminella nigrifrons,* a vector of corn stunt disease, and its congeners in the United States (Homoptera: Cicadellindae: Deltacephalinae), *Ann. Entomol. Soc. Am.* **60:**604–616.

74a. Kramer, J.P., and Whitcomb, R.F., 1968, A new species of *Baldulus* from gamagrass in Eastern United States with its possible implications in the corn stunt virus problem (Homoptera: Cicadellidae: Deltocephalinae), *Proc. Entomol. Soc. Wash.* **70:**88–92.

75. Kunkel, L.O., 1926, Studies on aster yellows, *Am. J. Bot.* **13:**646–705.

76. Kunkel, L.O., 1946, Leafhopper transmission of corn stunt, *Proc. Natl. Acad. Sci. USA* **32:**246–47.
77. Kunkel, L.O., 1948, Studies on a new corn virus disease, *Arch. Gesamte. Virusforsch.* **4:**24–46.
78. Laflèche, D., and Bové, J.M., 1970, Mycoplasmes dans les agrumes atteints de greening, de stubborn ou de maladies similaires, *Fruits* **25:**455–465.
79. Lee, I.M., Cartia, G., Calavan, E.C., and Kaloostian, G.H., 1973, Citrus stubborn disease organism cultured from beet leafhopper, *Calif. Agric.* **27:**14–15.
80. Liu, H.-Y., Gumpf, D.J., Oldfield, G.N., and Calavan, E.C., 1983, Transmission of *Spiroplasma citri* by *Circulifer tenellus, Phytopathology* **73:**582–585.
80a. Liv, H.-Y., Gumpf, D.J., Oldfield, G.N. and Calonan, E.C., 1983, The Relationship of *Spiroplasma citri* and *Circulifer tenellus*. Phytopathology **73:**585–590.
81. Madden, L.V., and Nault, L.R., 1983, Differential pathogenicity of corn stunting mollicutes to leafhopper vectors in *Dalbulus* and *Baldulus* species, *Phytopathology* **73:**1608–1614.
82. Maramorosch, K., 1955, The occurrence of two distinct types of corn stunt in Mexico, *Plant Dis. Rept.* **39:**896–898.
83. Markham, P.G., and Alivizatos, A.S., 1983, The transmission of corn stunt spiroplasma by natural and experimental vectors, in Gordon, D.T., Knoke, J.K., Nault, L.R., and Ritter, R.M. (eds): *Proc. Int. Maize Virus Dis. Colloq. Workshop 2–6 Aug. 1982,* Ohio State University, Ohio Agric. Res. Dev. Ctr., Wooster, pp. 56–61.
84. Markham, P.G., and Oldfield, G.N., 1983, Transmission techniques with vectors of plant and insect mycoplasmas and spiroplasmas, in Tully, J.G., and Razin, S. (eds): *Methods in Mycoplasmology,* II. Diagnostic Mycoplasmology, Academic Press, New York, pp. 261–267.
85. Markham, P.G., and Townsend, R., 1974, Transmission of *Spiroplasma citri* to plants, *Les colloques de l'Institut National de la Santé et de la Recherche Médicale* **33:**201–206.
86. Markham, P.G., and Townsend R., 1979, Experimental vectors of spiroplasmas, in Maramorosch, K., and Harris, K.F. (eds): Leafhopper-Vectors and Plant Disease Agents, Academic Press, New York, pp. 413–445.
87. Markham, P.G., Townsend, R., Bar-Joseph, M., Daniels, M.J., Plaskitt, A., and Meddins, B.M., 1974, Spiroplasmas are the causal agents of citrus little-leaf disease, *Ann. Appl. Biol.* **78:**49–57.
88. Markham, P.G., Townsend, R., Plaskitt, K., and Saglio, P., 1977, Transmission of corn stunt to dicotyledonous plants, *Plant Dis. Rept.* **61:**342–345.
89. Martin, H.H., Schilf, W., and Schiefer, H.G., 1980, Differentiation of Mycoplasmatales from bacterial protoplast L-forms by assay for penicillin binding proteins, *Arch. Microbiol.* **127:**297–299.
90. Martinez-Lopez, G., 1977, New maize virus disease in Colombia, in Williams, L.E., Gordon, D.T., Nault, L.R. (eds): *Proc. Int. Maize Virus Dis. Colloq. Workshop,* Ohio State University, Ohio Agric. Res. Dev. Ctr, Wooster, pp. 20–29.
91. McCoy, R.E., Caudwell, A., Chang, C.J., Chen, T.A., Chen, T.Y., Chiykowski, M.T., Cousin, M.T., Dale, J.L., DeLeew, G.T.N., Golino,

D.A., Hackett, K.J., Kirkpatrick, B.C., Marwitz, R., Petzold, H., Sinha, R.C., Sugiuva, M., Whitcomb, R.F., Yang, I.L., Zhu, B.M., and Seemiller, E., 1989, Plant Diseases associated with Mycoplasmas, in Whitcomb, R.F., and Tully, J.G. (eds): *The Mycoplasmas,* Volume V, Academic Press, San Diego, pp. 546–630.

92. Nast, J., 1972, Palearctic Auchennorhyncha (Homoptera), an annotated check list, Polish Academy of Sciences, Institute of Zoology, pp. 329–331. Polish Scientific Publishers, Warszawa, 550 pp.
93. Nault, L.R., 1980, Maize bushy stunt and corn stunt: A comparison of disease symptoms, pathogen host ranges and vectors, *Phytopathology* **70:**659–62.
94. Nault, L.R., 1985, Evolutionary relationships between maize leafhoppers and their plant hosts, in Nault, L.R., and Rodriquez, J.G. (eds): The Leafhoppers and Planthoppers, Wiley-Interscience, New York, pp. 309–330.
95. Nault, L.R., 1983, Origins of leafhopper vectors of maize pathogens in Mesoamerica, in Gordon, D.T., Knoke, J.K., Nault, L.R., and Ritter, R.M. (eds): Proc. Int. Maize Virus Dis. Colloq. Workshop 2–6 Aug. 1982, Ohio State University, Ohio Agric. Res. Dev. Ctr, Wooster, pp. 75–82.
96. Nault, L.R., and Bradfute, O.E., 1979, Corn stunt: Involvement of a complex of leafhopper-borne pathogens, in Maramorosch, K., and Harris, K. (eds): Leafhopper Vectors and Plant Disease Agents, Academic Press, New York, pp. 561–586.
97. Nault, L.R., and Delong, D.M. (eds): 1980, Evidence for co-evolution of leafhoppers in the genus *Dalbulus* (Cicadellidae: Homoptera) with maize and its ancestors, *Ann. Entomol. Soc. Am.* **73:**349–353.
98. Nault, L.R., and Knoke, J.K., 1981, Maize vectors. *So. Coop. Ser. Bull.* **247:**77–84.
99. Nault, L.R., Gordon, D.T., Robertson, D.C., and Bradfute, O.E., 1976, Host range of maize chlorotic dwarf virus, *Plant Dis. Rept.* **60:**374–377.
100. Nault, L.R., Madden, L.V., Styer, W.E., Triplehorn, B.W., Shambaugh, G.F., and Heady, S.E., 1984, Pathogenecity of corn stunt spiroplasma and bushy stunt mycoplasma to their vector, *Dalbulus longulus, Phytopathology* **74:**977–979.
101. Nault, L.R., Styler, W.E., Knoke, J.K., and Pitre, H.N., 1973, Semipersistent transmission of leafhopper-borne maize chlorotic dwarf virus, *J. Econ. Entomol.* **66:**1271–1273.
102. Nielsen M.W., 1968, The leafhopper vectors of phytopathogenic viruses (Homoptera, Cicadellidae) taxonomy, biology, and virus transmission, *Tech. Bull. U.S. Dept. Agri.* No. 1382.
103. Nielsen, M.W., 1979, Taxonomic relationships of leafhopper vectors of plant pathogens, in Maramorosch, K., and Harris, K.F. (eds): Leafhopper Vectors and Plant Disease Agents, Academic Press, New York, pp. 3–27.

103a. Nielson, M.W., 1985, Leafhopper systematics, in Nault, L.R., and Rodriguez, J.G. (eds): The Leafhoppers and Planthoppers, Volume 2, Wiley, New York, pp. 11–39.

104. Nielson, M.W., and Morgan, L.A., 1982, Developmental biology of the leafhopper, *Scaphytopius nitridus* (Homoptera: Cicadellidae) with notes on distribution, hosts and interspecific breeding, *Ann Entomol. Soc. Am.* **75:**350–352.

105. O'Hayer, K.W., Schultz, G.A., Eastman, C.E., Fletcher, J., 1984, Newly discovered plant hosts of *Spiroplasma citri, Plant Dis.* **68:**336–338.
106. O'Hayer, K.W., Schultz, G.A., Eastman, C.E., Fletcher. J., and Goodman, R.M., 1983, Transmission of *Spiroplasma citri* by the aster leafhopper *Macrosteles fascifrons* (Homoptera: Cicadellidae) *Ann Appl. Biol.* **102:**311–318.
107. Oldfield, G.N., 1987, Leafhopper vectors of citrus stubborn disease spiroplasma, in Wilson, M.R. and Nault, L.R. (eds), *Proc. of 2nd Int. Workshop on Leafhopper and Planthoppers of Econ. Importance,* CIE, London, pp. 151–159.
108. Oldfield, G.N., 1987, Ecological associations of *Spiroplasma citri* with insects, plants, and other plant mycoplasmas in the western United States, in Maramorosch, K., and Raychandhuri, S.P. (eds): Mycoplasma Disease of Crops: Basic and Applied Aspects, Springer-Verlag, New York, pp. 175–191.
109. Oldfield, G.N., and Calavan E.C., 1981, *Spiroplasma citri* non-rutaceous host, in Bové, J., and Vogel, R. (eds): Description and Illustration of Virus and Virus-like Diseases of Citrus: A Collection of Color Slides, 2nd ed., SETCO-IRFA, Paris, pp. 1–8.
110. Oldfield, G.N., Sullivan, D.A., and Calavan, E.C., 1984, Inoculativity of leafhopper vectors of stubborn disease in California, *Proc. 9th Conf. Int. Org. of Cit. Virol.,* pp. 125–130.
111. Oldfield, G.N., Sullivan, D.A., and Calavan, E.C., 1984, Aspects of transmission of the citrus stubborn agent by *Scaphytopius nitridus* (DeLong) *Proc. 9th Conf. Int. Org. of Cit. Virol.,* pp. 131–136.
112. Oldfield, G.N., Kaloostian, G.H., Pierce, H.D., Calavan, E.C., Granett, A.L., and Blue, R.L., 1976, Beet leafhopper transmits citrus stubborn disease, *Calif. Agric.* **30:**15.
113. Oldfield, G.N., Kaloostian, G.H., Pierce, H.D., Sullivan, D.A., Calavan, E.C., and Blue, R.L., 1977, New hosts of citrus stubborn disease, *Citrograph* **62:**309–12.
114. Oldfield, G.N., Kaloostian, G.H., Sullivan, D.A., Calavan, E.C., and Blue, R.L., 1978, Transmission of the citrus stubborn disease pathogen, *Spiroplasma citri,* to monocotyledonous plant, *Plant Dis. Rept.* **62:**758–760.
115. Oman, P.W., 1970, Taxonomy and nomenclature of the beet leafhopper, *Circulifer tenellus* (Homoptera: Cicadellidae), *Ann. Entomol. Soc. Am.* **63:**507–512.
116. Oman, P.W., 1948, Notes on the beet leafhopper, *Circulifer tenellus* (Baker), and its relatives (Homoptera: Cicadellidae), *J. Kansas Entomol. Soc.* **21:**10–14.
117. Pirone, T.P., 1972, Virus-like particles associated with a leafhopper transmitted disease of corn in Kentucky, *Plant Dis. Rpt.* **56:**652–656.
118. Pickens, E., Gerloff, R.K., and Burgdorfer, W., 1968, Spirochete from the rabbit tick, *Haemaphyslis leporispalustris* (Packard), *J. Bacteriol.* **95:**291–299.
119. Pitre, H.N., Combs, R.L., and Douglas, W.A., 1966, Gamagrass, *Tripsacum dactyloides:* A new host of *Dalbulus maidis,* vector of corn stunt virus, *Plant Dis. Rept.* **50:**570–571.

120. Poulson, D.F. and Sakaguchi, B., 1961, Nature of "sex-ratio" agent in *Drosophila, Science* **133:**1489–1490.
121. Power, A.G., 1987, Plant community diversity, Herbivore movement, and an insect-transmitted disease of maize, *Ecology* **69:**1658–1669.
122. Purcell, A.H., 1982, Evolution of the insect vector relationship, in Mount, M.S., and Lacy, G.H. (eds): *Pathogenic Procaryotes,* Volume 1, Academic Press, New York, pp. 121–156.
123. Purcell, A.H., 1982, Insect–vector relationships with procaryotic plant pathogens, *Ann. Rev. Phytopathol.* **20:**397–417.
124. Raju, B.C., Nyland, G., Backus, E.A., and McLean, D.L., 1981, Association of the spiroplasma with brittle root of horseradish, *Phytopathology* **71:**1067–1072.
124a. Raju, B.C., Purcell, A.H., and Nyland, G. 1984, Spiroplasmas from Plants with Aster Yellows Disease and X-Disease: Isolation and transmission by Leafhoppers. *Phytopathology* **74:**925–931.
125. Rana, G.L., Kaloostian, G.H., Oldfield, G.N., Granett, A.L., Calavan, E.C., Pierce, H.D., Lee, I.-M., and Gumpf, D.J., 1975, Acquisition of *Spiroplasma citri* through membranes by homopterous insects, *Phytopathology* **65:**1143–1145.
126. Razin, S., 1989, Molecular approach to mycoplasma phylogeny, in Whitcomb, R.F., and Tully, J.G. (eds): The Mycoplasmas, Volume V, Academic Press, San Diego, 653 p.
127. Rogers, M.J., Simmons, J., Walker, R.T., Weisburg, W.G., Woese, C.R., Tanner, R.S., Robinson, I.M., Stahl, D.A., Olsen, G., Leach, R.H., and Maniloff, J., 1985, Construction of the mycoplasma evolutionary tree from 5S rRNA sequence data, *Proc Natl. Acad. Sci. USA* **82:**1160–1164.
128. Romney, V.E., 1943, The beet leafhopper and its control on beets in Arizona and New Mexico, *U.S. Dept. Agri. Tech. Bull.* 855, 24 p.
129. Saglio, P., L'Hospital, M., Laflèche, D., Dupont, G., Bové, J.M., Tully, J.G., and Freundt, E.A., 1973, *Spiroplasma citri* gen. and sp. n.: A mycoplasma-like organism associated with "stubborn" disease of citrus, *Int. J. Syst. Bacteriol.* **23:**191–204.
130. Saglio, P., Laflèche, D., Bonissol, C., and Bové, J.M., 1971, Culture *in vitro* des mycoplasmes associes au "Stubborn" des agrumes et leur observation au microscope electronique, *C. R. Acad. Sci. (Paris)* **272:**1387–1390.
131. Saillard, C., Vignault, J.C., Bové, J.M., Raie, A., Tully, J.G., Williamson, D.L., Fos, A., Garnier, M., Gadeau, A., Carle, P., and Whitcomb, R.F., 1987, *Spiroplasma phoeniceum* sp. n.: A new plant pathogenic species from Syria, *Int. J. Syst. Bacteriol.* **37:**106–115.
132. Saillard, C., Vignault, J.C., Gadeau, A., Carle, P., Garnier, M., Fos, A., Bové, J.M., Tully, J.G., Whitcomb, R.F., 1984, Discovery of a new plant-pathogenic spiroplasma, *Israel J. Med. Sci.* **20:**1013–1015.
133. Spaar, D., Kleinhempel, H., Müller, H.M., Stanarius, A., and Schimmel, D., 1974, Culturing mycoplasmas from plants, *Coll. Syst. Nat. Sante Rech. Med.* **33:**207–213.
134. Stalheim, O.H.V., Ritchie, A.E., and Whitcomb, R.F., 1978, Cultivation,

serology, ultrastructure, and virus-like particles of spiroplasma 277F, *Curr. Microbiol.* **1:**365–370.
135. Stoner, W.N., and Gustin, R.D., 1967, Biology of *Graminella nigrifrons* (Homoptera: Cicadellidae), a vector of corn (maize) stunt virus, *Ann. Entomol. Soc. Am.* **60:**496–505.
136. Sullivan, D.A., Oldfield, G.N., Eastman, C., Fletcher, J., and Gumpf, D.J., 1987, The transmission of a citrus-infecting strain of *Spiroplasma citri* to horseradish, *Plant Dis.* **71:**469.
137. Swenson, K.G., 1971, Environmental biology of the leafhopper, *Scaphytopius delongi, Ann. Entomol. Soc. Am.* **64:**809–812.
138. Tsai, J.H., 1988, Bionomics of *Dalbulus maidis* (Delong and Wolcott), a vector of mollicutes and virus (Homoptera: Cicadellidae), in Maramorosch, K., and Raychudhuri, S.P. (eds): Mycoplasmalike Disease of Crops: Basic and Applied Aspects, Academic Press, New York, pp. 209–221.
139. Tully, J.G., Rose, D.L., Clark, E.A., Carle, P., Bové, J.M., Henegar, R.B., Whitcomb, R.F., Colflesh, D.E., and Wiliamson, D.L., 1987, Revised group classification of the genus *Spiroplasma* (class Mollicutes) with proposed new Groups XII to XXIII, *Int. J. Syst. Bacteriol.* **37:**357–364.
140. Tully, J.G., 1989, Class Mollicutes: New perspectives from plant and arthropod studies, in Whitcomb, R. F., and Tully, J.G. (eds): The Mycoplasmas, Volume V, Academic Press, San Diego, pp. 1–31.
141. Wallis, R.L., 1962, Spring migration of six-spotted leafhopper in western Great Plains, *J. Econ. Entomol.* **55:**871–874.
142. Whitcomb, R.F., 1989, *Spiroplasma kunkelii:* Biology and ecology, in Whitcomb, R.F., and Tully, J.G. (eds): The Mycoplasmas, Volume V, Academic Press, San Diego, pp. 488–544.
143. Whitcomb, R.F., 1988, Introduction: Mollicutes, hosts and evolution, in Hiruka, C. (ed): Tree Mycoplasmas and Mycoplasma Diseases, The University of Alberta Press, Edmonton, pp. 1–16.
144. Whitcomb, R.F., Tully, J.G., McCawley, P., and Rose, D.L., 1982, Application of growth inhibition test to *Spiroplasma* taxonomy, *Int. J. Syst. Bacteriol.* **32:**387–394.
145. Whitcomb, R.F., 1981, The biology of spiroplasmas. *Ann. Rev. Entomol.* **26:**397–425.
146. Whitcomb, R.F., 1980, The genus *Spiroplasma, Ann. Rev. Microbiol.* **34:**677–709.
147. Whitcomb, R.F., Tully, J.G., Bové, J.M., and Saglio, P., 1973, Spiroplasmas and acholeplasmas: Multiplication in insects, *Science* **182:**1251–1253.
148. Whitcomb, R.F., Chen, T.A., Williamson, D.L., Liao, C., Tully, J.G., Clark, T.B., Bové, J.M., Mouches, C., Rose, D.L., Coan, M.E., and Clark, T.B., 1986, *Spiroplasma kunkelii* sp. nov.: Characterization of the etiological agent of corn stunt disease, *Int. J. Syst. Bacteriol.* **36:**170–78.
149. Whiteside, J.O., Garnsey, S.M., and Timmer, L.W., 1988, Compendium of Citrus Diseases, The American Phytopathological Society Press, St. Paul, MN, 80 p.
150. Williamson, D.L., and Poulson, D.F., 1979, Sex ratio organisms (Spiroplasmas) of drosophila, in R.F., Whitcomb, and J.G. Tully (eds): The Mycoplasmas, Volume III, Academic Press, New York, London, pp. 175–208.
151. Williamson, D.L., Tully, J.G., and Whitcomb, R.F., 1989, The genus

Spiroplasma, in Whitcomb, R.F. and Tully, J.G. (eds): The Mycoplasmas, Volume V, Academic Press, New York, London, pp. 71–111.
152. Woese, C.R., Stackebrandt, E., Ludwig, W., Pater, B.J., Macke, T., and Hahn, C.M., 1985, What are mycoplasmas: The relationship of tempo and mode in bacterial evolution, *J. Mol. Evol.* **21**:305–316.
153. Young, D.A., Jr., 1955, Notes on the genus *Euscelidius Ribaut* in the United States, *Wash. Entomol. Soc. Proc.* **57**:78.
154. Young, D.A., Jr. Frazier, N.W., 1954, A study of the leafhopper genus *Circulifer* Zachvatkin (Homoptera: Cicadellidae), *Hilgardia* **23**:25–52.

11
Semipersistent Transmission of Viruses by Vectors with Special Emphasis on Citrus Tristeza Virus

Benjamin Raccah, Chester N. Roistacher, and Sebastiano Barbagallo

Introduction

Semipersistent viruses (SPV) have received less attention than either nonpersistent (NPV) (94, 95) or persistent viruses (PV) (129). The group of SPV includes some of the most economically important diseases in cultivated crops: viz., citrus tristeza virus (CTV) in citrus fruits, beet yellows virus (BYV) in beets, maize chlorotic dwarf virus (MCDV) in maize and rice tungro virus (RTV) in rice.

BYV transmission has been studied for almost 50 years (139, 140). The transmission characteristics were found to be similar to those of NPVs and PVs. Some of the differences between the various forms of transmission have been described in detail (128, 142), but there are still many missing links in our explanation of the specificity and dependence for this form of transmission. Therefore, an attempt is made in this chapter to update the present knowledge with details of some of the newer approaches to explain virus–vector interrelationships among SPVs.

Originally, only aphid-borne closteroviruses were included among SPVs; however, with time, several additional filamentous and spherical viruses were found to be transmitted semipersistently by aphids as well as by several other Homoptera. Special emphasis will be placed on CTV, epidemics of which have caused severe crop destruction. In the following sections, we will analyze various factors affecting the role of vectors in virus spread and evaluate the measures adopted for their suppression and control.

Citrus Tristeza Virus: A Special Case of a Devastating Disease

The tristeza disease of citrus is caused by CTV and transmitted solely by aphids. This virus is considered to cause one of the most destructive of all

---

Benjamin Raccah, Department of Virology, the Volcani Center, Bet Dagan, Israel.
Chester N. Roistacher, Department of Plant Pathology, University of California, Riveside, California 92521, USA.
Sebastiano Barbagallo, Department of Entomology, University of Catania, Via Valdisavoia 4, Catania, Italy.

known virus diseases afflicting tree crops. Its epidemiology has been reviewed recently (8, 11). CTV is detrimental to citrus in two ways:

Phloem Necrosis-Induced Bud Union Girdle. The phloem cells in the sour orange rootstock just below the rootstock are killed (this is termed *phloem necrosis*), which effectively girdles the tree in precisely the same manner as the mechanical girdle (118). This mode of destruction was demonstrated in tristeza epidemics worldwide, and over 50 million trees on sour orange rootstock have been killed in Brazil, Argentina, Venezuela, Florida, California, Spain, and Israel. The virus continues to be a threat to citrus in any area or country where sour orange is the predominant rootstock. Schneider (119, 120), has reviewed the pathological anatomy of citrus affected with CTV and has implicated the phloem as the area of virus replication and damage. It is also possible that toxins are involved in phloem necrosis, but this area of research has not been explored.

Stem Pitting. Certain CTV isolates can induce severe stem pitting beneath the bark on the scion and/or the rootstock. In severe cases, the affected branches lose their flexibility and can be readily snapped off at the new growth joints. Tree trunks are severely deformed and may show intense pitting. Leaves may be twisted and chlorotic, branches may show dieback, trees may be severely stunted or killed, and fruit size may be severely reduced (79, 104). The destructive potential of this stem-pitting form of CTV does not depend on the rootstock as does phloem necrosis form, since the virus attacks the scion or rootstock directly. Certain severe isolates of CTV have this ability to cause severe stem pitting or to reduce fruit size, although they can exist without causing symptoms in certain cultivars on tolerant rootstocks. For example, certain exceptionally severe strains of CTV exist in a symptomless form in Japanese Satsuma mandarins grafted on the highly tolerant trifoliate orange rootstock. However, if budwood or trees of such Satsumas are imported and established, the virus may be readily transmitted to susceptible hosts by *Toxoptera citricidus* (Kirk), or possibly other aphid vectors, and can ultimately decimate an industry of navel, Valencia, or other sweet oranges, regardless of rootstock (104). Similarly, severe stem-pitting CTV isolates can exist, without causing symptoms, in sweet orange or mandarin trees, but when they are transmitted by aphids to grapefruit they can induce severe stem pitting, smaller trees, and very small fruit. Certain virulent CTV isolates, when transmitted by vector, will be superimposed on the already existing CTV isolates and can be highly destructive to sweet orange scions (132). For example, in Florida and in Israel where sour orange is still a very important rootstock, newly introduced CTV isolates were highly destructive, whereas previous CTV isolates were only moderately destructive to the sour orange rootstock.

Viruses

Group of Viruses That Are Transmitted Semipersistently

Virus nomenclature is based primarily on physical and chemical properties such as structure, composition and nature of the nucleic acid, and the size of the virus proteins. Characteristics of a virus such as its dilution end-point and stability must be considered (73); biological features such as host reactions, the insect group that serves as its vector, and the mode of transmission are also used for classification.

Semipersistent transmission was mostly ascertained for RNA viruses (although the caulimoviruses, which are DNA viruses, exhibit certain semipersistent characteristics). Both spherical (36, 37, 86) and filamentous viruses have been found to be transmitted semipersistently (83).

A short description of these viruses is given herein. A more comprehensive description of the viruses can be found elsewhere (6, 9).

Closteroviruses

Beet yellows virus is the type member of the closteroviruses and the first virus for which semipersistent transmission was described (128). The group consists of very long, flexible particles ranging from ~750 to 2000 × 12 nm. The thermal inactivation point is 40 to 55°C, the longevity in sap is one to four days, and the dilution end-point ranges between 1/500 and 1/10,000. Bar-Joseph and Murant (9) tentatively grouped the viruses in three groups, based on their length (~700, 1300, and 1500 to 2000 nm). The molecular weight (MW) range varied with size (from 2.5×10^6 to 6.5×10^6). Their nucleic acid is single-stranded (ss) RNA. Murant et al. (83) pointed out that the longer closteroviruses are transmitted by aphids in a typical semipersistent manner. Some of the shorter viruses, such as heracleum latent virus (HLV), are known to require another closterovirus as a helper for vector transmission.

Much of the information on these viruses derives from two main members of this group, the BYV and the CTV. The BYV infects a wide range of plants in 15 botanical families (34). Viruses differing in strain severity were reported by Watson (141) and by Bennett (15). Sap inoculation also serves for strain differentiation. Cross-reaction was recorded among isolates that produce vein clearing, but not among the isolates that did not produce vein clearing. Strains of BYV were altered by passage through *Capsella* spp. The CTV primarily infects *Citrus* spp. and its related genera in the family Rutaceae, although its presence has been reported in a very few noncitrus hosts, i.e., *Passiflora* spp. Transmission is primarily by vector and bud propagation, although mechanical transmission is possible (40). Cross-protection is common and is used commercially to protect against severe stem pitting isolates.

Strains of tristeza were attenuated by passage of the virus through *Passiflora* spp. (107).

In addition to the typical closteroviruses, viruses that are structurally similar to them may differ by being transmitted by nonaphid vectors. These viruses will be designated in this chapter as "clostero-like viruses" (83). These include the grape leaf roll virus (GLRV) complex (25, 78) and the cucumber yellow cirus (CYV) (145).

Grape leafroll virus was found in at least three lengths, 700, 1400, and 2000 nm. It induces a syndrome of symptoms that includes stem pitting, leaf roll, and color change. It remains unclear as to what combination of virus components is necessary to produce certain symptoms in the host. Research is in progress with polyclonal and monoclonal antibodies and cDNA cloning to characterize the components of this virus complex. Several components of this complex are transmitted by mealybugs (38, 112, 133).

Yamashita et al. (145) described CYV as filamentous and 700 nm long; it was transmitted by the greenhouse whitefly *Trialeurodes vaporariorum* (West). A virus that resembles CYV was purified by Lot et al. (73) and appears to be similar to closteroviruses. The GLRV and the CYV might fit into the semipersistent mode of transmission, if all other characteristics are found to conform.

Maize Chlorotic Dwarf Virus Group

This group of Graminae viruses is transmitted semipersistently by leafhoppers. The nucleic acid is ss RNA of positive sense, with a molecular weight of $\sim 3.2 \times 10^6$. The particle molecular weight is $\sim 8.8 \times 10^6$, and its diameter is 30 nm (45). The host range is rather narrow. The MCDV is transmitted by two leafhopper species, *Graminiella nigrifrons* (Forbes) and the less efficient *Deltocephalus sonorus* (86).

It has been suggested that RTV (39) is sufficiently similar to MCDV to be classified in the same group (74). A polyhedral virus similar in size and shape to the MCDV was described for rice tungro-infected plants. However, a virus that is bacilliform in shape was also identified in the infected plant. Moreover, the bacilliform particles by themselves are sufficient to cause the disease syndrome, although the spherical particles aggravate the disease etiology (58).

Nongrouped Viruses That Are Transmitted Semipersistently

Filamentous Viruses

Cucumber Vein Yellowing Virus. This virus can be transmitted mechanically and by *Bemisia tabaci* (Genn.) in a semipersistent manner (88). Cucumber vein yellowing virus (CVYV) was purified and characterized 20 years later than CYV (122). The particle is slightly longer than CYV

(750 to 800 nm) and is reported to be the only filamentous virus that has a double stranded DNA nucleic acid (122).

Cowpea Mild Mottle Virus (CMMV). The virus was described in 1973, in leguminous plants. Its structure and size and the lack of knowledge of its potential vector, led Brunt and Kenten (21) to include it among the carlaviruses. However, Muniyappa and Reddy (30) subsequently showed it to be transmitted by *B. tabaci*. The virus is 600 to 650 × 13 nm in size. It has a wider host range than the typical carlavirus and has been shown to be transmitted efficiently through seed (2 to 90%). A serological affinity between CMMV and other carlaviruses was found by Iwaki et al. (64); however, their findings were not confirmed by other workers (1).

Spherical Viruses

Parsnip Yellow Fleck Virus (PYFV). The particle is spherical, ~30 nm in diameter, and contains a single strand of RNA of MW 3.3×10^6. The virus resembles picornaviruses in shape and size as well as in RNA and protein composition (54, 55). It is transmissible by sap and can be transmitted by the aphids *Cavariella aeqopodii* and *C. pastinacae* in a semipersistent manner. Transmission of the virus depends on the presence of the anthriscus yellows virus in the same host. The virus infects 11 plant species in the Umbelliferae family (54, 82).

Anthriscus Yellow Virus. This virus was also found in umbelliferous hosts; however, unlike PYFV, it is not sap-transmissible and is apparently found in the host's deeper tissue, since ultraviolet irradiation did not affect the virus source. The virus is transmitted in a semipersistent manner (36) and serves as a helper for the transmission of PYFV (37). This virus resembles the former with a shape and a RNA and protein composition similar to those of picornaviruses (54). In 1989, Hemida and Murant (54) reported partial purification of the virus (0.5 to 1.0 mg/kg leaves). The virus disintegrates after a few days at 5°C. Two kinds of particles were discerned, one with a buoyant density in cesium chloride of 1.54 g/cm^3 and the other with a density of 1.29 g/cm^3. The two kinds of particles contained four proteins within the MW range of 35 to 22×10^3.

Vectors

Various groups of sucking insects are known to transmit viruses semipersistently: leafhoppers (Cicadellidae), whiteflies (Aleyrodidae), and aphids (Aphididae) [homoptera] (Table 11.1). In recent reports (38, 112), The mealybugs (Pseudococcidae) were also found to transmit clostero-like viruses associated with the GLRV complex. However, the mode of transmission of these viruses has not yet been characterized.

TABLE 11.1. Vectors for several groups of semipersitent viruses.

Vectors	Viruses involved	Host range of vector species	References
Cicadellidae			
Nephotettix cincticeps	RTV	Rice and other Graminae	56,57
Graminiella nigrifrons	MCDV	Corn and other Graminae	62,86
Aleyrodidae			
Trialeurodes vaporariorum	CYV	Polyphagous, mainly Cucurbitaceae, Solanaceae, and Compositae	145
Bemisia tabaci	CMMV CYVV	Polyphagous, especially Malvaceae	64,80 88
Aphididae			
Acyrthosiphon pisum	BYV	Leguminosae	65
Aphis craccivora	CTV	Polyphagous, mainly Leguminosae	136
Aphis citricola (spiraecola)	CTV	Polyphagous, mostly Compositae	89,97
Aphis fabae	BYV	Polyphagous, mainly Leguminosae	65
Aphis gossypii	BYV CTV	Polyphagous, mainly Leguminosae and Cucurbitaceae	65 7,32 30,90
Aphis nasturtii	BYV	Polyphagous	65
Cavariella aegopodii	AYV/PYFV HLV	Umbelliferae	36,37,82 81
Cavariella pastinaceae	AYV/PYFV HLV	Umbelliferae	36,37,82 81
Cavariella theobaldii	AYV/PYFV HLV	Umbelliferae	36,37,82 81
Macrosiphum euphorbiae	BYV	Polyphagous	65
Myzus ascalonicus	BYV	Polyphagous	65
Myzus ornatus	BYV	Polyphagous	65
Myzus persicae	BYV CTV	Polyphagous	65 137
Toxoptera aurantii	CTV	Polyphagous, mainly Rutaceae	90
Toxoptera citricidus	CTV	Polyphagous, mainly Rutaceae	77
Uroleucon jaceae	CTV	Compositae, Cinareae	136
Pseudococcidae			
Planococcus ficus	GLRV[a]	Polyphagous, mainly grapes	112
Planococcus longispinus	GLRV[a]	Polphagous	133

[a] The mode of transmission for this complex has not been determined to date. The clostero-like structure of the virus suggests that it might be transmitted in a semipersistent manner.

Groups of Vectors

Each of the insect vector families mentioned above have specific morphological, biological, and ecological differences.

Cicadellidae

This family includes small hopping homoptera that live during their nymphal and adult stages on a variety of botanical families. They are found mostly on the underside of leaves and induce a typical chlorosis that is readily visible from the upper side of the infested leaf. They produce a number of generations each year and overwinter mostly as adults. Certain species are known to be monophagous and some are even oligophagous. Two species found to transmit semipersistent viruses are *Nephotettix cincticeps* (Uhl.) and *Graminiella nigrifrons* (Forbes), which are primary pests of rice and maize, respectively; occasionally, they are also found secondarily on other Graminaceae. The former was discerned in Southeast Asia and the latter in the southern United States.

Aleyrodidae

These are small sternorrhynchous homoptera that are distributed from equatorial regions to temperate regions. They are characterized by the presence of nymphal stages which are attached to the underside of leaves in a way that resembles scales. The adults resemble small flies, with bodies covered with a waxy white secretion. They include species that produce a number of generations each year and are mostly polyphagous. The nymphal stages are not involved in virus transmission and are frequently overlooked, even upon careful inspection, during the transport of vegetative material. The adult forms, which are responsible for virus transmission, are not very active fliers but can be carried long distances by the wind. The two species that are known as vectors of semipersistent viruses, *Trialeurodes vaporariorum* (West) and *Bemisia tabaci,* have a cosmopolitan distribution and are widespread pests of many crops. They are known vectors of many viruses, including those indicated in Table 11.1 (26).

Pseudococcidae

A group of a superfamily Coccoidea, mealybugs are characterized macroscopically by abundant waxy white secretions on their body; they are widespread mainly in subtropical and tropical regions. Certain species are polyphagous and are widely distributed throughout the world. They have limited active mobility and are distributed by transport of infested plant material. The first instar nymphs (crawlers) are more mobile and are often carried passively by the wind (125).

Aphididae

This is the largest family of insects involved in the transmission of semipersistent viruses. The importance of aphids in the transmission of viruses is exemplified by two major diseases: tristeza and beet yellows.

The biology of aphids is complicated. A comprehensive description is beyond the scope of this review and, for that information the reader is referred to Hille Ris Lambers (59) and Lees (70). The general scheme for aphid development is as follows: In tropical and subtropical regions, aphids mostly reproduce continuously through a succession of parthenogenetic generations during the spring and summer seasons; in the autumn, they produce one amphigonic generation and overwinter as fertilized eggs laid on the proper primary host.

Most aphid species are monophagous, or at most oligophagous. Among the few that are polyphagous, we find some of the most efficient vector species [e.g., *Myzus persicae* (Sulzer), *Aphis gossypii* (Glover), and *Aphis fabae* (Scop.)].

Two biological characteristics are believed to be the major reason for the success of aphids as vectors of numerous viruses: the short- and long-distance flights of the alates and their probing and feeding behavior on hosts and nonhosts. The importance of aphids as pests in the citrus grove was summarized by Barbagallo (2) and by Barbagallo and Patti (3).

The largest number of aphid species recorded for transmission of a semipersistent virus (>20) is for BYV (65); CTV is transmitted by seven species and Heracleum latent virus and beet yellow stunt virus by three species. There is less information on aphid transmission of other viruses. In many cases, one or two species only were reported, and those probably represent the species that were available for transmission rather than species resulting from a systematic search for potential vectors. For some viruses that are potentially semipersistent, there is no information on acquisition and inoculation times.

Biology of Vectors of Important Semipersistent Viruses

A number of different aphids are involved in the transmission of SPVs. In the following section, we will describe the biology and characteristics of the major aphid-vector species.

Aphid Vectors of Citrus Tristeza Virus

The literature mentions seven aphid species that are reported to transmit CTV. Most of these species colonize or temporarily reside on citrus, except for *Uroleucon* (*Dactynotus*) *jaceae* [L.], which infests certain *Compositae* and has been mentioned only once as a vector of CTV in India (136). The six other species are *Toxoptera aurantii, Toxoptera citricidus, Toxoptera aurantii, Aphis citricola* (*spiraecola*), *Aphis crac-*

civora (Koch), *Aphis gossypii,* and *Myzus persicae*. The vectorial efficiency of these species varies (30, 31, 89, 97, 107, 136, 137). All the above species were reported, in earlier studies, to feed on or to colonize citrus. Among the above-mentioned species, the most efficient vector of CTV is *T. citricidus*. There is a direct correlation between disease spread and the presence of aphids, wherever these aphids are prevalent (14, 27). A high vectorial capacity for spreading CTV was also attributed to the melon aphid (*A. gossypii*), where *T. citricidus* was absent. *Aphis gossypii* can assume a unique role in the transmission of certain CTV isolates in the absence of *T. citricidus* (7, 111). Another species that might contribute to CTV epidemics is the green citrus aphid *Aphis citricola* Van Der Goot. Its low transmission efficiency could be offset by its sometimes large populations found on citrus.

The role of other aphid species designated as vectors appears to be secondary and of limited importance. When ELISA was used to detect CTV in aphids that were allowed to feed on infected source plants, several additional aphid species were considered able to acquire CTV from infected citrus plant tissue, viz., *Aphis jacobaeae* (Schrank), *Aphis nerii* (Fonscolombe), *Hyalopterus pruni* (Geoff.), and *Macrosiphum rosae* (L.) (27). Of these, only *A. nerii* has been reported on citrus in Israel, and only once (18); the other species were never recorded in any direct conjunction with citrus. Among the species of aphids known to be vectors of CTV, but not of critical importance, are the green peach aphid *M. persicae* (137) and *A. craccivora* (136). Although these two species can be seen now and then in a number of groves in various parts of the world, they seem to be insignificant with respect to the spread of tristeza disease.

The Brown Citrus Aphid (Toxoptera citricidus)

As indicated above, this brown citrus aphid is the most efficient vector of CTV. The species is widespread throughout Africa, the southern Sahara, all of South America, Australia and New Zealand, India, Sri Lanka, and all countries of the Far East. It has not been recorded in the Mediterranean region and in the Near East, but citrus specialists fear its accidental introduction into regions where citrus is grown. The Sahara desert seem to have served as a natural barrier to the spread of the aphid (126). A similar, though partial barrier exists in Central America. This explains the efforts made by plant quarantine agencies to confine it or otherwise to reduce the possibility of its casual or accidental introduction.

The aphid maintains a continuous parthenogenetic development, and each generation, under favorable conditions, lives for less than one week. The host range is more restricted than that of *T. aurantii*. The various species of the genus *Citrus* are the preferred hosts for this species. However, it can also be found on other plants in the Rutaceae and even on plants of other botanical families, such as the genera *Passiflora, Calodendron, Mangifera,* and *Anacardium,* especially during seasons when the aphid population is high.

The Black Citrus Aphid (Toxoptera aurantii)

This aphid has a wider geographic distribution than *T. citricidus,* which extends from the subtropical to temperate regions, between latitudes 45° north and south. It is also seen in cooler regions inside heated greenhouses. Its biology resembles that of *T. citricidus,* with a developmental cycle of typical anholocyclic parthenogenesis. The duration of development, under optimal conditions (22 to 25°C and 60 to 70% RH), less than one week. This aphid is polyphagous but mostly reproduces on *Citrus* species. Aphid colonies can be seen on numerous plants in various botanical families, such as the genera *Coffea, Theobroma, Cola, Camellia, Rhamnus, Viburnum,* and *Ficus. Toxoptera aurantii* has been recorded once as a vector of CTV (90).

*The Green Citrus Aphid (*Aphis citricola *van der Goot, spiraecola Patch)*

This nearctic species has spread during the last 40 years to the Mediterranean region, Australia, and Africa. It undergoes anholocyclic development, which includes a series of parthenogenetic generations in tropical and subtropical climates. Limited holocyclic development has been noted in temperate or cold regions. The green citrus aphid is a polyphagous species, which infests a variety of plants, mainly Compositae, Rutacee, and Rosaceae families. Among citrus, oranges, mandarins, and grapefruit are affected, with limited infestation of lemons. This species was recorded as a vector of CTV in several countries (30, 89, 96, 146).

The Melon (Cotton) Aphid (Aphis gossypii)

This species is of cosmopolitan distribution, although it is more common in tropical and subtropical regions. The aphid is polyphagous, infesting hundred of plants from many dicotyledonous families and serving as a vector for more than 50 different plant viruses (65). The species undergoes anholocyclic development, mainly but in rare cases and under certain conditions, holocycle development has occurred. The regions infested by *A. gossypii* resemble those reported for the green citrus aphid. Transmission of CTV by *A. gossypii* is of great vectorial importance in countries that do not harbor *T. citricidus* (31, 90, 96, 105, 108). Changes in the transmissibility of CTV isolates by the melon aphid (7, 111) are the reason for increased interest in this species in Europe and the United States. *Aphis gossypii* is one of the species that has been reported to transmit the BYV (65).

APHID VECTORS OF BYV

Twenty-two aphid species have been reported to transmit BYV (65). However, only two species are considered epidemiologically important, *M. persicae* and *A. fabae.* Others are considered less important and, therefore, will not be described in detail.

The Green Peach Aphid (Myzus persicae)

This aphid species is cosmopolitan and probably the most notable vector of virus diseases for the largest number of plant species. It is extraordinarily polyphagous, colonizing dicotyledons (although, in certain cases, it has also been recorded on monocotyledons). Among its host plants of main economic importance are the following: *Prunus persicae* (peach), *Solanum tuberosum* (potato), *Beta vulgaris* (beet), *Nicotiana tabacum* (tobacco), *Brassica oleraceae* (cabbage), and *Citrus* spp. (e.g., orange). The biological development of the insect is holocyclic, especially in temperate or cold climates, where eggs are laid on *Prunus*. However, the species is frequently anholocyclic, especially in tropical or warm climates. In one and the same area, holocyclic and anholocyclic reproduction may coexist. *Myzus persicae,* the principal vector of BYV also transmits beet yellows stunt virus (34) and carnation necrotic fleck virus (63). Attempts to transmit CTV by this species have never been successful (97), except for one unconfirmed report (137).

The Black Broad Bean Aphid (Aphis Fabae)

This is also a cosmopolitan and polyphagous species. It prefers herbaceous cultivated hosts, especially Leguminosae (*Vicia baba, Phaseolus vulgaris, Vigna sinensis,* and *B. vulgaris*). Its biological development can be holocyclic or anholocyclic, especially in temperate and cooler climates. The primary host is *Evonymus europaeus*. The incidence of this species on beet has been related to its role in spread of BYV (35).

Additional aphid species were recorded for transmission of BYV (65); the majority are aphids that infest beets as casual feeders. The actual capability of certain species to transmit BYV is not known but should be resolved by more accurate and careful studies. Among these aphids are the following: *Acyrthosiphon pisum* (Harris), *A. gossypii, Aphis nasturtii* Kalt., *Macrosiphum euphorbiae* (Thomas), *Myzus ascalonicus* (Donc.), *Myzus ornatus* (Laing), and *Rhopalosiphoninus staphylae* (34).

Aphid Vectors for Other Semipersistent Viruses

Among other closteroviruses, carnation necrotic fleck virus (CNFV) was reported to be transmitted by *M. persicae* (63). Duffus (34) established semipersistent transmission for beet yellow stunt virus by three aphid species—*Myzus euphorbiae, M. persicae,* and *Nasonovia lactucae.*

The heracleum latent closteroviruses are transmitted semipersistently by three aphid species: *Cavariella aegopodii* (Scop.), *Cavariella theobaldii* (Gill. and Bragg), and *Cavariella pastinacae* (L.) (13). These three species also transmit two spherical viruses, the anthriscus yellow virus (AYV) and the parsnip yellow fleck virus (PYFV) (36, 37). *Cavariella aegopodii* (the willow/carrot aphid) is cosmopolitan in distribution, reproduces on several willow species (*Salix* spp.) that serve as primary hosts, and has many herbaceous plants of the Umbelliferae family as secondary hosts. Anholocyclic development on the latter hosts may occur frequently, especially in mild temperate regions. *Cavariella theobaldii* and *C. pastinaceae* also have *Salix* as primary host plant but have *Heracleum sphondylium* and *Pastinaca sativa* as their main secondary hosts; both these species may be found occasionally on other Umbelliferae where they are pests of only minor interest.

Virus–Vector Relationships

Semipersistent Transmission of Plant Viruses

Watson and Roberts (144) were the first to use measurable criteria for classifying modes of transmission of plant viruses by vectors. The terms used were *nonpersistent* and *persistent,* on the basis of the duration of retention of virus by the viruliferous vector (ranging from minutes to days, accordingly). Deviations from these two modes required an intermediate mode, which was defined by Sylvester (128) as *semipersistent.*

In the years since, several alternative classifications have been proposed on the basis of new findings. Thus, the term *stylet-borne* was suggested, based on the findings of Van Hoof (135) and Bradley and Ganong (19). Experiments carried out by Sylvester and Bradley (131) with BYV suggested that this virus differs from the so-called nonpersistent viruses, but other workers maintained that semipersistent viruses should be included among the stylet-borne viruses (42). More recently, a process of regurgitation, which was discovered in feeding aphids (49) and leafhoppers (51), led to the development of the "ingestion–egestion" concept (47, 48), whereby virus is believed to be retained momentarily in the foregut lumen and later egested in the process of feeding on subsequent hosts. In this case, both the NPVs and SPVs should be regarded as noncirculative, since they do not enter the hemocoel and the salivary system of the vector. However, the PVs that pass through the gut wall and circulate in the hemolymph to the salivary system are circulative.

In this chapter, we prefer to retain the original persistence-based terminology. Duration of virus retention is measurable and has a definite biological meaning. The term *noncirculative* may also be applied, since it includes both nonpersistent and semipersistent virus subcategories.

Watson (139, 140) was the first to conduct transmission experiments with BYV, the type member of the closteroviruses. Her early experiments were carried out before Sylvester (128) defined the intermediate semipersistent mode. Although she was aware of some irregularities in the transmission of BYV, she was forced at that time to include it among the persistent viruses. It was two years before Sylvester summarized Watson's and his own transmission experiments and defined the third and intermediate mode of transmission, namely, semipersistent (128, 142).

About the time that Watson described the transmission of BYV by the green peach aphid, Meneghini (77) described the first association of the tristeza disease with *T. citricidus*. Costa and Grant (27) presented timed acquisition and inoculation feeding data that, 15 years later, were found to agree with the semipersistent definition of Sylvester (128). Their findings were questioned by Reuterma and Price (100) who claimed transmission of the virus by the same species within seconds. However, the latter results were never confirmed in other laboratories, and confirmation is needed before any classifications are changed. Semipersistent transmission of CTV by the melon aphid was reported by Norman et al. (92) and confirmed by Raccah et al. (96). The contradiction between the findings of Reuterma and Price (100) and those of other workers led Lim and Hagedorn (71) to assume that CTV should be categorized among the so-called bimodally transmitted viruses. However, in the more than 16 years since the work of Reuterma and Price was published, their results have not been confirmed; therefore; we do not favor the use of the term *bimodal transmission* for CTV.

Characteristics of Semipersistent Transmission

The following features were found to conform with semipersistent transmission:

1. The duration of retention of the virus in the vector (usually one to four days) is intermediate: longer than in nonpersistent viruses but shorter than in persistent viruses.
2. A positive correlation exists between the duration of acquisition and the probability of transmission of BYV. (In transmission of NPVs, there is a negative correlation between the duration of acquisition and the probability of transmission.) This implies a certain amount of accumulation of virus within the insect.
3. Preacquisition fasting does not increase the probability of transmission of BYV (unlike NPVs).
4. Transmission can be achieved as soon as the vector is passed from the source of virus to the test plant; thus, no discernible latent period can be recorded (unlike persistent viruses).
5. Virus is retained by the vector at the same rate throughout a series of one-hour feeding periods on test plants (see Ref. 142, Tables 5–7).
6. Virus is lost if the vector molts between acquisition and its inoculation feeding (unlike persistent viruses).

Semipersistent transmission was reported also for CVYV and CMMV by *Bemisia tabaci,* for CYV by *Trialeurodes vaporariorum,* and MCDV by leafhoppers (86).

Specificity

Transmission of plant viruses is specific: viruses are transmitted by only one group of insects, even if other insects are allowed to feed on the same virus-source tissue. Apparently, a biological affinity exists between the virus and its insect vector. The authors of comprehensive reviews (83, 93, 94, 130) have discussed various aspects of virus–vector relationships. Therefore, we will limit ourselves here to descriptions of phenomena that are found among semipersistent viruses.

Certain SPVs are transmitted by a number of vector species but not by others, even within the same genus (e.g., for CTV, *Aphis gossypii* and *A. citricola* but not *A. craccivora*) (10, 97, 106). The number of species involved in transmitting a particular virus often represents the vector species tested but not the full range of vectors. For those few viruses for which comprehensive studies have been done, we know that the vector range varies greatly from one SPV to the next. Thus, BYV is transmitted by many aphid species, whereas CTV is transmitted by seven species, with only three or four confirmed in more than one laboratory. For other SPVs, between one and three aphid species are involved [e.g., *M. persicae* for carnation necrotic fleck virus (63) and *M. euphorbiae, N. lactucae,* and *R. staphylae* for beet yellow stunt (34)]. A clear explanation for the affinity between these viruses and their vectors, among closteroviruses, is still lacking; however, a possible mechanism can be inferred from other virus–vector relations.

Localization of Semipersistent Viruses in Their Vectors

Little is known about where SPVs are carried in their vectors. The only experimental evidence for an association between the viruses and the vector's foregut is available from two spherical viruses: PYFV, which is present in parsnip (85), and MCDV in maize (24). Both were found in plant cells embedded in a dense granular matrix material. The first was the aphid-borne PYFV, which was localized by electron microscopy in the heads of *Cavariella aegopodii* in which virus-like particles (VLPs) were seen behind the cibarial pump (84). More recently, VLPs resembling MCDV in purified preparations and in MCDV-infected plants (50) (23 to 31 nm in diameter) were observed adhering to the cuticular lining of the precibarium, cibarium (sucking pump), pharynx, and fore-esophagus in viruliferous *Graminiella nigrifrons* leafhoppers carrying MCDV (24). No VLP were observed in the alimentary canal beyond the cardiac valve, nor in the maxillary salivary duct. The authors did not see VLPs in the hemocoel or in any other internal organ of the leafhoppers. Childress and Harris (24) demonstrated the impossibility of casual acquisition of the particles by feeding leafhoppers, since no VLPs were found in the foregut of the nonvector leafhopper species *Dalbulus maidis* (DeLong and Wolcott) which was treated the same as the vector *G. nigrifrons*. No VLPs were detected in leafhoppers that were allowed a postacquisition feeding on healthy plants (24), a fact that is in agreement with the decreased retention of semipersistent viruses after a postacquisition feeding. Both for the aphid-borne PYFV (84) and for the leafhopper-borne MCDV, the VLPs occurred in thin layers or small to large clusters embedded in slightly stained matrix material (M-material) or in a densely stained substrate, which, in turn, were apparently attached by M-material to the cuticula (Ref. 84 for PYFV in aphids; Ref. 24 for MCDV in leafhoppers). It is thought that the VLP are virions of the viruses involved and that the M-material is similar in both cases, as well as in nematodes. The recent study by Childress and Harris (24) presents the first experimental description of a suspected site for retention of a semipersistent virus in its leafhopper vector. In other cases, the presumed presence of virus on the mouthparts was deduced from biological tests.

Helper-Assisted Transmission Among Semipersistent Viruses

The mechanism of transmission of SPVs by their vectors is still unclear. Singh and Chakraborty (124) claimed to obtain transmission of CTV by aphids feeding through membranes on crude infected citrus extract. However, attempts by us and others to repeat the work with either crude extracts or purified virus were not successful (Raccah and Bar-Joseph, unpublished results; Raccah and Garnsey, unpublished results). The infectivity of partially purified virus was proved for at least two viruses: BYV is mechanically inoculable with difficulty. CTV is inoculable by slashing (40). This indicates that the virus particles are infective but also might imply that a component needed for aphid transmission is missing in the purified virus. Such a case was proved to exist among the nonpersistent potyviruses (83, 93). Attempts to extract and test such a component have not been successful to date (99). These failures do not exclude the possibility that a helper is produced, but the appropriate extraction condition or experimental set-up, or both, have not been found. Dependence of HLV on heracleum virus 6 (HV6) was recently reported, but the nature of the dependency is still under investigation (81).

A more successful outcome was attained for semipersistent icosahedral viruses. For these viruses, whether aphid-borne or leafhopper-borne, a dependence mechanism was discovered that seems to follow the general scheme of helper-assisted transmission as reported for potyviruses and caulimoviruses (52).

Dependence Among Semipersistent Viruses

Dependence among semipersistent viruses has been ascertained for three systems: (a) dependence of the rice tungro baciliform virus (RTBV) on the rice tungro spherical virus (RTSV); (b) dependence of purified MCDV on a factor acquired from virus-infected plants (62); and (c) dependence of PYFV on the presence of AYV for its transmission (36, 37).

A tentative member of the MCDV group, RTSV is transmitted semipersistently by its leafhopper vector. Findings in biological studies imply that RTSV produces a helper factor that helps in the transmission of RTBV (56, 57). This was deduced from a series of experiments in which transmission of RTBV was possible only if the plants were coinfected with both components of the rice tungro disease or if the leafhoppers were given an acquisition access feeding on plants infected with RTSV prior to the acquisition feeding on the plants infected with RTBV.

A dependence of purified virus on a factor found in virus-infected plants was reported recently by Hunt et al. (62). They made use of a pair of MCDV strains, differing in symptoms, to test the dependence of purified virus on the factors produced in the plants in the presence of virus. The results imply that the factor involved is a virus-coded protein, although the possibility that the virion itself is involved cannot be excluded.

The third system resembles the second. Experiments based on the early findings of Murant and Goold (82) described the interesting dependency of the semipersistent PYFV on the semipersistent AYV by the aphid *Cavariella aeqopodii*. The PYFV is sap transmissible whereas the AYV is not; therefore, mechanical inoculation from a mixture of the two viruses resulted in the presence of PYFV only. When the PYFV was present alone, aphids failed to transmit it, but transmissibility was restored when AYV was reintroduced into plants previously inoculated with PYFV.

The acquisition sequence was important. Successful transmission of PYFV was obtained if the aphids were fed first on the helper virus (AYV) before being allowed to acquire the helped virus (PYFV). No transmission was obtained if the sequence were reversed (first on PYFV and second on AYV), which is similar to the case of potyviruses [see reviews of Pirone (93) and Murant et al. (83)].

It has not yet been possible to determine if the AYV produces a helper that assists in the PYFV transmission or if the AYV particles themselves are needed to effect PYFV transmission. Some progress was achieved recently in the process of AYV purification (54).

Epidemiology and Control

Epidemiology of CTV and Its Vectors

Citrus tristeza is one of the best-known examples of a major epidemic virus disease that is spread primarily by aphid vectors. However, little is known about the factors that affect its epidemiology: primarily the movement of its aphid vectors, the flight patterns of aphid species that serve as vectors, and the interrelationships of the aphid vectors, CTV, and the plant. Efficient transmission depends on the following: factors that affect vector activity (flight, colonization, of host preference); factors that affect the transmissibility of the virus isolates; and environmental and agrotechnical factors that affect virus multiplication, aphid population, and the conditions for infection.

Factors That Affect Vector Activity

Semipersistent viruses are retained by the aphid for less than 48 hours. This has an obvious effect on the spread of the virus in the field. Cases in which BYV is progressively and rapidly reduced from the margins of infected regions were reported by Duffus (35) and by Russell (115). A distance of one to two miles from an infected source was considered an effective barrier to virus spread.

The ability to fly makes alates the primary aphid form that transports the virus under field conditions. Secondary spread may be also caused by wingless aphids walking from the canopy of one tree to the next. Passive transport by the wind may occur, but it is of negligible importance, since only few wingless aphids have been found trapped in suction traps exposed close to citrus groves (Raccah, unpublished observation). Since they are more restless, alates tend to stay for shorter periods on one host compared to wingless aphids. However, alates do stay for longer periods after a long flight (68) or when they larviposit. Citrus tristeza virus is transmitted in a semipersistent manner by *T. citricidus* (27, 60) and by *A. gossypii* (92, 96). Like other SPVs, CTV transmission rates will be low if the duration of acquisition time is short (>30) and high if acquisition time is long (24 hours) (106, Table 3).

Since alates are the principal vectors for virus spread under field conditions, what factors are needed for successful transmission of CTV? Two sources of viruliferous alates can be considered to be involved in spread, as follows.

The first source relates to alate forms that arise in the progeny of feeding on noncitrus or virus-free hosts, or both. These alates are virus-free when they first land in a citrus grove. They become viruliferous only after completing an acquisition feeding on an infected citrus leaf. The probability of a visiting nonviruliferous alate becoming infective depends on the duration of flight that preceded landing. After a long flight, the aphids tend to stay long enough to acquire CTV. The probability for inoculation then depends on the feeding preference of the alates. Some of the viruliferous alates might desert the citrus plot before inoculation occurred (this might happen with polyphagous aphids such as *A. gossypii* and *A. citricola*).

The second source of viruliferous alates is from populations of aphids reproducing in the infected citrus grove. The proportion of viruliferous alates in this population depend on the number of alates that acquired the virus during the "teneral period" (the period after the last nymphal stage and before first flight). As mentioned previously, SPVs are not retained in the aphid if the vector molts between the acquisition feeding and inoculation feedings. Thus, only alates that have acquired the virus after the projected molt can be considered infective. These alates, either after a short flight in the same grove or after a long flight to more distant groves, are capable of inoculating virus immediately upon landing. The number of alates produced and the proportion of those that migrate are related to their conditions of development (59, 70). The development of large aphid populations results in crowding which is a strong incentive for migration. If crowding occurs when the reproduction host is infected, some of the migrating alates will carry the virus.

The conditions may often exist on the reproduction host that, for *T. citricidus* or *A. citricola* (and sometime also for *A. gossypii*), is citrus, the primary host for CTV. For *M. persicae*, the reproductive host is beet, the primary host for BYV.

Unfortunately, most of our knowledge on the flight behavior of aphids is on species that are poor vectors or are not vectors of CTV. Very little is known about possible alternative hosts and the flight behavior of *T. citricidus*, an important vector of CTV. There is a high probability that this species will continue to migrate north from South America through Central America into Mexico and the United States, and from central Africa into the Mediterranean region. Research on the flight behavior of this species is needed.

Factors That Affect Virus Transmissibility

Certain aphid species may fail to transmit some CTV isolates or transmit others at a very high rate. This was shown for *A. gossypii* in Israel by Bar-Joseph and Loebenstein (7) and in California by Roistacher et al. (111). They clearly demonstrated that efficiency of transmission was not related to any property of the vector *A. gossypii* but rather to the intrinsic property of the CTV isolate. Aphids collected from different parts of a particular region would consistently transmit CTV isolates at low, intermediate, or high rates. In most cases, the aphid sources used were not collected from citrus but rather from melon or other crops, yet they had the same ability to transmit virus. An explanation for the mechanism by which the virus changes from poorly transmissible (less than 5%) to highly transmissible (100%) forms has been offered by Bar-Joseph (4). He proposed that highly transmissible and poorly transmissible CTV isolates may coexist in the same tree and that possibly by cross-protection, the more transmissible isolates are suppressed. After a period of time, this protection breaks down and the more transmissible isolates are more readily transmitted by *A. gossypii* to non-infected citrus trees. Raccah et al. (98) have shown that CTV isolates differing in transmissibility coexisted in the same tree. Highly transmissible isolates were recovered after aphid transmission from trees that had been thought to be infected by poorly transmissible isolates only.

Why one aphid species such as *T. citricidus* is a more efficient vector than other species is not understood. However, the potential of *A. gossypii* and, possibly, *A. citricola* to inoculate highly transmissible isolates of CTV at high levels of efficiency has serious epidemiological implications. A recent serious outbreak of destruction of sweet orange trees on sour orange rootstock in Israel by a seedling yellows isolate of CTV transmitted by *A. gossypii* may be cited as an example (8, 15). All studies indicate that *T. citricidus* will transmit CTV at very high rates of efficiency wherever it is found. It is an extremely efficient vector for CTV. Concern that this aphid species may be introduced into areas where it is not now present, that is, in the Mediterranean basin or in North America, has been expressed by many workers (46, 69, 106, 111, 126). Climatic conditions for growth of citrus are quite similar worldwide, and *T. citricidus* is currently found in countries throughout the world with a wide range of temperatures and humidities. There is no known reason why this efficient vector cannot exist in areas and countries where it is not now present. The possibility of importing this vector, with its potential for transmitting such highly destructive CTV isolates as the Capao Bonito CTV (79), or the recently reported isolates in Peru (104), is serious. The importance of enforcing strict quarantine laws and regulations cannot be overemphasized.

The severity of the CTV isolate and its transmission rate by *A. gossypii* are apparently correlated. The severity of CTV isolates can be measured quantitatively (41). Roistacher and Bar-Joseph (105) showed that two severe seedling yellows and stem pitting isolates of CTV were transmitted by *A. gossypii* at rates of 79% and 92%, respectively, from sweet orange to grapefruit, compared with only 14% for a non-seedling yellows and non-stem pitting isolate. Severe strains were transmitted more efficiently by *A. gossypii* than by the milder reacting strain from grapefruit to grapefruit, grapefruit to Mexican lime, or lemon to Mexican lime. The acquisition and receptor hosts are important considerations in the transmissibility of CTV. Table 11.2 shows the relative rates of transmission by *A. gossypii* of three CTV isolates in three citrus acquisition hosts to three

TABLE 11.2. The effect of the source plant and virus isolate on the efficiency of transmission of CTV by *Aphis gossyipii*.

Acquisition host	Receptor host	Rate of transmission (%)
Sweet orange	Mexican lime	100
Sweet orange	Grapefruit	65
Grapefruit	Mexican lime	21
Sweet orange	Lemon (1)	28
Lemon	Lemon	0

Adapted from Roistacher and Bar-Joseph (105).

receptor hosts (105). High efficiency (100%) was obtained in transmission from sweet orange to Mexican lime, in contrast to no transmission from lemon to lemon.

When given a choice, populations of *A. gossypii* were found to occur in equal number feeding on young leaves of sweet orange or grapefruit and showed a significant preference for these cultivars over young lemon leaves (105).

Whereas the range of noncitrus hosts for *T. citricidus* is limited, *A gossypii* has a wide noncitrus feeding range. In studies in which ELISA was used for testing, CTV could not be detected in nine herbaceous species subjected to transmission feedings by *A. gossypii* in 170 transmission tests. However, CTV was transmitted to various *Passiflora* spp. from infected sweet orange seedlings by *A. gossypii,* and the virus remained in the *Passiflora* spp. for the life of the plant (109). Thus, CTV isolates may vary in their ability to be transmitted by *A. gossypii* from sweet orange to *Passiflora* spp. One CTV isolate was transmitted from a sweet orange acquisition host to *P. gracilis* at 58% efficiency compared to 0 to 17% efficiency for three other CTV isolates in their three sweet orange acquisition hosts (107).

Environmental and Agrotechnical Factors

Temperature has a direct effect on aphid population growth and activity. It also affects the vigor of citrus growth flushing and the content of virus in acquisition sources (7). Other factors that affect virus transmission are the variety of host plants (Table 1 in Ref. 105) and pruning practices. Severe pruning of citrus during the winter months will induce abundant succulent growth in the spring, resulting in a large population increase of aphids. In Florida, a typical practice is machine hedging; this results in uniform flush growth in the spring. If hedged in only one direction, tree canopies may touch and could theoretically create optimum conditions for short-distance vector spread (R.K. Yokomi, personal communication). Klas (67) and Schwarz (121) showed that the increase in populations of *T. citricidus* in yellow traps was directly associated with the growth flush. Except for the work of Dickson et al. (32) on the spread of CTV in citrus by *A. gossypii,* no other definitive studies have been done with aphid flight in relation to the spread of CTV.

The Development of Citrus Tristeza Virus Epidemics

Once CTV is newly established in an area and a vector is present to transmit it with reasonable efficiency, it may take about four years for 25% of the citrus trees to become infected (11, 102, 103). After this period, the virus quickly becomes endemic and the infection curve rises rapidly. Thus, tristeza eradication programs face a difficult task in preventing complete dissemination of the virus that results in epidemics. Rapid decline of citrus on sour orange rootstock may be a function of virulence of the CTV isolates present, their transmissibility, and the prevailing temperatures. Although endemic in Florida since 1956 and present in Israel since 1970, tristeza has not induced the total destruction of the industry on sour orange rootstock as it did in South America in the 1930s and 1940s. However, the introduction or possible development of more virulent isolates is now destroying many thousands of trees in both these areas.

The very rapid spread of CTV by *T. citricidus* in South America has been studied only by observation of declining trees on sour orange rootstock, and specific reports are few. Bennett and Costa (14) reported on the unpublished observations of Moreira who studied the decline of 2606 trees over a six-year period in Bahianinha sweet orange on sour orange rootstock at the Limeira Experiment Station in Brazil. He reported 11.8% infection in November 1942, 19.5% infection approximately one year later, and 55.1, 72.5, and 100% infection, respectively, in three successive years. Five years after the initial observation, all the trees were dead.

Once introduced, *T. citricidus* spreads very rapidly. Geraud, in a personal communication, reports as follows: The rapid movement of *T. citricidus* from one area to another was observed in Venezuela in 1976–1978. *T. citricidus* was first found in the southern Lake Maracaibo region of western Venezuela in early 1956, apparently spreading from Colombia. At about the same time, the aphid was found localized near the Brazilian border in southeastern Venezuela. Within a two year period the vector moved throughout Venezuela and could be detected almost everywhere in the country where citrus was grown. It is our belief that the vector was carried long distances by the winds and found citrus and their relatives to feed on and establish colonies. However, movement of plant stock occurred intensively at that time due to increased citrus production, which may have helped the dispersal of the aphid. The first dramatic disease outbreak of tristeza were observed within two years of colonization by *T. citricidus* (1980) of given citrus growing areas, in the north central region of Venezuela. There, evidence was found of scattered small endemic foci of diseased trees that had survived for about 20 years, previous to the colonization of the area by this vector." Since 1980, over 6 million trees on sour orange rootstock have been reported killed in Venezuela (76).

A similar picture of tristeza spread is given by Ducharme et al. (33) for Argentina. Tristeza was first noticed in Misiones Province in 1937, and by 1942, all the groves on sour orange rootstocks were dead. It was reported in West Corrientes in about 1932 and was widespread by 1936. It was first reported in Santa Lucia in 1938 and moved into Goya (50 miles away) and north to Corrientes (75 miles from Bella Vista). All the trees were dead by 1942. The writers discuss why it may have spread so rapidly into all three areas. Entre Rios lies 250 miles southeast of West Corrientes and 400 miles south of Misiones. Here, there were natural barriers not in citrus. The writers suggest that the disease was spread by unrestricted movement of plants, and by aphids. They point out that there is a string of wild and dooryard citrus trees along the Uruguay River where aphids may land, feed, and breed. Some of the foci of CTV inoculum may have been established through propagation of dooryard citrus and citrus introduction by settlers.

Principles of the Approaches to Control

Control of virus epidemics in plants is based on understanding the parameters for the development of epidemic situations. With vector-borne viruses, an epidemic will develop if there are sources of virus in host tissue from which vectors can readily acquire the virus, a population of active and efficient vectors, and a sufficient number of target plants that are likely to become infected (12, 95). All the above parameters are affected by environmental conditions.

Control measures are singularly or in combination aimed at interfering with processes leading to a rapid spread of virus. Control can be attained (a) by reducing the virus at its source by quarantine by eliminating alternative host plants for the virus, or by suppressing and rouging; (b) by interfering with vector transmission; and (c) by rendering the target crop resistant to virus infection.

Approaches for controlling virus sources in vegetatively propagated perennial crops differ from those adopted for seed-borne annual crops. The difference is illustrated in this chapter for two SPVs: CTV in citrus and BYV in beets.

Reduction of Sources of Citrus Tristeza Virus in Citrus

Two major approaches for reduction of the CTV sources are by *quarantine,* when the virus or a particular virus strain is still absent in the citrus growing region, or *detection–eradication* when the virus is present but in a relatively limited amount.

Quarantine. This is perhaps the most important but most difficult method for control. In most citrus-growing areas of the Mediterranean region, especially in the countries of the Near East where the sour orange is still the predominant rootstock, importation of tristeza-infected budwood or trees in the presence of melon aphids presents an immanent and potential danger. The tristeza epidemic in Spain in the 1960s through the 1970s and the recent outbreak of a more virulent destructive tristeza in Israel (8, 15) are probably the result of unauthorized importation of tristeza-infected citrus budwood from other countries. There are many strains of tristeza throughout the world, and some can be extremely destructive to citrus when inadvertently introduced to an area (41, 79, 104). Some CTV strains do not cause symptoms in satsuma mandarins, and, if infected budwood is imported and propagated, nearby sweet orange trees can become infected via aphid transmission. The virus is readily spread by vector to other sweet orange groves within a country. This was recently shown to be the cause of the serious stem pitting infestation and subsequent decline of almost all the navel orange trees in Peru (104).

Effective quarantine involves intensive efforts to educate growers and the public about the destructive potential of tristeza (as well as other viruses). The efforts and costs involved in publicity and education, combined with stricter quarantine laws backed up by heavy fines, might be very helpful in preventing introduction of CTV. This is especially important in areas where sour orange is still the predominant rootstock. It is also important in preventing the introduction of new, more virulent CTV strains.

The importance of establishing an efficient, early-warning detection program and rigidly enforcing quarantine laws cannot be overemphasized for citrus-growing areas where tristeza is not endemic. Not only does CTV kill citrus on sour orange rootstock, but certain virulent isolates also can attack scions directly to reduce fruit size and yields and to induce severe stem pitting, dieback, and tree death (104). Rigid quarantine laws are the primary defense against invasion by severe reacting isolates, and a program for cross-protection and selection is the only long-range solution, once severe isolates are established.

Detection and Eradication. The recent use of the enzyme-linked immunosorbent assay (ELISA) for detecting CTV has permitted intensive and extensive surveys of large areas of citrus in a relatively short period of time. Such surveys conducted in Israel, California, Spain, and Italy have uncovered many CTV-infected trees. When these surveys are backed by legislation permitting the destruction of the infected trees, they can be effective in limiting the spread of the virus. This has been especially true in the central valley of California, which is still one of the few remaining areas relatively free of tristeza and where tristeza remains under surveillance and control.

The program for detecting and eradicating tristeza in Israel is a remarkable example of the use of ELISA, combined with the latest computer technology. It has compiled a comprehensive record of the presence of CTV in almost every citrus grove in the country (16). Wherever the virus was found, the trees were destroyed. However, in recent years, virus spread could not be controlled in some areas and the disease became established. Groves surrounding infected areas are monitored and watched. A program using ELISA for detection, followed by eradication of infected trees, is highly recommended in those areas where sour orange is still the predominant rootstock and where data on the presence or absence of CTV are not available.

Reduction of Sources of Beet Yellows Virus in Beets

For BYV, which is not seed-borne, most of the virus is brought into the plot by aphids that acquired it elsewhere. Heathcote and Cockbain (53) found that most of the virus was carried over a relatively short distance. As an SPV, the virus is retained for longer periods in its vectors than are nonpersistent viruses and, therefore, the vectors are likely to carry them longer distances. Loss of virus is probably the result of repeated attempts to feed on intermediate plants on the way to the plot; thus, barriers or cover crops such as barley or mustard provide protection from yellows to sugar beet plants (35).

Eliminating overwintering beets that harbor the virus greatly reduces the probability of infection in neighboring crops (35). Another approach is based on the use of virus-free material, by careful testing and selecting of propagation material [as for carnation necrotic fleck virus in Israel (Loebenstein, personal communication)].

Control of Spread by Interference with Virus Transmission

Since, by definition, all SPVs are vector-borne, a logical approach to the control of their spread is to reduce vector numbers, vector activity, and vector capacity to acquire and inoculate virus. Attempts to use insecticides to control vectors of yellows were made in Europe (101) and in England (61). Insecticides did reduce and delay infection, but they were tested on a limited, noncommercial scale.

The likelihood of controlling vectors for SPVs is greater than for NPVs, since the former are transmitted within a minimum of 30 minutes and reach a reasonable rate of transmission only after two hours (60, 91, 96, 142). Insecticides will have some effect, especially on colonizing species; however, many alates are likely to take off to avoid the insecticide and later attempt to feed, and thus, inoculate, elsewhere.

Bradley et al. (20) discovered the effect of waxy compounds on the transmission of NPVs. Field studies were also carried out by Vanderveken and Semal (134) and Russell (116). Although mineral oils were used in the past as an efficient insecticide against scales in the citrus grove, we are not familiar with any work in which mineral oil was used to prevent the spread of CTV in citrus.

In recent years, new fast-acting pyrethroids can affect transmission due to a so-called knock-down effect. These were tested alone (44) or in combination with mineral oil (43) against BYV transmission. The results were promising, although a convenient mode of application must be developed in order to ensure good coverage of leaves.

Applications of mineral oils using high volume and pressure sprayers were suggested by Simons (123), and found effective. In Israel, the use of a sprayer to apply the liquid at a high volume and pressure improved the performance of an oil plus pyrethroids combination against cucurbits infected by the NPV zucchini yellow mosaic virus (Raccah, Mor, and

Dubitsky, unpublished results). The above procedure could be considered for the prevention of SPVs in areas of intensive activity of aphids that are likely to carry the virus. The potential use of pyrethroids in a grove is doubtful, since this class of insecticides might affect beneficial insects adversely.

Other approaches to control are based on interfering with the landing behavior of aphids. This can be achieved by covering the plants to be protected with coarse white nets that camouflage the plants, to hide them from landing aphids. Bar-Joseph and Fraenkel (5) used the same principle to prevent *A. citricola* from landing on citrus seedlings. The method was tested on a limited scale and might be considered in places where virus-free trees are to be grown in the presence of CTV.

For perennials, infection in seedlings can be avoided if the nurseries are in virus-free regions. Budwood sources are maintained in the Negev in Israel, far from tristeza sources. A similar practice for young trees in containers is followed in Spain at the quarantine station for AVASA (87), and in California at the Lindcove Field Station (22). Costa et al. (29) have used mulches to reduce aphid colonization in outdoor nurseries in Brazil. These practices allow the nurseries to maintain virus-free seedlings before protective isolates are introduced.

Finally, hedging and pruning have a considerable effect on the vigor and time of growth flushes. New flushes are a major attraction to aphids and lead to a great increase in aphid colonization. Similarly, the proximity of crops that are heavily colonized by potential vectors might affect the number of aphids that reach the crops. This is true for Cruciferae and Leguminosae as sources of *M. persicae* and *A. fabae,* vectors of BYV, and cotton and cucurbits or rosaceous plants as a source of *A. gossypii* and *A. citricola*. Obviously, it is difficult to control aphids in plots that belong to different owners, but a regional policy of aphid control might help all the growers.

Use of Resistance Mechanisms

This is probably the cheapest and easiest measure to adopt, provided it works. Unfortunately, the development of resistant varieties is laborious and often the breeder is not always victorious in the race against the appearance of new strains of virus. Approaches for resistance to the virus for perennial trees differ from those for herbaceous annual crops. The long process of producing resistant varieties in annuals is even longer than in perenniels. However, the prospect of using cross-protection is greater in perennials, since the protective strains will remain in the trees for years. Moreover, "immunization" does work and is practical.

Citrus Rootstocks Resistant to Citrus Tristeza Virus

Once tristeza becomes endemic and trees on sour orange rootstock are destroyed, the next procedure is to find a resistant rootstock tolerant to the CTV bud–union phloem necrosis. The rootstock must also be evaluated for its effect on fruit quality and size, tree size, *Phytophthora* resistance, etc. The destruction of trees on sour orange rootstock by CTV opened up an entirely new field in horticulture for the development of citrus rootstocks. Many fine, tristeza-tolerant rootstocks have been developed by horticulturists and research is continuing. Conditions in different areas of the world favor different tristeza-tolerant rootstocks. For example, the rough lemon is preferred in South Africa, the trifoliate orange in Australia and in Japan, the Rangpur lime in Brazil, the citrange in California and in Spain, and the Cleopatra mandarin in Venezuela and other tropical regions. Once tristeza becomes established, a research program is needed to test and evaluate various tristeza-tolerant rootstocks specific to that country or to a region within that country.

The new tristeza-tolerant rootstocks that replaced the tristeza-intolerant sour orange revitalized the citrus industry throughout the world. However, new strains of stem pitting CTV are currently spreading and have the potential to destroy grapefruit and sweet orange scions directly, regardless of their rootstocks. Control of these destructive strains presents a more difficult problem that cannot be resolved by using tolerant rootstocks.

Tristeza Cross-Protection

Citrus tristeza virus is destructive to citrus in two ways: by inducing a phloem necrosis near the bud–union on susceptible scions on sour orange rootstock (sweet–sour destruction) or by inducing intense stem pitting in certain scions such as Mexican lime, grapefruit tangelo, and sweet orange, regardless of rootstock. The presence of severe, CTV stem pitting strains results in lower yields, smaller fruit size and, ultimately, tree decline and even death. Cross-protection is currently being used in some countries as an effective method for controlling the damaging effects of tristeza. First, CTV isolates must be found that are protective, and then they must be intensively tested. Of the many millions of trees affected by the sweet–sour decline of citrus, few if any surviving trees have been reported as sources of protective inoculum. However, Wallace and Drake (138) in California found that recovered shoots from seedling yellows tristeza-infected plants did have some protective ability when they were used as inoculum in sweet–sour decline field tests. In time, this protection broke down, apparently as new destructive isolates spread into the experimental plot. Similarly, Stubbs (127) also tested some protective isolates against sweet–sour decline in Australia, but protection there also broke down in time. Studies are continuing in Florida, Israel, and Brazil on testing protective isolates against sweet–sour CTV-induced decline.

Researchers have been more successful in finding protective isolates against the severe stem pitting forms of tristeza. Costa and Muller (29) have reviewed the work done on cross-protection in Brazil. They collected budwood from over 70 surviving Mexican lime trees throughout Brazil that contained mild CTV isolates. After many years of testing and research, two isolates were selected as superior for cross-protection. These are now in use to protect the Pera sweet orange against the severe stem pitting strains currently found in Brazilian citrus. These protective isolates have proved remarkably successful in Brazil. Finding protective isolates of CTV against stem pitting strains affecting grapefruit has been much more difficult.

Another approach to finding protective isolates was outlined by Wallace and Drake (138) who used budwood from recovered shoots of seedling yellows-infected greenhouse plants as protective inoculum. This technique was employed by Roistacher and Bar-Joseph (107) and Roistacher et al. (109) to discover a number of isolates that were highly protective when challenged under greenhouse conditions but that could not be tested in the field in California. A different approach was adopted by Yokomi et al. (147) who used aphid vectors to select CTV-protective mild isolates.

One other method of selecting protective isolates was recently developed by Roistacher et al. (109, 110) who used *Passiflora* spp. as a filtering host for CTV. Severe stem pitting isolates of CTV present in sweet orange holding plants were first vector-inoculated into *Passiflora* spp. by *A. gossypii* and then transmitted back to Mexican lime seedlings by aphids. The stem pitting factor appeared to be filtered out by the *Passiflora*. Six attenuated and highly protective isolates were derived by this means. When used as protective inoculum in sweet orange or grapefruit seedlings and challenged with severe CTV-stem pitting or seedlings yellows isolates (either by vector or buds or both), all six isolates were shown to be highly effective in cross-protecting against yellows or pitting. Again, these promising results were obtained in greenhouse tests and could not be tested in the field in California.

The production of grapefruit, limes, and sweet oranges may be difficult in countries where destructive stem pitting isolates of CTV are present and readily transmitted by aphid vectors. Continued intensive research will be needed to develop protective CTV isolates with proven protective ability under field conditions in these countries.

Resistance to Beet Yellows Virus in Beets and to Rice Tungro Virus in Rice

In most studies, breeding for resistance did not distinguish between BYV and the persistent beet western yellows virus (BWYV). Successful breeding for resistance or tolerance was reported by a number of workers both in the United States and Europe (75, 101, 114). However, Russell (117), who did much of the breeding, finds this approach to be only a partial answer to the epidemiology of yellows. Partial resistance to RTV and leafhopper colonization was reported by Khush (66).

Conclusion and Summary

Although more than 30 years have passed since semipersistent transmission was defined by Sylvester, little has been added to our knowledge of the precise mechanism of transmission and how it differs from nonpersistent and persistent transmission by vectors.

The duration of virus retention by the vector defines the persistence that, for SPVs, is one to four days. There is a positive correlation between the duration of acquisition and inoculation feedings and the probability of transmission. No latent period is needed, and the vector is capable of transmitting the virus immediately after leaving the virus source plant. Upon molting, the vector ceases to be viruliferous, presumably as a result of its shedding the intima. Transmission data and electron microscopic observation of the foreguts of leafhoppers allowed a postacquisition feeding seem to indicate that virions are indeed washed down the alimentary canal after prolonged feeding on an healthy host (24, 48, 51).

Many SPVs are propagated in the phloem of the host plant cells (49, 113). Therefore, active retention of virions in the vector's foregut is necessary to prevent rapid flushing of the virions by fluids ingested from the phloem. Overall, the characteristics of semipersistent virus transmission seem compatible with an internal vector retention site (24, 47–49, 94). The basic difference between nonpersistent and semipersistent transmission may be the tenacity with which virus can be held in the transmissible state and its titer at retention sites in the foreguts of aphid and leafhopper vectors (24, 72).

Viruses that are transmitted semipersistently belong to the closteroviruses, to the maize chlorotic dwarf virus group that infects monocots, and to such ungrouped viruses as CVYV, AYV, and PYFV. Viruses that belong to this latter group are filamentous and spherical. They infect a variety of botanical families and are widespread.

All the known vectors of semipersistent viruses belong to the order Homoptera. The largest number of viruses and those of the utmost economic importance are transmitted by aphids. However, the viruses of corn and rice that greatly affect yield are transmitted semipersistently by leafhoppers.

A greater specificity between vectors and viruses is encountered among SPVs. Fewer vector species are known to transmit semipersistent viruses than are known for nonpersistent ones. The reason for this specificity is still unknown. Moreover, transmissibility by vectors of certain virus isolates varies. We still do not fully understand what makes certain isolates more transmissible than others. The epidemics of tristeza in Israel and in the United States have been attributed to a change in the transmissibility of the virus isolates.

In addition to differences in transmissibility, certain viruses and virus isolates may depend on other viruses or virus products for their vector transmission. The term *helper* was coined to describe a virus or a virus product that assists in the transmission of another virus. Helper-assisted transmission has been studied intensively in the nonpersistent potyviruses, in which a virus gene was found to code for a polypeptide responsible for the association of the virus with aphid stylets (17). A similar association was discovered more than 10 years ago for the AYV–PYFV complex by Murant et al. (84), although the nature of the association is still not known. Another helper-dependent transmission was recently reported for spherical leafhopper-borne virus (62). It will be a challenge to determine if helper-assisted transmission also exists among closteroviruses. Attempts are now being made also to use clones for the characterization of the CTV genome (113) (Bar-Joseph, personal communication). If successful, these attempts might shed light on this important issue. One outcome of the above attempts might be the development of resistant mechanisms aimed at transmission of the virus by the vector rather than replication of the virus in the host.

The semipersistent nature of transmission has direct implications in epidemiology. Transport of virus by vectors depends on the duration of its retention in the vector. This, along with more precise and detailed data on flight patterns of major vector species, will provide the missing links in the puzzle of the epidemiology of several important SPV diseases. Attempts to relate BYV incidence to environment have been successful in some years but not in others (143).

However, virus-dependent changes are no less important than vector-dependent ones. Rapid decline and destruction of trees in Peru (and possibly in Israel) may be due to illegal importation of budwood containing more virulent strains from other areas.

Control of virus diseases, especially those causing severe epidemics like tristeza, in areas where disease is not prevalent is achieved primarily by quarantine, and by eliminating virus-infected plants in places where the virus exists in a low population of trees or is "dormant" (i.e., non-transmissible, or low-transmissible and symptomless). This can be monitored by the use of such serological methods as ELISA. The use of monoclonal antibodies and specific cDNA probes might facilitate the distinction between various levels of severity of CTV isolates in the future.

Once sweet–sour decline has become established, it is not possible to control it; then, the only recourse is to use resistant rootstocks. However, severe stem pitting CTV can be controlled by cross-protection.

Very little research has been carried out on preventing SPVs in herbaceous crops and trees by affecting vector capacity. Earlier findings on the effect of mineral oils on BYV and the more recent ones on the use of pyrethroids should be tested for their efficacy in preventing transmission of viruses in citrus, beet, carnation, and other crops affected by SPVs.

Finally, a better understanding of such SPV diseases as tristeza and beet yellows, which can cause severe epidemics, is important to many nations and societies. Collaboration, exchange of knowledge, and learning from the mistakes of others might prevent disasters caused by the introduction of severe CTV isolates. The control of such a major disease as tristeza may depend on political and economic forces that can dictate possible strategies. The choice of an effective control strategy is not always based on the ideal, that is, the best scientific information available but is usually a compromise among all the political and economical forces within the industry and the country. Unfortunately, if a wrong strategy is adopted, there may not always be a way to correct a poor decision, and an industry might be seriously affected. In the case of the tristeza disease, with its potential for epidemic, the industry could be debilitated or even destroyed.

Appendix

Identification of Main Aphid Vectors of CTV and BYV

1. Head deeply excavated at center of frontal profile. 1st and 7th abdominal segments without marginal tubercles (microscopic observation). Antennae subequal to or longer than body length. Aphids light green or sometimes pinkish, siphunculi mostly uniform in color (at least in apterae and nymphs) or brown at its distal part only 2

   –Head frontal profile straight or slightly sinuous. 1st and 7th abdominal segments with well-developed tubercules (seen well under a microscope). Antennae significantly shorter than body length (their distal parts do not reach or exceed only slightly the base of siphunculi). Dark colors of aphids vary: deep green or yellowish with siphunculi totally pigmented (blackish or brown) 3

2. Head with smooth cuticle and frontal depression diverging laterally. Cornicles slender and long, cylindrical and slightly distally attenuated, directed in parallel or slightly divergent posteriorly (under the microscope, hexagonal reticulation at the aphid's distal part is visible). Cauda with 10 to 12 hairs. Alatae without distinctive dorsal pigmentation on abdomen. Body length, 2.4–3.8 mm. .... *Macrosiphum euphorbiae*

   –Head with roughish cuticle; marked frontal excavation with convergent sides. Cornicles slightly swollen at their distal halves, no apical reticulation; typically convergent orientation toward the cauda. The cauda with hairs. Dorsal alate abdomen has a sclerified patch, olive-green to brown. Body length, 1.4–2.5 mm .... *Myzus persicae*

3. A stridulatory apparatus (visible only under a microscope) consists of a series of fine serrated cuticular cells on the abdomen, posterior to siphuncular bases, and an internal line; spine-like setae on the hind tibiae. Aphids dark brown (lighter and brighter in larval forms). Posteriorly directed antennae slightly exceed the siphuncular base. Distal part of the 6th antennal segment generally four times longer than the basal part of the same segment (visible only under a microscope) 4

   –Absence of a stridulatory apparatus. Aphids different colors (black, bluish-brown, mottled green, yellowish, or apple green). Antennae shorter than above (if aimed posteriorly, they do not exceed the siphuncular base). Distal part of 6th antennal segment 2.5 to 3.5 times longer than basal part 5

4. The 3rd antennomere of apterae (and usually the fourth) is uniformly pale. The 3rd antennomere of alatae is black with anterior, double bifurcated wings, wing pterostigma is pale brown. Body length, 1.8–2.8 mm. .... *Toxoptera citricidus*

   –Apterae III and IV antennal segment brown at the apex. Alatae pale III antennal segment brown at the brown apex. Anterior wings with a single bifurcation in median innervation. Pterstigma black. Body length 1.6–2.2 mm .... *Toxoptera aurantii*

5. Femoral and dorsal hairs slender, longer than the minimum thickness of the femur near the trochanterofemoral suture (seen under the microscope). Cauda has 7 to 20 hairs, rarely less than 8 to 9. ... 6

   –Anterior femoral hairs very short, not more than 1/2 to 1/3 of the minimum diameter of the femur near the trochantero femoral suture. Cauda has 5 to 7 hairs, usually 3 on each side 7

6. The 8th abdominal segment has generally 4 dorsal hairs (seen under the microscope). Cauda has 10 to 20 hairs, tongue-shaped; no distinct constriction at the basal third. Dorsum of apterae with stripped segmental sclerification; is blackish, with a white waxy, transversally striped secretion. Body length, 1.6–2.5 mm .... *Aphis fabae*

   –The 8th abdominal segment has only two dorsal hairs. Cauda with 7–16 setae distinctly constricted at the third basal part. Dorsum in apteare practically without pigmentation. Apple green or yellowish green (particularly in summer forms). Body length, 1.3–2.2 .... *Aphis citricola*

7. Abdomen has a strong dorsal reticulated sclerification in the apterae so live aphids appear brightly black. Cauda black, same color pigment as the siphunculi. Color is dull black. Body length, 1.6–2.4 mm .... *Aphis craccivora*

–Abdomen not dorsally sclerified (possible small stripes on abdominal segments posterior to the 5th segment). Cauda pale, less pigmented than siphunculi. Color in live aphids varies considerably from brown yellowish to green-bluish or mottled green. Body length, 1.1–1.9 mm .... *Aphis gossypii*

The key is prepared for the tentative identification of species in the field using a magnifying glass. For more accurate identification, it is suggested that specimens be mounted on microscopic slides. The procedures for specimens preparation are given elsewhere (e.g., Stroyan, ref. 126). Space limitations allowed us to include only the major vector species for transmission of two major closteroviruses, BYV and CTV.

Acknowledgments. The authors are grateful to Dr. R. Yokomi, from the Horticultural Research Laboratory, USDA, Orlando, Florida for the critical reading of the manuscript, and to Dr. Kerry Harris, from Texas A & M University, for prepublished results. This is Publication No. 2462-E, 1988 series of the Agricultural Research Organization, Bet Dagan, Israel.

References

1. Adams, A.N., and Barbara, D.J., 1982, The use of F(ab')2-based ELISA to detect serological relationships among carlaviruses, *Ann. Appl. Biol.* **101:**495–500.
2. Barbagallo, S., 1966, L'Afidofauna degli agrumi in Sicilia, *Entomologia* **2:**201–260.
3. Barbagallo, S., and Patti, I., 1986, The citrus aphids: Behavior, integrated pest control in citrus groves, in Cavalloro, R. and Di Martino, E. (eds): *Integrated Pest Control in Citrus Groves, Proc. Expert's Meeting Acireale, 26–29 Mar. 1985* A.A. Balkema, Rotterdam, & Boston, pp. 67–75.
4. Bar-Joesph, M., 1978, Cross-protection incompleteness: A possible cause for natural spread of citrus tristeza virus after a prolonged lag in Israel, *Phytopathology* **68:**1110–1111.
5. Bar-Joseph, M., and Fraenkel, H., 1983, Spraying citrus plants with Kaolin suspensions reduces colonization by the spiraea aphid (*Aphis citricola* van der Goot), *Crop Prot.* **2:**371–374.
6. Bar-Joseph, M., Garnsey, S.M., and Gonsalves, D., 1979, The closteroviruses: A distinct group of elongated plant viruses, *Adv. Virus Res.* **25:**93–168.
7. Bar-Joseph, M., and Loebenstein, G., 1973, Effects of strain, source plant, and temperature on the transmissibility of citrus tristeza virus by the melon aphid, *Phytopathology* **63:**716–720.
8. Bar-Joseph, M., Marcus, R., and Lee, R.F., 1989, The continuous challenge of citrus tristeza virus control, *Annu. Rev. Phytopathol.* **XX:**in press.
9. Bar-Joesph, M., and Murant, A.F., 1982, Closterovirus group, *CMI/AAB Description of Plant Viruses, CMI/AAB* No. 260.
10. Bar-Joesph, M., Raccah, B., and Loebenstein, G., 1977, Evaluation of the

main variables that affect citrus tristeza virus transmission by aphids, *Proc. Int. Soc. Citricult.* **3**:958–961.

11. Bar-Joseph, M., Roistacher, C.N., and Garnsey, S.M., 1983, The epidemiology and control of citrus tristeza disease, in Plumb, R.T., and Thresh, J.M. (eds): Plant Virus Epidemiology, Blackwell, Oxford, pp. 61–72.
12. Bar-Joseph, M., Roistacher, C.N., Garnsey, S.M., and Gumpf, D.J., 1981, A review on tristeza, an ongoing threat to citriculture, *Proc. Int. Soc. Citricult.* **1**:419–423.
13. Bem, F., and Murant, A.F., 1979, Host range, purification and serological properties of heracleum latent virus, *Ann. Appl. Biol.* **92**:243–256.
14. Bennett, C.W., and Costa, A.S., 1949, Tristeza disease of citrus, *J. Agric. Res.* **78**:203–237.
15. Bennett, C.W., 1960, Sugar beet yellows disease in the United States, *Tech. Bull. U.S. Dept. Agric.* **1218**:1–63.
16. Ben Ze'ev, I.S., Bar-Joseph, M., Nitzan, Y. and Marcus, R. 1989, A severe citrus tristeza virus isolate causing the collapse of trees of sour orange before virus is detectable through the canopy, *Ann. Appl. Biol.* **114**:293–300.
17. Berger, P.H., and Pirone, T.P., 1986, The effect of helper component on the uptake and localization of potyviruses in *Myzus persicae, Virology* **153**:256–261.
18. Bodenheimer, F.S., and Swirski, E., 1957, The Aphidoidea of the Middle East, Weizmann Science Press, Jerusalem, 378 p.
19. Bradley, R.H.E., and Ganong, R.Y., 1955a, Evidence that potato virus Y is carried near the tip of the stylets of the aphid vector *Myzus persicae, Can. J. Microbiol.* **1**:775–782.
20. Bradley, R.H.E., Wade, C.V., and Wood, F.A., 1962, Aphid tranmission of potato virus Y inhibited by oils, *Virology* **18**:327–328.
21. Brunt, A.A., and Kenten, R.H., 1974, Cowpea mold mottle virus, *CMI/AAB Description of Plant Viruses,* No. 140.
22. Calavan, E.C., Mather, A.M., and McEachern, E.H., 1978, Registration, certification and indexing of citrus trees, in Reuther, W., Calavan E.C., and Carman, G.E. (eds): Citrus Industry, Volume 4, University of California, Riverside, Ca., pp. 185–222.
23. Cambra, M., Hermoso de Mendoza, A., Moreno, P., and Navarro, L., 1982, Use of enzyme-linked immunosorbent assay (ELISA) for detection of citrus tristeza virus (CTV) in different aphid species, *Proc. Int. Soc. Citricult.* **1**:444–448.
24. Childress, S.A., and Harris, K.F., 1989, Localization of virus-like particles in the foreguts of viruliferous *Graminiella nigrifrons* leafhoppers carrying the semipersistent maize chlorotic dwarf virus, *J. Gen. Virol.* **70**:247–251.
25. Conti, M., Milne, R.G., Luisoni, E., and Boccardo, G., 1980, A closterovirus from stem-pitting diseased grapevine, *Phytopathology* **70**:394–399.
26. Costa, A.S., 1969, Whiteflies as virus vectors, in Maramorosch, K. (ed): Viruses, Vectors and Vegetation, Interscience, New York, pp. 95–119.
27. Costa, A.S., and Grant, T.J., 1951, Studies on transmission of the tristeza virus by the vector, *Aphis citricidus, Phytopathology.* **41**:105–113.
28. Costa, A.S., and Muller, G.W., 1980, Tristeza controlled by cross protection. A US-Brazil cooperative success, *Plant Dis.* **64**:538–541.
29. Costa, C.L., Muller, G.W., Costa, A.S., and Sobrinho, J., 1974, Reducing

tristeza infection of citrus seedlings by repelling *Toxoptera citricidus* with rice husk mulch, in Weathers, L.G. and Cohen, M. (eds): *Proc. 6th Int. Conf. Organization of Citrus Virologists,* University of California Press, Berkeley, pp. 97–100.

30. Davino, M., and Patti, I., 1986, Preliminary results of citrus tristeza virus transmission by aphids in Italy, in Cavalloro, R. and Di Martino, E. (eds): *Integrated Pest Control in Citrus Groves. Proc. Expert's Meeting Acireale, 26–29 Mar. 1985* A.A. Balkema, Rotterdam, & Boston, pp. 305–309.
31. Dickson, R.C., Flock, R.A., and Johnson, M.McD. 1951, Insect transmission of citrus quick decline virus, *J. Econ. Entomol.* **44:**172–176.
32. Dickson, R.C., Johnson, M.M., Flock, R.A., and Laird, E.F., 1956, Flying aphid populations in southern California citrus groves and their relation to the transmission of the tristeza virus, *Phytopathology* **46:**204–210.
33. Ducharme, E.P., Knorr, L.C., and Speroni, H.A., 1951, Observations on the spread of tristeza in Argentina, *Fla. Citrus Mag.* **13:**10–14.
34. Duffus, J.E., 1972, Beet yellow stunt, a potentially destructive virus disease of sugar beet and lettuce, *Phytopathology* **62:**161–179.
35. Duffus, J.E., 1973, The yellowing virus diseases of beet, *Adv. Virus Res.* **18:**347–386.
36. Elnagar, S., and Murant, A.F., 1976a, Relations of the semipersistent viruses, parsnip yellow fleck and anthriscus viruses, with their vector, *Cavariella aegopodii, Ann. Appl. Biol.* **84:**153–167.
37. Elnagar, S., and Murant, A.F., 1976b, The role of the helper virus, anthriscus yellow, in the transmission of parsnip yellow fleck virus by the aphid *Cavariella aegopodii, Ann. Appl. Biol.* **84:**169–181.
38. Engelbrecht, D.J., and Kasdorf, G.G.F., 1985, Association of a closterovirus with grapevines indexing positive for grapevine leafroll disease and evidence for its natural spread in grapevine, *Phytopathol. Medit.* **24:**101–105.
39. Galvez, G.E., 1968, Purification and characterization of rice tungro virus by analytical density-gradient centrifugation, *Virology* **35:**418–426.
40. Garnsey, S.M., Gonsalves, D., and Purcifull, D.E., 1977, Mechanical transmission of citrus tristeza virus, *Phytopathology* **67:**965–968.
41. Garnsey, S.M., Gumpf, D.J., Roistacher, C.N., Civerolo, E.L., Lee, R.F., Yokomi, R.K., and Bar-Joseph, M., 1987, Towards a standardized evaluation of the biological properties of citrus tristeza virus, *Phytophylactica* **19:**151–157.
42. Garrett, R.G., 1973, Non-persistent aphid-borne viruses, in Gibbs, A.J. (ed): Viruses and Invertebrates, Elsevier North-Holland, Amsterdam, London, pp. 476–492.
43. Gibson, R.W., and Rice, A.D., 1986, The combined use of mineral oils and pyrethroids to control plant viruses transmitted non- and semi-persistently by *Myzus persicae, Ann. Appl. Biol.* **109:**465–472.
44. Gibson, R.W., Rice, A.D., and Sawicki, R.M., 1982, Effects of the pyrethroid deltamethrin on acquisition and inoculation of viruses by *Myzus persicae, Ann. Appl. Biol.* **100:**49–54.
45. Gingery, R.E., 1976, Properties of maize chlorotic dwarf virus and ints ribonucleic acid, *Virology* **73:**311–318.
46. Granhall, I., 1961, The danger of introducing tristeza and its most efficient vector into the Mediterranean area, in Price, W.C. (ed): *Proc. 2nd Conf. Intl.*

Organization Citrus Virologists, University of Florida Press, Gainsville, pp. 113–115.
47. Harris, K.F., 1977, An ingestion-egestion hypothesis of non-circulative virus transmission, in Harris, K.F., and Maramorosch, K. (eds): Aphids as Virus Vectors, Academic Press, New York, London, pp. 165–220.
48. Harris, K.F., 1979, Leafhoppers and aphids as biological vectors: Vector–virus relationships, in Maramorosch, K., and Harris, K.F. (eds): Leafhopper Vectors and Plant Disease Agents, Academic Press, New York, pp. 217–308.
49. Harris, K.F., and Bath, J.E., 1973, Regurgitation by *Myzus persicae* during membrane feeding: its likely function in transmission of nonpersistent plant viruses, *Ann. Entomol. Soc. Am.* **66:**793–796.
50. Harris, K.F., and Childress, S.A., 1983, Cytology of maize chlorotic dwarf virus infection in corn, *Int. J. Trop. Dis.* **1:**135–140.
51. Harris, K.F., Treur, B., Tsai, J., and Toler, R., 1981, Observations on leafhopper ingestion–egestion and vector feeding behavior in noncirculative transmission by leafhoppers, *J. Econ. Entomol.* **74:**446–453.
52. Harrison, B.D., and Murant, A.F., 1984, Involvement of virus-coded proteins in transmission of plant viruses by vectors, in Mayo, M.A., and Harrap, K.A. (eds): Vectors in Virus Biology, Academic Press, New York, pp. 1–36.
53. Heathcote, G.D., and Cockbain, A.J. 1966, Aphids from mangold clamps and their importance as vectors of beet viruses, *Ann. Appl. Biol.* **57:**321–336.
54. Hemida, S.K., Murant, A.N., and Duncan, G.H., 1989, Purification and some particle properties of anthriscus yellows virus, a phloem-limited semi-persistent aphid-borne virus. *Ann. Appl. Biol.* **114:**71–86.
55. Hemida, S.K., and Murant, A.N., 1989, Some properties of parsnip yellow fleck virus. *Ann. Appl. Biol.* **114:**87–100.
56. Hibino, H., 1983, Relations of rice bacilliform and rice tungro spherical viruses with their vector *Nephotettix virescens, Ann. Phytopathol. Soc. Japan* **49:**545–553.
57. Hibino, H., and Cabauatan, P.Q., 1987, Infectivity neutralization of rice tungro-associated viruses acquired by vector leafhoppers, *Phytopathology* **77:**473–476.
58. Hibino, H., Roechan, M., and Sudarisman, S., 1978, Association of two types of virus particles with penyakit habang (tungro disease of rice) in Indonesia, *Phytopathology* **68:**1412–1416.
59. Hille Ris Lambers, D., 1966, Polymorphism in Aphididae. *Ann. Rev. Entomol.* **11:**47–78.
60. Hughes, W.A., and Lister, C.A., 1953, Lime dieback in the Gold Coast, a virus disease of the lime, *Citrus aurantifolia* (Christmann) Swingle, *J. Hortic. Sci.* **28:**131–140.
61. Hull, R., and Heathcote, G.D., 1967, Experiments on the time of application of insecticide to decrease the spread of yellowing viruses of sugar beet, 1954–1966, *Ann. Appl. Biol.* **60:**469–478.
62. Hunt, R.E., Nault, L.R., and Gingery, R.E., 1988, Evidence for infectivity of maize chlorotic dwarf virus and for helper component in its leafhopper transmission, *Phytopathology* **78:**499–504.
63. Inouye, T., and Mitsuhata, K., 1973, Carnation necrotic fleck virus, *Ber. Ohara, Inst. Landwirtsch, Biol. Okayama Univ.* **15:**195–206.

64. Iwaki, M., Thongmeearkom, P., Prommin, M., Honda, Y., and Hibi, T., 1982, Whitefly transmission and some properties of cowpea mild mottle virus, *Plant Dis.* **66**:365–368.
65. Kennedy, J.S., Day, M.F., and Eastop, V.F., 1962, A Conspectus of Aphids as Vectors of Plant Diseases. Commonwealth Institute of Entomology, London, p. 114.
66. Khush, G.S., 1980, Breeding for multiple disease and insect resistance in rice, in Harris, M.K. (ed): Biology and Breeding for Resistance to Arthropods and Pathogens in Agricultural Plants, *Texas Agric. Exp. Sta.* MP-1451, pp. 341–354.
67. Klas, F.E., 1980, Population densities and spatial patterns of the aphid *Toxoptera citricida* Kirk, in Calavan, E.C., Garnsey, S.M., and Timmer, L.W. (eds): *Proc. 8th Conf. Int. Organization Citrus Virologists,* University of California Press, Riverside, pp. 83–87.
68. Kring, J.B., 1972, Flight behavior of aphids, *Annu. Rev. Entomol.* **17**:461–492.
69. Lee, R.F., Brlansky, S.M., Garnsey, S.M., and Yokomi, R.K., 1987, Traits of citrus tristeza virus important for mild strain cross protection of citrus: The Florida approach, *Phytophylactica* **19**:215–218.
70. Lees, A.D., 1966, The control of polymorphism in aphids, *Adv. Insect Physiol.* **3**:207–277.
71. Lim, W.L., and Hagendorn, J.M., 1977, Bimodal transmission of plant viruses, in Harris, K.F., and Maramorosch, K., (eds): Aphids as Virus Vectors, Academic Press, New York, London, pp. 237–252.
72. Lopez-Abella, D., Bradley, R.H.E., and Harris, K.F., 1988, Correlation between stylet paths made during superficial probing and the ability of aphids to transmit nonpersistent viruses, in Harris, K.F. (ed): *Advances in Disease Vector Research,* Volume 5, Springer-Verlag, New York, pp. 251–285.
73. Lot, H., Delecolle, B., and Lecoq, H., 1983, A whitefly transmitted virus causing muskmelon yellows in France, *Acta Hortic.* **127**:175–182.
74. Matthews, R.E.F., 1982, Classification and nomenclature of viruses. Fourth Report of the International Committee on Taxonomy of Viruses. *Interviology* **17**:1–199.
75. McFarlane, J.S., Skoyen, I.O., and Lewellen, R.T., 1969, Development of sugar beet breeding lines and varietes resistent to yellows, *J. Am. Soc. Sugar Beet Technol.* **15**:347–360.
76. Mendt, R., Plaza, G., Boscan, R., Matinez, J., and Lastra, R., 1984, Spread of citrus tristeza virus and evaluation of tolerant rootstocks in Venezuela, in Garnsey, S.M., Timmer, L.W., and Dodds, J.A. (eds): *Proc. 9th Conf. Intrntl. Organization Citrus Virologists,* University of California Press, Riverside, pp. 95–99.
77. Meneghini, M., 1946, Sobre a natureza e transmissibilidade da doenca "tristeza" dos citrus, *Biologico* **12**:285–287.
78. Milne, R.G., Conti, M., Lesemann, D.E., Stellmach, G., Tanne, E., and Cohen, J., 1984, Closterovirus-like particles of two types associated with diseased grapevines, *Phytopathol. Z.* **110**:360–368.
79. Muller, G.W., Rodriguez, O., and Costa, A.S., 1968, A tristeza virus complex severe to sweet orange varietes, in Childs, J.F.L. (ed): *Proc. 4th Conf. Intrntl. Organization Conf. Citrus Virologists,* University of Florida Press, Gainsville, pp. 64–71.

80. Maniyappa, V., and Reddy, D.V.R., 1983, Transmission of cowpoea mild mottle virus by *Bemisia tabaci* in a non-persistent manner, *Plant Dis.* **67:**391–393.
81. Murant, A.F., and Duncan, G.H., 1984, Nature of the dependence of heracleum latent virus on heracleum virus 6 for transmission by the aphid *Cavariella theobaldi, Proc. 4th Int. Congr. Virol. Sendai, Japan, 1984.* **W 45–3:**328.
82. Murant, A.F., and Goold, R.A., 1968, Purification, properties and transmission of parsnip yellow fleck, a semi-persistent aphid-borne virus, *Ann. Appl. Biol.* **62:**123–137.
83. Murant, A.F., Raccah, B., and Pirone, T.P., 1988, Transmission by vectors, in Milne, R.G. (ed): The Filamentous Viruses, Volume 4, Plenum Publishing, New York, pp. 237–273.
84. Murant, A.F., Roberts, I.E., and Elnagar, S. 1976, Association of virus-like particles with the foregut of the aphid *Cavariella aeqopodii* transmitting the semipersistent viruses anthriscus yellows and parsnip yellow fleck, *J. Gen. Virol.* **31:**47–57.
85. Murant, A.F., and Roberts, I.M., 1977, Virus-like particles in phloem tissue of chervil (*Anthriscus cerefolium*) infected with anthriscus yellows virus, *Ann. Appl. Biol.* **85:**403–406.
86. Nault, L.R., Styer, W.E., Knoke, J.K., and Pitre, H.N., 1973, Semipersistent transmission of leafhopper-borne maize chlorotic dwarf virus, *J. Econ. Entomol.* **66:**1271–1273.
87. Navarro, L., Juarez, J., Pina, J.A., and Ballester, J.F., 1984, The citrus quarantine station in Spain, in Garnsey, S.M., Timmer, L.W., and Dodds, J.A. (eds): *Proc. 9th Conf. Int. Organization Citrus Virologists,* University of California Press, Riverside, pp. 365–370.
88. Nitzany, F., and Cohen, S., 1960, A whitefly transmitted virus of cucurbits in Israel, *Phytopathol. Med.* **1:**44–46.
89. Norman, P.A., and Grant, T.J., 1954, Preliminary studies of aphid transmission of tristeza virus in Florida, *Citrus Industry* **35:**10–12.
90. Norman, P.A., and Grant, T.J., 1956, Transmission of tristeza virus by aphids in Florida, *Proc. Fla. St. Hort. Soc.* **69:**38–42.
91. Norman, P.A., Sutton, R.A., and Burditt, A.K., Jr., 1968, Factors affecting transmission of tristeza virus by melon aphids, *J. Econ. Entomol.* **16:**238–242.
92. Norman, P.A., Sutton, R.A., and Selhime, A.G., 1972, Further evidence that tristeza virus is transmitted semipersistently by the melon aphid, *J. Econ. Entomol.* **65:**593–594.
93. Pirone, T.P., 1977, Accessory factors in nonpersistent virus transmission, in Harris, K.F., and Maramorosch, K. (eds): Aphids as Virus Vectors, Academic Press, New York, London: pp. 221–236.
94. Pirone, T.P., and Harris, K.F., 1977, Nonpersistent transmission of plant viruses by aphids, *Ann. Rev. Phytopathol.* **15:**55–73.
95. Raccah, B., 1986, Non-persistent viruses: Epidemiology and control, *Adv. Virus Res.* **31:**387–429.
96. Raccah, B., Loebenstein, G., and Bar-Joseph, M., 1976, Transmission of citrus tristeza virus by the melon aphid, *Phytopathology* **66:**1102–1104.
97. Raccah, B., Loebenstein, G., Bar-Joseph, M., and Oren, Y., 1976, Transmission of tristeza by aphids prevalent on citrus, and operation of the tristeza

suppression programme in Israel, in Calavan, E.C. (ed): *Proc. 7th Conf. Int. Organization Citrus Virologists,* University of California Press, Riverside, pp. 47–49.

98. Raccah, B., Loebenstein, G., and Singer, S., 1980, aphid-transmissibility variants of citrus tristeza virus in infected citrus trees, *Phytopathology* **70:**89–93.
99. Raccah, B., and Singer, S., 1987, The incidence and vectorial potential of the aphids which transmit citrus tristeza virus in Israel, *Phytophylactica* **19:**173–177.
100. Reuterma, M., and Price, W.C., 1972, Evidence that tristeza virus is stylet-borne, *FAO Plant Prot. Bull.* **20:**111–114.
101. Rietberg, H., and Hijner, J.A., 1956, Die Bekámpfung der Vergilbungskrankheit der Rüben in den Niederlanden, *Zucker* **9:**483–485.
102. Roistacher, C.N., 1981a, A blueprint for a disaster—Part 1: The destructive potential for seedling yellows, *Citrograph* **67:**28–32.
103. Roistacher, C.N., 1981b, A blueprint for a disaster—Part 2: The destructive potential of seedling yellows, *Citrograph* **67:**48–53.
104. Roistacher, C.N., 1988, Observations on the decline of sweet orange trees in coastal Peru caused by stem pitting tristeza, *FAO Plant Protectron Bull.* **36:**19–26.
105. Roistacher, C.N., and Bar-Joesph, M., 1984, Transmission of tristeza and seedling yellow tristeza virus by *Aphis gossypii* from sweet orange, grapefruit and lemon to mexican lime, grapefruit and lemon, in Garnsey, S.M., Timmer, L.W., and Dodds, J.A. (eds): *Proc. 9th Conf. Int. Organization Citrus Virologists,* University of California Press, Riverside, pp. 9–18.
106. Roistacher, C.N., and Bar-Joesph, M., 1987a, Aphid transmission of citrus tristeza virus: A review, *Phytophilactaca* **19:**163–167.
107. Roistacher, C.N., and Bar-Joseph, M., 1987b, Transmission of citrus tristeza virus (CTV) by *Aphis gossypii* and by graft inoculation to and from *Passiflora* Spp., *Phytophylactica* **19:**179–182.
108. Roistacher, C.N., Bar-Joseph, M., and Gumpf, D.J., 1984, Transmission of tristeza and seedling yellows tristeza virus by small populations of *Aphis gossypii, Plant Dis.* **68:**494–496.
109. Roistacher, C.N., Dodds, J.A., and Bash, J.A., 1987, Means of obtaining and testing protective strains of seedling yellows and stem pitting tristeza virus: A preliminary report, *Phytophylactica* **19:**199–203.
110. Roistacher, C.N., Dodds, J.A., and Bash, J.A., 1988, Cross-protection against citrus tristeza seedling yellows (CTV-SY) and stem pitting (CTV-SP) viruses by protective isolates developed in greenhouse plants, in Timmer, L.W., Garnsey, S.M., and Navarro, L. (eds): *Proc. 10th Conf. Int. Organization Citrus Virologists,* University of California, Riverside, pp. 91–100.
111. Roistacher, C.N., Nauer, E.M., Kishaba, A., and Calavan, E.C., 1980, Transmission of citrus tristeza virus by *Aphis gossypii* reflecting changes in virus transmissibility in California, in Calavan, E.C., Garnsey, S.M., and Timmer, L.W. (eds): *8th Proc. Conf. Int. Organization Citrus Virologists,* University of California Press, Riverside, pp. 76–82.
112. Rosciglione, B., Castellano, M.A., Martelli, G.P., Savino, V., and Cannizzaro, G., 1983, Mealybug transmission of grapevine virus A, *Vitis* **22:**331–347.

113. Rosner, A., Ginzburg, I., and Bar-Joseph, M., 1983, Molecular cloning of complementary DNA sequences of citrus tristeza virus RNA, *J. Gen. Virol.* **64:**1757–1763.
114. Russell, G.E., 1964, Breeding for tolerance to beet yellows virus and beet mild yellowing virus in sugar beet, *Ann. Appl. Biol.* **53:**377–388.
115. Russell, G.E., 1965, The host range of some English isolates of beet yellowing viruses, *Ann. Appl. Biol.* **55:**245–252.
116. Russell, G.E., 1970, Effects of mineral oil on *Myzus persicae* (Sulz.) and its transmission of beet yellows virus, *Bull. Entomol. Res.* **59:**691–694.
117. Russell, G.E., 1972, Inherited resistance to virus yellows in sugar beet, *Proc. R. Soc. Ser. B* **181:**267–279.
118. Schneider, H., 1954, Anatomy of bark bud union, trunk and roots of quick-decline-affected sweet orange trees on sour orange rootstock, *Hilgardia* **22:**567–601.
119. Schneider, H., 1959, The anatomy of tristeza-virus-infected-citrus, in Wallace, J.M. (ed): Citrus Virus Diseases, University of California Press, Riverside, pp. 73–84.
120. Schneider, H., 1969, Pathological anatomies of citrus affected by virus diseases and by apparent-inherited disorders and their use in diagnosis, in *Proc. Int. Soc. Citricult.* **3:**1489–1494.
121. Schwartz, R.E., 1965, Aphid-borne diseases of citrus and their vectors in South Africa. B. Flight activities of citrus aphids, *S. Afr. J. Agric. Sci.* **8:**931–940.
122. Sela, I., Assouline, I., Tanne, E., Cohen, S., and Marco, S., 1980, Isolation and characterization of a rod-shaped, whitefly-transmissible DNA-containing plant virus, *Phytopathology* **70:**226–228.
123. Simons, J.N., 1982, Use of oil sprays and reflective surfaces for control of insect-transmitted plant viruses, in Maramorosch, K., and Harris, K.F. (eds): Vectors and Plant Diseases: Approaches to Control, Academic Press, New York, pp. 71–93.
124. Singh, A.B., and Chakraborty, N.K., 1978, Transmission of citrus tristeza virus by a membrane-feeding technique, *Hort. Res.* **17:**83–86.
125. Strickland, A.H., 1950, The dispersal of Pseudococcidae [Hemiptera-Homoptera] by air currents in the Gold Coast, *Proc. R. Soc. Ser. A. (London)* **25:**1–9.
126. Stroyan, H.L.G., 1961, Identification of aphids living on citrus, *FAO Plant Protein Bull.* **9:**45–65.
127. Stubbs, L.L., 1964, Transmission of protective inoculations with viruses of the citrus tristeza complex, *Austr. J. Agric. Res.* **15:**752–770.
128. Sylvester, E.S., 1956, Beet mosaic and beet yellows virus transmission by the green peach aphid, *J. Am. Soc. Sugar Beet Technol.* **9:**57–61.
129. Sylvester, E.S., 1980, Circulative and propagative virus transmission by aphids, *Ann. Rev. Entomol.* **25:**257–286.
130. Sylvester, E.S., 1985, Multiple acquisition of viruses and vector-dependent prokaryotes: Consequences on transmission, *Ann. Rev. Entomol.* **30:**71–88.
131. Sylvester, E.S., and Bradley, R.H.E., 1962, Effect of treating stylets of aphids with formalin on the transmission of sugar beet yellows virus, *Virology* **17:**381–386.
132. Tamaki, S., Roistacher, C.N., and Dodds, J.A., 1983, California field isolates

of citrus tristeza virus (CTV) have little cross-protective ability against severe seedling yellows strains of CTV, *Phytopathology* **73:**962.
133. Tanne, E., Ben Dov, Y., and Raccah, B., 1987, Transmission of clostero-like particles by mealybugs (*Pseudococcidae*) in Israel, *Proc. Int. Council Meeting for Virus Diseases of Grapevine. (Kiryat Anavim, Israel, 1987),* pp. 15–16.
134. Vanderveken, J., and Semal, J., 1966, Aphid transmission of beet yellows virus inhibited by mineral oils, *Phytopathology* **56:**1210–1211.
135. Van Hoof, H.A., 1957, On the mechanism of transmission of some plant viruses, *Proc. Kon. Akad. Wetensch. Ser. C.* **60:**314–317.
136. Varma, P.M., Rao, D.G., and Capoor, S.P., 1965, Transmission of tristeza virus by *Aphis craccivora* (Koch) and *Dactynotus jacea* (L.), *Indian J. Entomol.* **1:**67–71.
137. Varma, P.M., Rao, D.G., and Vasudeva, R.S., 1960, Additional vectors of tristeza disease of citrus in India, *Curr. Sci.* **29:**359.
138. Wallace, J.M., and Drake, R.J., 1974, Field performance of tristeza susceptible citrus trees carrying virus derived from plants that recovered from seedling yellows, in Weathers, L.G., and Cohen, M. (eds): *Proc. 6th Conf. Int. Organization Citrus Virologists,* University of California Press, Berkeley, pp. 67–74.
139. Watson, M.A., 1940, Studies on the transmission of sugar beet yellows virus by the aphid *Myzus persicae., Proc. R. Soc. Ser. B (London)* **128:**535–552.
140. Watson, M.A., 1946, The transmission of beet mosaic and beet yellows viruses by aphides: A comparative study of a non-persistent and a persistent virus having host plants and vectors in common, *Proc. R. Soc. Ser. B (London)* **133:**200–219.
141. Watson, M.A., 1951, Beet yellows and other yellowing virus diseases of sugar beet, *Report of the Rothamsted Exp. Sta.,* pp. 157–167.
142. Watson, M.A., 1972, Transmission of plant viruses by aphids, in Kado, C.I., Agrawal, H.O. (eds): Techniques in Plant Virology, Van Nostrand, New York, pp. 131–167.
143. Watson, M.A., Heathcote, G.D., Lauckner, F.B., and Sowray, P.A., 1975, The use of weather data and counts of aphids in the field to predict the incidence of yellowing viruses of sugar-beet crops in England in relation to the use of insecticides, *Ann. Appl. Biol.* **81:** 181-198
144. Watson, M.A., and Roberts, F.M., 1939, A comparative study of the transmission of Hyoscyamus virus 3, potato virus Y and cucumber virus I by the vectors *Myzus persicae* (Sulz.), *M. circumflexus* (Buckton) and *Macrosophum gei* (Koch), *Proc. R. Soc. Ser. B (London)* **127:**543–576.
145. Yamashita, S., Doi, Y., Yora, K., and Yoshino, M., 1979, Cucumber yellows virus: Its transmission by the greenhouse whitefly *Trialeurodes vaporariorum* (Westwood), and the yellowing disease of cucumber and muskmelon caused by the virus, *Ann. Phytopathol. Soc. Japan* **45:**484–496.
146. Yokomi, R.K., and Garnsey, S.M., 1987, Transmission of citrus tristeza virus by *Aphis gossypii* and *Aphis citricola* in Florida, *Phytophylactica* **19:**169–172.
147. Yokomi, R.K., Garnsey, S.M., Lee, R.F., and Cohen, M., 1987, Use of vectors for screening protecting effects of mild CTV isolates in Florida, *Phytophylactica* **19:**183–185.

Index

Accessory glands, 102
Acertagallia, as *Spiroplasma citri* vector, 278
Acetylcholine, as neurotransmitter, 29
Acyrthosiphon pisum
  as beet yellows virus vector, 311
  as pea enation mosaic virus vector, 194-195
Adiopokinetic hormone, 44-45
Adipokinetic hormone-like peptide, 41
Adrenalin, diuretic effects, 34
Adrenocorticotropic hormone, 49
Aedes aegypti
  as *Borrelia burgdorferi* vector, 140–141
  diuretic factors, 34
  midgut
    anatomy, 32, 35
    endocrine cells, 36
    host-parasite interactions in, 49
    secretory granules, 38
  yolk deposition, 34
Aedes atropalpus
  as *Borrelia burgdorferi* vector, 140–141
  midgut endocrine cells, 36
  yolk deposition, 34
Aedes head peptide, 45, 46
Aedes sierrensis, aggregation pheromones, 116
Aedes triseriatus, as *Borrelia burgdorferi* vector, 140–141
Agallia constricta, as rice dwarf virus vector, 198
Age determination, of sandflies, 102–104
Alenquer virus, 101
Aleyrodidae, as semipersistent virus vectors, 305, 306, 307
Alfalfa dwarf disease
  epidemiology, 257, 258
  pathogens, 244
  symptoms, 245–246
Allele frequency variation, of sandflies, 113–114
Almond leaf scorch, 244
  epidemiology, 257, 258
  symptoms, 245–246
Amblyomma, aggregation pheromones, 116
Amblyomma americanum, as *Borrelia burgdorferi* vector, 139–140
Amino acids, as neurotransmitters, 29
Anopheles albimanus, rDNA, 14
Anopheles arabiensis
  rDNA, 6, 13
    evolution, 21, 22
    intergenic spacer structures, 11, 19–20
    structure, 7, 8
  as malaria vector, 1
  phylogenetic relationships, 4
  restriction fragment length polymorphisms, 17
Anopheles bwambae
  as malaria vector, 1
  phylogenetic relationships, 3
Anopheles darlingi, nulliparous biting activity, 103

Anopheles freeborni, diuretic factors, 34
Anopheles gambiae
DNA probe, 114
gametocyte infectivity, 68
as malaria vector, 1
phylogenetic relationships, 3–4
range, 3
Anopheles gambiae species complex
definition, 1
rDNA, 1–28
applied uses, 15–20
characteristics, 4–6
evolution, 21–22
genome, 6–7
intergenic spacer structures, 8–12, 16, 18, 19–20, 22, 23
intervening sequences, 12–14
restriction fragment length polymorphisms, 16–18, 23
structure, 6–8
hybridization, 2–4
phylogenetic relationships, 3–4
species identification, 1–2
Anopheles maculipennis, crossing experiments, 108
Anopheles melas
rDNA, 7, 13
evolution, 22
intergenic spacer structures, 11–12
structure, 7, 8
as malaria vector, 1
phylogenetic relationships, 3
range, 3
restriction fragment length polymorphisms, 17
Anopheles merus
rDNA, 6, 7, 13
evolution, 22
intergenic spacer structures, 11–12
structure, 7, 8
as malaria vector, 1
phylogenetic relationships, 3–4
range, 3
restriction fragment length polymorphisms, 17
Anopheles quadriannulatus
rDNA, 6, 7, 13
evolution, 21–22
intergenic spacer structures, 11, 20
structure, 7, 8
as malaria vector, 1
phylogenetic relationships, 3
restriction fragment length polymorphisms, 17
Anopheles stephensi, midgut immunoreactive cells, 44
Anthriscus yellows virus
semipersistent transmission, 193, 328, 329
helper virus in, 329
parsnip yellow fleck virus dependence in, 315, 316
RNA, 305
vectors, 311
Antibodies, gametocytemia and, 73
Aphids. *See also* names of specific species and genera
as beet yellows virus vector, 330–332
black citrus. *See Toxoptera aurantii*
brown citrus. *See Toxoptera citricidus*
as citrus tristeza virus vector, 301–302, 308–310, 330–332
as closterovirus vectors, 301
green citrus. *See Aphis citricola*
green peach. *See Myzus persicae*
melon. *See Aphis gossypii*
as rice virus vector, 210, 211, 213
as semipersistent virus vector, 305, 306, 308
control, 324–325
factors affecting, 317–320
sheath saliva, 251
Aphis citricola
as citrus tristeza virus vector, 308–309, 310
control, 325
as potato virus Y vector, 194
Aphis craccivora
as citrus tristeza virus vector, 308–309

as potato virus Y vector, 194
Aphis fabae
as beet yellows virus vector, 310, 311
control, 325
Aphis gossypii
as citrus tristeza virus vector, 309, 310, 317
as cucumber mosaic virus vector, 193–194
as potato virus Y vector, 194
transmission efficiency, 318, 319–320
Aphis jacobaeae, as citrus tristeza virus vector, 309
Aphis nasturtii, as beet yellows virus vector, 311
Aphis nerii, as citrus tristeza virus vector, 309
Apophylis, as rice yellow mottle virus vector, 229
Apple decline, 154
Arabis mosaic virus
antigenic variants, 179
detection in virus, 200
dispersal, 180
geographical distribution, 162, 164
hosts, 153, 154, 155
natural spread, 165
physicochemical properties, 158
serology, 161, 163
vectors, 165, 173, 174, 175, 178–179
Arbia virus, 98
Arboledas virus, 96–97, 101
Argas, as spirochete vector, 127, 129
Argas persica, as *Borrelia anserina* vector, 132
Arracacha virus A
geographical distribution, 162
natural spread, 165
physicochemical properties, 159
Arracacha virus B
geographical distribution, 162
natural spread, 165
physicochemical properties, 158–159
Artichoke Italian latent virus
geographical distribution, 162
host, 152
natural spread, 165
physicochemical properties, 159
vectors, 165, 199–200
Artichoke vein banding virus
geographical distribution, 162
host, 152
natural spread, 165
physicochemical properties, 158
Artichoke yellow ringspot virus
geographical distribution, 162
host, 152
natural spread, 165
physicochemical properties, 159
Aster yellows mycoplasma-like organism, 271, 279
Atriplex halimus, attraction index, 106
Aulacorthum solani
as subterranean clover red leaf virus vector, 197
as tobacco mosaic virus vector, 192

Bacteria
endophytic, 259
epiphytic, 258–259
as frost damage cause, 258
xylem-inhabiting, 243–266
epidemiology, 257–259
group I, 244
group II, 244, 256
pathogenicity, 245–246
strain differences, 246–247
vectors, 244, 245, 250–252
xylem environment and, 247–250
Baldulus tripsici
host, 269
as *Spiroplasma kunkelii* vector, 269, 282
Barley stripe mosaic virus, 201
Barley yellow dwarf virus, 195, 196, 197, 221
Barley yellow striate mosaic virus, 198
Beet curly top virus, 198–199
Beetles
as bacteria vector, 251
bean leaf, 202

Beetles (*cont.*)
as plant virus vectors, 201–202
as rice virus vector, 210, 211, 213, 214, 229
Beet leafhopper transmitted virescence agent, 279
Beet necrotic yellow vein virus, 201
Beet yellows stunt virus, 311
Beet yellows virus
control, 324
semipersistent transmission, 301, 312–313, 314, 315
vectors, 308, 310–311
identification, 330–332
Bemisia tabaci
as cowpea mild mottle virus vector, 305
as cucumber vein yellowing virus vector, 304
as semipersistent virus vector, 304, 305, 307
as tomato yellow leaf curl virus vector, 202
Biogenic amines
diuretic effects, 34
as neurotransmitter, 29–30
Black line disease, 155
Blood feeding
by *Lutzomyia,* 95, 97
by *Phlebotomus,* 95
by sandflies, 102
Blowfly, insulin-like peptide, 40, 47–48
Blueberry leaf mottle virus
geographical distribution, 162
grapevine decline and, 153, 154
natural spread, 165
physicochemical properties, 159
serology, 161
Blue tongue virus, 98
Bombyx mori, peptide hormones, 39–40
Boophilus, as *Borrrelia theileri* vector, 133–134
Borrelia anserina, behavior in vector, 132
Borrelia burgdorferi
behavior in vectors, 134–146
midgut penetration, 134–136, 139
in nonspecific vectors, 139–141
salivary gland transmission, 138–139
transovarial transmission, 137–138
vectors, 127
detection, 139–140
vesicle formation, 141–144, 145
Borrelia crocidurae, behavior in vector, 132, 133
Borrelia duttonii
behavior in vector, 130
transmission, 132
Borrelia recurrentis
behavior in vector, 130
transmission, 131
Borrelia theileri, behavior in vector, 133–134
Borrelia vincenti, behavior in vector, 131
Borrelioses. *See also* specific *Borrelia* species
bovine, 133–134
spirochete-vector relationships in, 127–150
behavior in vectors, 130–146
Bouquet disease, 153
Brain, of mosquito, 31
immunoreactive cells, 42, 43
vertebrate-like peptide content, 40
Brittle root disease. *See* Horseradish brittle root disease
Broad bean yellow band virus, 168
Brome mosaic virus, 200–201
Bud yellow virus, 303

Cacopsylla pyricola, xylem feeding, 252
Calliphora, immunoreactive peptides, 40, 41. *See also* Blowfly
Calliphora erythrocephala, rDNA, 4, 7
Cardical cell, 32, 33
Carlaviruses, 305
Carnation necrotic fleck virus, 311
Cassava green mottle virus, 159, 163, 165

Cavariella aegopodii
as heracleum virus vector, 311
hosts, 311–312
as parsnip yellow fleck virus vector, 305, 316
as semipersistent viruses vector, 311–312
Cavariella pastinacae
as heracleum virus vector, 311
hosts, 312
as parsnip yellow fleck virus vector, 305, 316
Cavariella theobaldii
as heracleum virus vector, 311
hosts, 312
Ceratoma trifurcata, as plant virus vector, 202
Cercopidae, xylem feeding by, 250
Cercropin, 107
Cereal tillering disease, 216
Chaetocnema, as rice yellow mottle virus vector, 229
Changuinola virus, 97
Cherry leafroll virus
as flat apple cause, 154
geographical distribution, 162, 164
host, 155
natural spread, 165
physicochemical properties, 159
as plant death cause, 178
serology, 161
as union necrosis and decline cause, 154
Cherry leafspot virus, 155
Cherry rasp-leaf virus
geographical distribution, 162
hosts, 154, 155
natural spread, 165
physicochemical properties, 158
vectors, 165
Chicory yellow mottle virus
geographical distribution, 162
host, 153
natural spread, 165
physicochemical properties, 159
Children, gametocytemia in, 68, 69, 72
Chironomus, rDNA, 4, 5
Chloroquine, 61
Chlorosis, 275
Chytridiomycetes, 201
Cicadae, xylem feeding by, 250
Cicadellinae. *See also* Leafhopper(s)
as semipersistent virus vector, 305, 306, 307
as spiroplasma vector, 269
Cicadellinae. *See also* Leafhopper(s)
foregut, 256–257
as *Xylella fastidiosa* vector, 243, 254–257
as *Xylella* vector, 256–257
xylem feeding by, 243–244, 250, 251–252
Cicadulina mbila
as maize streak virus vector, 197, 199
as *Spiroplasma kunkelii* vector, 282
Circadian rhythm, of gametocytemia, 65
Circulifer tenellus, 276–277
as beet curly top virus vector, 198, 279
as beet leafhopper transmitted virescence agent, 279
geographical distribution, 269
host, 269
as mycoplasma-like organism vector, 279, 280
as *Spiroplasma citri* vector, 269
Citrus blight, 244
Citrus stubborn disease
causal agent, 274, 275
symptoms, 275
vectors, 267
Citrus tristeza virus, 301–302
bimodal transmission, 313
control, 322–323, 329
epidemics, 320–322
epidemiology, 316–317
host range, 320
hosts, 303–304
as phloem necrosis-induced bud girdle cause, 302, 326
semipersistent transmission, 313, 314, 315
as stem pitting cause, 302, 326–327
vectors, 301–302, 308–310
acquisition time, 317
factors affecting, 317, 318

Citrus tristeza virus (*cont.*)
identification, 330–332
transmission efficiency, 318–320
Clavibacter xyli, 245, 256
Climate control cabinet, 93, 94
Clostero-like virus, 304
Closteroviruses
aphid-borne, 301
hosts, 303–304
physicochemical properties, 303
Coccoidea, 307
Cockroach, secretory granules, 38
Cocoa necrosis virus, 159, 161, 163, 165
Colladonus montanus, as *Spiroplasma citri* vector, 278
Corn stunt disease
causal agent, 280, 281
complex, 282
vectors, 267
Corpora allata, 30
Corpora cardiaca, 40
Corpus cardiacum, 33, 41
Cowpea mild mottle virus, 305, 313
Cowpea mosaic virus, 201
Crimson clover latent virus, 159, 162, 165
Crossing characteristics, of sandflies, 108–110
Cucumber mosaic virus, 193–194
Cucumber necrosis virus, 201
Cucumber vein yellowing virus, 304–305
helper virus, 328, 329
Cucumber yellows virus, 304
Culicoides, as blue tongue virus vector, 98
Culex pipiens, neurotransmitters, 32
Curly top virus, 197–198. *See also* Beet curly top virus
Cycas necrotic stunt virus, 159, 163, 165

Dalbulus, as *Spiroplasma kunkelii* vector, 270
Dalbulus elimatus
host, 270
as *Spiroplasma kunkelii* vector, 281
Dalbulus gelbus, as *Spiroplasma kunkelii* vector, 282
Dalbulus guevari
host, 270
as *Spiroplasma kunkelii* vector, 282
Dalbulus guzmani
host, 270
as *Spiroplasma kunkelii* vector, 282
Dalbulus longulus
host, 270
as *Spiroplasma kunkelii* vector, 282
Dalbulus maidis
host, 270
as maize rayado fino virus vector, 199, 283
as *Spiroplasma kunkelii* vector, 281
Dalbulus quinquenotatus
host, 270
as *Spiroplasma kunkelii* vector, 282
Dalbulus tripsacoides
host, 270
as *Spiroplasma kunkelii* vector, 282
Deerfly, as Lyme disease vector, 140, 146
Deltocephalineae, 269
Deltocephalus sonorus, as maize chlorotic dwarf virus vector, 304
Deoxyribonucleic acid (DNA)
of bud yellows virus, 303
of cucumber vein yellowing virus, 305
of insect-borne rice viruses, 216
of maize streak virus, 199
of rice dwarf virus, 218
Deoxyribonucleic acid probe
plant virus detection applications, 191, 197
taxonomic applications
Anopheles gambiae, 15–19
sandflies, 114–115
rDeoxyribonucleic acid (rDNA), of *Anopheles gambiae* species complex, 1–28
applied uses, 15–20
characteristics, 4–6
evolution, 21–22
genome, 6–7

intergenic spacer structure, 8–12, 16, 18, 19–20, 22, 23
intervening sequences, 12–14
restriction fragment length polymorphisms, 16–18, 23
structure, 6–8
Dermacentor albipictus, as *Borrelia burgdorferi* vector, 139
Dermacentor parumapertus, as *Borrelia burgdorferi* vector, 139
Dermacentor variabilis, as *Borrelia burgdorferi* vector, 139, 139–140
Diptera
daily growth layers, 104
rDNA, 4–6
midgut endocrine cells, 36
pancreatic polypeptide immunoreactivity, 41
Diterpenoid, 110–111
Diuretic factor, 34
Dopamine, 32, 34
Drosophila
adipokinetic hormone-like peptide content, 41
allele frequency variation, 113
DNA probe, 114–115
rDNA, 4–5
peptide immunoreactivity, 40
as sigma virus vector, 100
song production, 111
Drosophila sex-ratio organism, 267

Ecdysteroid, 33–34
Egg development neurosecretory hormone, 33, 34
Electron microscopy, for plant virus detection, 191, 195, 198–200, 201
Electrophoresis, sandfly taxonomy applications, 113–114
Empoasca, as *Spiroplasma citri* vector, 278
Empoasca fabae, phloem feeding by, 251
Endocrine cell, of mosquito midgut, 35–38
Enzyme-linked immunosorbent assay (ELISA)
for plant virus detection, 191, 202, 323
in aphids, 193, 195, 196, 197, 309
in planthoppers, 198–199
for spiroplasma detection, 274
Eriophyes tulipae
as brome mosaic virus vector, 200–201
as wheat streak mosaic virus vector, 200
Erythroneura variabilis, as *Spiroplasma citri* vector, 278
Escherichia coli, sugar-related inhibition of, 107
Eucelis, xylem feeding by, 252
Eupteryx atropunctata, as tobacco mosaic virus vector, 192
Euscelidius plebejus, as *Spiroplasma citri* vector, 278
Euscelidius variegatus
geographical distribution, 270–271
hosts, 271
as *Spiroplasma citri* vector, 278
as *Spiroplasma kunkelii* vector, 281, 282
Euscelidius varigatus, xylem feeding by, 252
Euscelis plebejus, as *Spiroplasma citri* vector, 276
Eutettix tenellus, as curly top virus vector, 197–198
Exitianus exitiosus, as *Spiroplasma kunkelii* vector, 272, 282
Exocytosis, 38

Farnasene, 110
Fijivirus, 199, 215, 216
Flat apple disease, 154
Follicular relic, 102–103
Foregut
of leafhopper, 249, 256–257
of mosquito, 35
in plant virus transmission, 193
in semipersistent transmission, 314
Xylella colonization, 253–254

Frost damage, bacteria-related, 258
Fungi, as plant virus vectors, 201

Gall, rice virus-related, 216–217
Gametocyte
  in malaria transmission, 65
  patency measurement, 61
  presence/absence determination, 64–65
Gametocytemia
  asexual parasitemia and, 66, 84
  gametocyte density in, 61, 62, 63–64, 65
    age factors, 71
    in children, 65, 68, 69, 72
    circadian rhythm, 65
    high-density, 66–67
  infectiousness, 59, 66
    age factors, 68–71, 79–84
    contingency table analysis, 70–71
    immune response, 71–73, 78–79
    reservoir of infection, 67–71, 81–83, 84
    transitions from, 84–86
  infectiousness versus, 65–67, 78
  mathematical models of malaria and, 73–86
  patterns, 61–65, 77–78
    antibodies and, 73
    during remission, 63
  quantitative measurement, 64
Gametocytogenesis, 60–61
Gammae, of *Borrelia burgdorferi*, 130, 131, 142, 143, 144, 146
Gamma globulin, 72–73
Gastrin, 48
Genome
  Anopheles gambiae, 6–7
  nepoviruses, 151, 156, 179, 180
  tobraviruses, 151, 176
Glossina, rDNA, 4–5, 9
Glossina morsitans morsitans, midgut immunoreactive cells, 44
Glucagon, hoverfly tissue content, 40
Graminella nigrifrons
  as maize chlorotic dwarf virus vector, 198, 304, 314
  as semipersistent virus vector, 307
  as *Spiroplasma kunkelii* vector, 272, 282
Graminella sonora, as *Spiroplasma citri* harborer, 278
Granule, secretory, 30, 38
Grape leaf roll virus complex, 304
Grapevine Bulgarian latent virus
  geographical distribution, 162
  host, 153
  natural spread, 165
  physicochemical properties, 159
  serology, 161
Grapevine chrome mosaic virus, 153
  geographical distribution, 162
  natural spread, 165
  physicochemical properties, 159
  serology, 161
Grapevine decline, 153, 154
Grapevine degeneration, 153
Grapevine fanleaf virus
  dispersal, 180
  geographical distribution, 162, 164
  grapevine decline and, 153
  grapevine degeneration and, 153
  host, 179
  natural spread, 165
  physicochemical properties, 158
  serology, 160–161, 163
  vectors, 151, 165, 170, 175
Grapevine yellow vein virus, 165
Graphocephala atropunctata as *Xylella fastidiosa* vector, 250, 254, 258
  xylem feeding by, 250

Haemaphysalis leporispalustris, as *Borrelia burgdorferi* vector, 139, 140, 146
Harpobittacus australis, pheromone gland, 110
Helper virus, in semipersistent transmission, 329
Hemiptera, as vectors, 250
Heracleum latent virus, 303
  semipersistent transmission, 315
  vectors, 308, 311–312

Hibiscus latent ringspot virus, 159, 163, 165
Hoja blanca virus. *See also* Rice hoja blanca virus
  Echinochloa, 223
Homaladisca coagulata, xylem feeding by, 253
Homoptera
  as closterovirus vectors, 301
  as xylem-inhabiting bacteria vectors, 243–266
Honeydew, as sandfly food source, 104–105
Hopperburn, 251
Hordnia circellata, as *Spiroplasma citri* vector, 278
Hormonal factors, in host-parasite interactions, 49–50
Horsefly, as Lyme disease vector, 140, 141, 146
Horseradish brittle root disease, 271, 274–275, 276
Host-parasite relationships, hormonal interactions, 49–50
Hoverfly, peptides, 40
Hyalopterus pruni, as citrus tristeza virus vector, 309
Hydrocarbon, cuticular, 111–112
Hyperomyzus lactucae, as lettuce necrotic yellows virus vector, 197

Icerya purchasei, 105
Icosahedral virus, 315
Immune response, in gametocytemia, 71–73, 78–79
Immunoreactivity, neuropeptide-like, 39–46, 47–49
  adiopokinetic hormone-like, 44–45
  phe-met-arg-phe-amide-like, 40–41
Inclusion body, 217, 219, 225
Insulin-like peptide, 40, 47–48
Isoenzymes, of sandflies, 113–114
Ixodes dammini, as *Borrelia burgdorferi* vector, 127, 134, 135, 136–137, 139, 140
Ixodes dentatus, as *Borrelia burgdorferi* vector, 139, 140, 146
Ixodes neotomae, as *Borrelia burgdorferi* vector, 139, 140
Ixodes pacificus, as *Borrelia burgdorferi* vector, 127, 134, 138, 139,
Ixodes persulcatus, as *Borrelia burgdorferi* vector, 127, 139
Ixodes ricinus, as *Borrelia burgdorferi* vector, 127, 134, 137–138, 139
Ixodes scapularis, as *Borrelia burgdorferi* vector, 127, 139, 140

Juvenile hormones, 30

Lachnus roboris, honeydew production by, 104
La Crosse virus, 100
Lamellar body, endocrine, 38–39
Laodelphax striatellus
  as barley yellow striate virus vector, 198
  as rice virus vector, 211, 213
    rice black-streaked dwarf virus, 212
    rice stripe virus, 212, 225–226
Laverania, gametocyte development in, 61
Leafhopper(s). *See also* specific species and genera
  beet. *See Circulifer tenellus*
  Deltacephaline, 252
  foregut, 249, 256–257
  mouthparts, 249
  as Pierce's disease vector, 243–244
  as rice dwarf virus vector, 210, 213–214
  semipersistent transmission by, 305, 306, 307
  sharpshooter. *See* Cicadellinae
  as spiroplasma vector, 267–299
    as artificial vector, 273
    as experimental vector, 273
    as natural vector, 273
    spiroplasma-leafhopper relationships, 283–288

Leafhopper(s) (*cont.*)
transmission terminology, 272–274
Typhlocybine, 251
Leafhopper-spiroplasma relationships, 283–288
inoculativity, 284
transmission patterns, 284–285
Leishmania, vectors, 102–103, 106–107
Leishmania braziliensis braziliensis, vectors, 112
Leishmania donovani, vectors, 104
Leishmania infantum, vectors, 112
Leishmania major, vectors, 107, 114
Lettuce necrotic yellows virus, 197
Leucokinin, 34
Leucomyosuppressin, 46
Leucophaea maderae, leucomyosuppressin, 46
Leucosulfakinin, 46
Lice, spirochete behavior in, 130, 131
LL5 cell line, 97
Longidorus
geographical distribution, 168–169, 170–171
as nepoviruses vector
virus association, 177
virus retention, 174–175, 180
virus transmission mechanism, 178
Longidorus apulus
as artichoke Italian latent virus vector, 165
geographical distribution, 169
Longidorus attenuatus
as raspberry ringspot virus vector, 172
as tomato black ringspot virus vector, 174
Longidorus diadecturus
as mulberry ringspot virus vector, 171
as peach rosette mosaic virus vector, 165, 170
Longidorus elongatus
as artichoke Italian latent virus vector, 199–200
morphometry, 170
as peach rosette mosaic virus vector, 170
as raspberry ringspot virus vector, 165, 172, 173, 174, 175, 199
as tomato black ringspot virus vector, 174, 199
virus detection in, 199–200
virus retention site, 174–175, 176, 177
Longidorus fasciatus, as artichoke Italian latent virus vector, 165
Longidorus macrosoma, as raspberry ringspot virus vector, 165, 172, 173, 176–177
Longidorus martini, as mulberry ringspot virus vector, 165, 171
Longidorus profundorum, morphometry, 170
Longidorus veneacola
geographical distribution, 169
morphometry, 170
Lucerne Australian latent virus, 159, 163, 165
Lucerne Australian symptomless virus, 157, 158, 163, 165
Luteoviruses, 195, 221
Lutzomyia anthophora
blood feeding by, 97
colonization, 92
as Rio Grande virus vector, 97
Lutzomyia braziliensis, isoenzymes, 114
Lutzomyia cruciata
colonization, 92
follicular relic, 102
Lutzomyia flaviscutellata
climate control for, 93
follicular relics, 103
gonotropic cycles, 100
isoenzymes, 113
oviposition, 96
as Pacui virus vector, 98, 99
Lutzomyia furcata
colonization, 92
follicular relics, 103
Lutzomyia gomezi, as Arboledas virus vector, 96–97
Lutzomyia intermedia, colonization, 92

Lutzomyia longipalpis
antennae, 116
blood feeding by, 95
as blue tongue virus vector, 98
as changuinola virus vector, 97
chromosomes, 112–113
colonization, 96
crossing experiments, 108–110
diterpenoid production, 110–111
DNA probe, 115
intrathoracic viral inoculation, 97
LL5 cell line, 97
oviposition, 96
as Pacui virus vector, 98
as Rift Valley fever virus vector, 98
sex pheromones, 117–118
song patterns, 111
as vesiculovirus vector, 101
Lutzomyia papatasi, sugar feeding, 105
Lutzomyia shannoni, colonization, 92
Lutzomyia spinicrassa, chromosomes, 113
Lutzomyia trapidoi
colonization, 92
as vesiculovirus vector, 97
Lutzomyia umbratilis
colonization, 92
as phlebovirus vector, 101
Lutzomyia walkeri, colonization, 92
Lutzomyia ylephiletor, isoenzymes, 113
Lyme disease, causal agent. *See* *Borrelia burgdorferi*
Lymnaea stagnalis, lamellar bodies, 39

Macrogametocyte, 65
Macrosiphum euphorbiae
as barley yellow dwarf virus vector, 195
as beet yellows virus vector, 311
as pea seed-borne mosaic virus vector, 193
Macrosiphum rosae, as citrus tristeza virus vector, 309
Macrosteles fascifrons
as aster yellows vector, 271, 279
geographical distribution, 271
host range, 271
as oat blue dwarf virus vector, 198
as spiroplasma vector, 271, 277, 278
Macrosteles quadrilineatus, 277
Macrosteles severini
geographical distribution, 271
as *Spiroplasma citri* vector, 277
Macrosteles sexnotatus, as *Spiroplasma kunkelii* vector, 282
Maize bushy stunt mycoplasma-like organism, 281, 282
Maize chlorotic dwarf virus
differentiation, 281
location in vector, 314, 315
semipersistent transmission, 198, 301, 313, 315, 316
vectors, 283
Maize chlorotic dwarf virus group, 304
Maize rayado fino virus, 199, 281, 282–283
Maize rough dwarf virus, 216
Maize streak virus, 197, 199
Maize stripe virus, 199, 225
Malaria
mathematical models, 73–86, 87
age factors, 79–84
gametocytemia patterns, 77–78
gametocytogenesis, 77
historical background, 73–74
immune response, 78–79
infectiousness versus gametocytemia, 78
transitions out of infectiousness, 84–86
transmission model, 59–60
mosquito infection level, 49
Manduca sexta, peptide immunoreactivity, 40. *See also* Tobacco hornworm
Maraba virus, 101
Mealybug. *See* Pseudococcidae
Medial neurosecretory cell, 33–34
Membrane-feeding, by sandflies, 97, 116–117
Merozoite, 60

Metopolophium dithodium, as rice giallume virus vector, 221
Miaze stripe virus, 225
Midgut, of mosquito
  endocrine system, 30
    anatomy, 32, 35, 36–37
    endocrine cells, 35–38, 46–47
    neuropeptide-like immunoreactivity, 41–46, 47–49
    subcellular organization, 38–39
  host-parasite interactions in, 49–50
Milaparvata lugens, as rice virus vector, 211, 213
Mineral oil, for aphid control, 324–325, 329
Mite, virus detection in, 200–201
Mollicute, 268, 269
Mollusc, peptide immunoreacivity, 40
Mosaic viruses. *See* specific mosaic viruses
Mosaic virus yellow bud, 155
Mosquito. *See also* specific species and genera
  aggregation pheromones, 116
  alimentary tract, 35
  as *Borrelia burgdoferi* vector, 140–141, 146
  brain, 31, 40, 42, 43
  gametocyte infection, 66, 67
  gametocytogenesis, 60–61
  hindgut, 35
  as La Crosse virus vector, 100
  midgut endocrine system
    anatomy, 32, 35, 46–47
    endocrine cells, 35–38, 46–47
    neuroendocrine system comparison, 46–50
    neuropeptide-like immunoreactivity, 41–46, 47–49
    subcellular organization, 38–39
  neural system, 29
  neuroendocrine system
    anatomy, 32–33, 47
    midgut endocrine system comparison, 46–50
    neuropeptide-like immunoreactivity, 47–48
    neurotransmitters, 29–30
    regulatory factors, 33–35
  as San Angelo virus vector, 100
  venereal virus transmission by, 101
M particle, 157, 158
Mulberry black ringspot virus, 159
Mulberry ringspot virus
  geographical distribution, 163
  natural spread, 165
  physicochemical properties, 159
  vectors, 165, 171
Musca domestica, adipokinetic hormone-like peptide content, 41
Mycetocyte, symbiote-virus association, 213–214
Mycoplasma-like organism, 268
  aster yellows, 271, 279
  maize bushy stunt, 281, 282
  orange, 213
  spiroplasma relationship, 268
  vectors, 278, 279, 280
  virescence-type, 279, 280
  yellow dwarf, 213
Myrobalan latent ringspot virus, 159, 162, 165
Myzus ascalonicus, as beet yellows virus vector, 311
Myzus euphorbiae, as beet yellow stunt virus vector, 311
Myzus ornatus, as beet yellows virus vector, 311
Myzus persicae
  as barley yellow dwarf virus vector, 195
  as beet yellows stunt virus vector, 311
  as beet yellows virus vector, 310, 311
  as carnation necrotic fleck virus vector, 311
  as citrus tristeza virus vector, 309
  control, 325
  as potato virus vector, 192–193, 194
  as tobacco mosaic virus vector, 192, 193
  as tobacco vein mottling virus vector, 194

Nasonovia lactucae, as beet yellow stunt virus vector, 311
Necrosis
  artichoke yellow ringspot, 152
  black line disease, 155
  cucumber, 201
  potato black ringspot, 153
  rice, 212
  rice stripe, 212
  tobacco, 201
  tomato top, 153, 159, 162, 165
  union, 154
Nematocera, perisympathetic organs, 32
Nematodes, as plant virus vectors. *See also* Nepoviruses; Tobraviruses; names of specific species and genera
  as passive carriers, 178
  transmission efficiency, 178–179
  virus detection in, 199–200
  virus-vector associations, 172–177
    nematode-virus interaction, 174–177
    variation in transmission, 173–174
Neoaliturus haematoceps
  geographical distribution, 271
  as *Spiroplasma citri* vector, 277
Nephotettix spp., as yellow dwarf mycoplasma agent vector, 213
Nephotettix bakeri
  as rice grassy stunt virus vector, 222
  as rice ragged stunt virus vector, 224
Nephotettix cincticeps
  geographical distribution, 212
  overwintering, 218
  plant colonization by, 212–213
  as rice bunchy stunt virus vector, 217–218
  as rice dwarf virus vector, 198, 218–219
  as rice gall dwarf virus vector, 220
  as rice transitory yellowing virus vector, 226
  as rice tungro viruses vector, 227
  as semipersistent virus vector, 307
  xylem feeding by, 252
Nephotettix lugens
  as rice grassy stunt virus vector, 222
  as rice ragged stunt virus vector, 224
Nephotettix malayanus
  as rice gall dwarf virus vector, 220
  as rice tungro viruses vector, 227
Nephotettix muiri, as rice grassy stunt virus vector, 222
Nephotettix nigropictus, 213
  detection, 199
  as rice dwarf virus vector, 218
  as rice gall dwarf virus vector, 220
  as rice transitory yellowing virus vector, 226
  as rice tungro viruses vector, 227, 228
Nephotettix parvus, as rice tungro viruses vector, 227
Nephotettix virescens
  geographical distribution, 212
  as rice bunchy stunt virus vector, 217–218
  as rice dwarf virus vector, 218
  as rice gall dwarf virus vector, 220
  as rice transitory yellowing virus vector, 226, 226
  as rice tungro viruses vector, 214–215, 227, 228
Nepoviruses, 151–189
  diseases induced by, 152–153
  dispersion, 180
  genome, 151, 156, 179, 180
  grouping
    geographical, 156, 162–164
    by natural spread, 165–166
    by physicochemical properties, 156, 157–160
    serological, 156, 160–161, 163
  host range, 152, 179, 180
  protein coat polypepetides, 157, 158–159, 160
  RNA, 151, 156, 157, 160, 176, 179
  vectors, 168, 168–172
  virus-vector associations, 172–177
Nervous system, of mosquito
  divisions, 31
  ganglia, 31

Nervous system, of mosquito (*cont.*)
neuropeptide-like immunoreactivity, 41–46
neurotransmitters, 30
Neural system, of mosquito, 29
Neuroendocrine regulatory factors, 33–34
Neuroendocrine system, of mosquito
anatomy, 32–33, 47
midgut endocrine system comparison, 46–50
neuropeptide-like immunoreactivity, 47–48
neurotransmitters, 29–30
regulatory factors, 33–35
Neurosecretory cell, 32–33
medial, 33–34
Neurotransmitters. *See also* specific neurotransmitters
of mosquito, 29–30
Nilaparvata lugens
as barley yellow striate virus vector, 198
detection, 199
as rice grassy stunt virus vector, 212
as rice ragged stunt virus vector, 212
Nitrogen-fixing bacteria, 258
Nonpersistent viruses, 301, 312, 324
Noradrenaline, 34
Nucleic acid hybridization, for virus detection, 199, 202

Oat blue dwarf virus, 198
Octopamine, 32, 34
Odontostyle, as virus retention site, 175, 176–177, 178
Olive latent ringspot virus
geographical distribution, 162
host, 155
natural spread, 165
physicochemical properties, 158
protein coat polypeptides, 160
Ollarianus strictus, as *Spiroplasma citri* harborer, 278, 278–279
Olphidium brassicae, tobacco necrosis virus relationship, 201
Olphidium cucurbitacearum, cucumber necrosis virus relationship, 201
Onchiostyle, 178
Oncometopia, 256
Onion mosaic virus, 200
Onion yellow dwarf virus, 200
Orange leaf mycoplasma agent, 213
Orbivirus, 97
Ornithodoros, as spirochete vector, 127, 128–129
Ornithodoros erraticus, as spirochete vector, 132, 133
Ornithodoros hermsi, as spirochete vector, 132
Ornithodoros moubata, as spirochete vector, 130
gut material regurgitation by, 139
spirochete behavior in, 131–132
transovarial transmission by, 133
Ornithodorus tartakovskyi, as spirochete vector, 132
Ornithodorus tholozani, as spirochete vector, 132, 133
Ornithodorus turicata, as spirochete vector, 132
Ornithodorus verrucosus, as spirochete vector, 132
Orosius argentatus, xylem feeding by, 252
Oviposition, by sandfly, 95–96

Pacui virus, 98, 99
Pagaronia, 256
Pancreatic peptide, 41, 42, 44
Pangola stunt virus, 216
Paracrine secretion, 48
Paralongidorus, geographical distribution, 169
Paratrichodorus
geographical distribution, 171
as tobacco rattle virus vector, 168, 172
as tobravirus vector, 164
Paratrichodorus allius, as tobacco rattle virus vector, 167

Paratrichodorus anemones, as
tobacco rattle virus vector, 167
Paratrichodorus christiei, as tobacco
rattle virus vector, 167, 168
Paratrichodorus minor
geographical distribution, 171
as tobacco rattle virus vector, 167
Paratrichodorus nanus, as tobacco
rattle virus vector, 167
Paratrichodorus pachydermus
geographical distribution, 171
as tobacco rattle virus vector, 167,
172, 176, 177
Paratrichodorus porosus
geographical distribution, 171
as tobacco rattle virus vector, 167
Paratrichodorus teres, as tobacco
rattle virus vector, 167
Paratrichodorus tunisiensis
geographical distribution, 171
as tobacco rattle virus vector, 167
Parsnip yellow fleck virus
helper virus, 329
location in vector, 314, 315
semipersistent transmission, 305,
315, 316, 328, 329
vectors, 311
Peach rosette mosaic virus
geographical distribution, 162
grapevine decline and, 154
hosts, 153, 155
natural spread, 165
physicochemical properties, 159
vectors, 165, 165, 170
Peach willow leaf rosette virus, 155
Pea early-browning virus
geographical distribution, 167, 167,
171
host range, 156, 167
serology, 168
vectors, 167, 168
Pea enation mosaic virus, 194–195
Pear psylla. *See Cacopsylla pyricola*
Pea seed-borne mosaic virus, 193
Pediculus humanus
as *Borrelia recurrentis* vector, 131
as relapsing fever spirochete vector,
130
Penyakit habang. *See* Tungro
Penyakit merah. *See* Tungro
Pepper ringspot virus, 156, 167
Peptides
ecdysteroidogenic activity, 33–34
gonadotropic acitivity, 34
myotropic, 34
as neurotransmitter, 29–30
Perisympathetic organ, 32
Persistent viruses, 301
Pheromone, of sandflies, 116–118
aggregation, 116–117
sex, 117–118
tergal spots and, 108, 109, 110–111
Philaenus spumarius, xylem feeding,
250
Phlebotomus argentipes
chromosomes, 113
as *Leishmania donovani* vector, 104
sugar feeding, 104
Phlebotomus ariasi
blood feeding by, 95
colonization, 92
cuticular hydrocarbons, 112
as *Leishmania* vector, 102–103, 112
ovaricular stalk dilations, 102–103
sugar feeding, 104–105
Phlebotomus bergeroti, spermathecae,
115
Phlebotomus colabaensis,
chromosomes, 113
Phlebotomus duboscqi
colonization, 92
gonotropic cycles, 100
oviposition, 96
Phlebotomus langeroni,
colonization, 92
Phlebotomus longicuspis,
spermathecae, 115
Phlebotomus major, spermathecae,
116
Phlebotomus martini, colonization, 92
Phlebotomus papatasi
age determination, 104
aggregation pheromones, 116–117
characterization, 107–108
chromosomes, 113
climate control for, 93

Phlebotomus papatasi (*cont.*)
culture, 95
DNA probe, 114
as *Leishmania major* vector, 114
as Pacui virus vector, 98
plant feeding, 105–106
as sandfly fever virus vector, 96, 97–98
spermathecae, 115
as Toscana virus vector, 98
Phlebotomus pedifer
blood feeding by, 95
colonization, 92
Phlebotomus perfiliewi
colonization, 92
cuticular hydrocarbons, 112
isoenzymes, 113
spermathecae, 115
Phlebotomus perniciosus
accessory glands, 102
anatomy, 107
as Arbia virus vector, 98
chromosomes, 113
colonization, 92
isoenzymes, 113
spermathecae, 115
as Toscana virus vector, 98, 100–101
Phloem-feeding insects, 251
Phony disease of peach, 244
causal agent, 246–247, 256
symptoms, 245–246
Phytoreovirus, 218, 220
Picornavirus, 305
Pierce's disease of grape
causal agent, 243
host range, 245
pathogenesis, 246–247
transmission efficiency, 256
epidemiology, 257–258
symptoms, 245–246
vectors, 243–244
Plant feeding, by sandflies, 105–106
Planthopper, as rice virus vector, 210, 211
Plant viruses. *See also* specific plant viruses
detection in vectors, 191–208
in aphids, 192–197
in beetles, 201–202
by electron microscopy, 191, 195, 198–200, 201
in fungi, 201
in leafhoppers, 197–199
in mites, 200–201
in nematodes, 199–200
by nucleic acid hybridization, 191
in planthoppers, 197–199
by radioactive isotope tagging, 192
by serology, 191, 193–194, 198, 199
in thrips, 202
by vector bioassay, 191
in whiteflies, 202
Plasmodiophoromycetes, 201
Plasmodium falciparum
gametocyte infectivity, 66, 73
gametocytemia and, 60–61, 67
in neurosyphilis, 62
patterns, 62–65
trophozoite density, 62
Plasmodium knowlesi, gametocyte development in, 65
Plasmodium malariae, gametocyte development in, 61
Plasmodium ovale, gametocyte development in, 61
Plasmodium vivax, gametocyte development in, 61
Plum lead scald, 244
Polymyxa betae, as beet necrotic yellow vein virus vector, 201
Polymyxa graminis
as barley yellow mosaic virus vector, 201
as rice necrosis virus vector, 212
Potato black ringspot virus
geographical distribution, 162
host, 153
natural spread, 165
physicochemical properties, 158
serology, 161
Potato leaf roll virus, 195
Potato virus, 159
Potato virus U, 153, 163, 165
Potato virus X, 192, 200
Potato virus Y, 192-193, 194

Potato yellow dwarf virus, 198
Precibarium, Xylella colonization, 254–256, 257
Precipitin ring test, 198
Pregnancy, gametocytemia during, 63
Proctolin, 29
Prohormone, secretory granule secretion by, 38
Prosopsis farcta, 106
Protein coat polypeptides, of nepoviruses, 157, 158–159, 160
Prothoracic gland, ecdysteroid secretion, 30
Prothoraciocotropic hormone, 39
Pruning
  growth flush effects, 325
  virus transmission effects, 320
Psammomys obesus, 105
Pseudo-aucuba disease, 153
Pseudococcidae, as semipersistent virus vector, 305, 306, 307
Pseudomonas aeruginosa, 107
Pseudomonas solanacearum, 244, 245
Psychodopygus carrerai, isoenzymes, 113
Psychodopygus complexus
  cuticular hydrocarbons, 111–112
  isoenzymes, 113
Psychodopygus wellcomei
  cuticular hydrocarbons, 111–112
  isoenzymes, 113
  nulliparous biting activity, 103
Psychodopygus yucumensis, isoenzymes, 113, 114
Pyrethroid, 324, 329

Quarantine, for citrus tristeza virus control, 322–323

Raspberry ringspot virus, 152, 154
  antigenic variants, 179
  detection in vector, 199
  geographical distribution, 162, 162
  natural spread, 165
  physicochemical properties, 158
  as plant death cause, 178
  serology, 161, 163
  vectors, 165, 172, 173, 173, 174, 175, 176–177
Ratoon stunt disease of sugar cane, 245
Recilia dorsalis
  as orange leaf mycoplasma agent vector, 213
  as rice dwarf virus vector, 218, 219
  as rice gall dwarf virus vector, 220
  as rice tungro viruses vector, 227, 228
Regenerative cell, 36–38
Relapsing fever, vectors, 130
Restriction fragment length polymorphisms, 16–18, 23
Rhabdovirus, 97
Rhipicephalus evertsi, as *Borrelia theileri* vector, 133
Rhipicephalus sanguineus, as *Borrelia burgdorferi* vector, 139
Rhopalosiphoninus staphylae, as beet yellows virus vector, 311
Rhopalosiphum padi
  as barley yellow dwarf virus vector, 195, 196, 197
  as rice giallume virus vector, 221
Ribonucleic acid (RNA)
  of anthriscus yellow virus, 305
  of insect-borne rice viruses, 215–216
  of nepoviruses, 151, 156, 157, 160, 176, 179
  of parsnip yellow fleck virus, 305
  of rice black-streaked dwarf virus, 215, 216
  of rice dwarf virus, 215, 216, 218
  of rice gall dwarf virus, 215, 216, 220
  of rice grassy stunt virus, 215, 216, 221
  of rice hoja blanca virus, 215
  of rice ragged stunt virus, 215, 216, 224
  of rice stripe virus, 215, 224–225
  of rice transitory yellowing virus, 216
  of rice yellow mottle virus, 216

Ribonucleic acid (RNA) (*cont.*)
of tobraviruses, 151, 166, 168, 176, 179–180
Ribonucleic acid polymerase, 1, 2
rRibonucleic acid (rRNA), dipteran, 4–5
Rice black-streaked dwarf virus
disease symptoms, 216–217
geographical distribution, 216
morphology, 216
outbreaks, 209, 216
RNA, 215, 216
vectors, 210, 211, 212, 217
Rice bunchy stunt virus, 210, 212, 217–218
Rice dwarf virus, 218–219
detection in vector, 198, 199
outbreaks, 209, 218
RNA, 215, 216, 218
transovarial transmission, 213–214
vectors, 198, 210, 211, 212, 218–219
Rice gall dwarf virus
outbreaks, 210
RNA, 215, 216, 220
vectors, 199, 210, 211, 220
Rice giallume virus
geographical distribution, 211
vectors, 210, 211, 213, 221
Rice grassy stunt virus, 221–222
outbreaks, 209
RNA, 215, 216
vectors, 199, 210, 212, 222
Rice hoja blanca virus
first outbreak, 209, 222–223
geographical distribution, 211
RNA, 215
vectors, 210, 211, 212, 223
Rice necrosis virus, 212
Rice ragged stunt virus, 223–224
outbreaks, 209–210, 223
RNA, 215
vectors, 199, 210, 211, 212, 224
Rice stripe necrosis virus, 212
Rice stripe virus, 222, 224–226
detection in vectors, 199
outbreaks, 209, 224
RNA, 215
vectors, 210, 211, 225–226
detection, 199
overwintering, 212
Rice transitory yellowing virus
geographical distribution, 212
outbreaks, 209, 226
RNA, 216
vectors, 210, 211, 226, 226
Rice tungro bacilliform virus. *See also* Tungro
rice tungro spherical virus dependence, 315–316
semipersistent transmission, 214, 215
vectors, 210, 227–228
Rice tungro spherical virus, 228–229. *See also* Tungro
semipersistent transmission, 214–215
vectors, 210, 211, 227–228
Rice tungro virus(es), 301
Rice viruses, insect-borne, 209–241
geographical distribution, 209–212
morphology, 216
nucleic acids and proteins, 215–216
vector species, 212–213
virus-vector interactions, 213–215
Rice waika virus. *See* Rice tungro spherical virus, 227
Rice wilted stunt virus, 222
Rice yellow mottle virus
geographical distribution, 211, 212
outbreaks, 209, 228–229
RNA, 216
semipersistent transmission, 214
vectors, 210, 211, 213, 229
Rift Valley fever virus, 98
Ringspot viruses. *See* specific ringspot viruses
Rio Grande virus, 97
Rubus Chinese seed-borne virus
geographical distribution, 163
natural spread, 165
nucleoproteins, 157, 158
serology, 161

Saliva, sheath, 251
Salivary gland, *Borrelia burgdorferi* transmission in, 138–139

San Angelo virus, 100
Sandfly, phlebotomine, 91–126. *See also* specific species and genera
  age determination, 102–104
  as Arboledas vector, 101
  chemically-mediated behavior, 116–118
  colonization, 92–96
    adults, 94, 95–96
    climatic control, 92–93
    equipment, 94
    food, 95
    immature stages, 93, 95
  as *Leishmania* vectors, 91, 106–107
  as Maraba virus vector, 101
  as orbivirus vector, 97
  as rhabdovirus vector, 97
  species group and complexes, 107–116
    chromosomal analysis, 112–113
    crossing characteristics, 108–110
    cuticular hydrocarbons, 111–112
    DNA probes, 114–115
    electrophoresis analysis, 113–114
    morphological investigation, 115–116
    semiochemical identification, 110–111
    song patterns, 111
  sugar feeding, 104–107
  as virus reservoir/vector, 96–102
    experimental infection, 96–97
    new virus isolation, 101
    sandfly susceptibility, 97–100
    transovarial transmission, 96, 100–101
Sandfly fever virus, 96, 97–98
Sap, xylem, 247, 248–250
Sarcophaga bullata, rDNA, 4, 6
Satsuma dwarf virus
  geographical distribution, 163
  natural spread, 165
  physicochemical properties, 158
  protein coat polypeptides, 158, 160
Scaphytopius acutus, 272
Scaphytopius acutus delongi, as *Spiroplasma citri* vector, 271–272, 277–278
Scaphytopius californiensis, as *Spiroplasma citri* vector, 272, 277–278
Scaphytopius nitridus, as *Spiroplasma citri* vector, 271, 277–278
Sciara coprophila, rDNA, 4
Semipersistent viruses, 301
  control, 317–327, 329
  description, 303–305
  epidemiology, 316–317
  filamentous, 304–305, 328
  helper viruses, 328–329
  as noncirculative viruses, 312
  spherical, 305, 328
  transmission time, 324
  vectors, 305–312, 328
    identification, 330–332
  virus-vector relationship, 312–316
Sergentomyia, colonization, 92
Serotonin, 34
Sesselia pussilla, as rice yellow mottle virus vector, 229
Shigella sonnei, 106–107
Sigma virus, 100
Simulium damnosum, DNA probe, 114
Sitobion avenae
  as barley yellow dwarf virus vector, 195, 196, 197
  as rice giallume virus vector, 221
Slaterivirus, 160
Sogatodes cubanus, as rice hoja blanca virus vector, 223
Sogatodes orizicola
  as rice hoja blanca virus vector, 223
  as rice virus vector, 211, 212, 213
Somatostatin, 40, 48
Southern bean mosaic virus, 193, 202
Spermathecae, of sandflies, 115–116
Spirochete-vector relationships, in borrelioses, 127–150. *See also* specific *Borrelia* species
  behavior in vectors, 130–146
Spiroplasma
  Drosophila sex-ratio organism, 267
  hosts, 267
  leafhopper vectors, 267–299
  taxonomy, 268–269
  transmission terminology, 272–274

Spiroplasma citri
as citrus stubborn disease vector, 268
detection in vectors, 284
diseases caused by, 274–276
harboring organisms, 278–279
as horseradish brittle root disease causal agent, 276
hosts, 275–276
vectors, 269, 276, 276–278
artificial transmission in, 278
experimental transmission in, 286
Spiroplasma kunkelii
associated pathogens, 282–283
as corn stunt disease vector, 268
detection in vectors, 284
diseases caused by, 280–281
hosts, 280–281
vectors, 271–272, 281–282, 287
Spiroplasma phoeniceum
classification, 268, 283
experimental transmission, 287
hosts, 283
as mycoplasma-like disease vector, 280
Spiroplasmataceae, 268
Spissistilus festinus, as *Spiroplasma citri* vector, 278
Spittlebug. *See also* Cercopidae
meadow. *See Philaenus spumarius*
as Pierce's disease vector, 243–244
xylem feeding by, 250
Staphylococcus aureus, 106–107
Stem pitting and decline, of cherry, 155
Stirellus bicolor, as *Spiroplasma kunkelii* vector, 272, 282
Stomoxys calcitrans, pheromones, 111
Strawberry latent ringspot virus, 153
geographical distribution, 162, 164
hosts, 155
natural spread, 165
as peach willow leaf rosette cause, 155
physicochemical properties, 158
serology, 161
vectors, 165, 173–174, 175
Strawberry latent virus, 154
Strawberry leafroll virus, 178–179
Streptococcus, 106–107
Stylet, in semipersistent transmission, 193, 312
Subterranean clover red leaf virus, 197
Sugar beet curly top virus, 279. *See also* Beet curly top virus
Sugar feeding, by sandflies, 104–107
Sumatra disease, 244, 245, 257
Symbiote-virus association, in mycetocyte, 213–214

Tabanus nigrovittatus, as *Borrelia burgdorferi* vector, 141
Tenui virus group, 221, 223
Tergal spot, 108, 109, 110–111
Terthron albovittata, as rice stripe virus vector, 225
Tetranychus urticae, as plant virus vector, 200
Thrip, as plant virus vector, 202
Tick. *See also* specific species and genera
aggregation pheromones, 116
spirochete behavior in, 130–146
Tobacco hornworm, insulin-like peptide, 40, 47–48
Tobacco mosaic virus
detection in vectors, 192, 193
in mites, 200
vectors, 202
Tobacco necrosis virus, 201
Tobacco rattle virus
detection in virus, 200
geographical distribution, 167, 171
host range, 156, 167
vectors, 167, 168, 171, 172
retention site, 179–180
virus-vector association, 176
Tobacco ringspot virus, 153, 200
geographical distribution, 162, 164
grapevine decline and, 154
natural spread, 165
physicochemical properties, 158
protein coat polypeptides, 160
serology, 161, 161, 163
vectors, 170

Tobacco severe etch virus, 193, 194
Tobacco vein mottling virus, 194
Tobraviruses
  classification, 166, 168
  diseases induced by, 156
  dispersion, 180
  gene pool, 168
  genome, 151, 176
  nematode vectors, 171
  physicochemical properties, 164, 166
  RNA, 151, 166, 168, 176, 179–180
  serological grouping, 168
Tomato black ring virus, 152
  detection in vector, 199
  geographical distribution, 162, 164
  natural spread, 165
  physicochemical properties, 159
  pseudorecombinant variant, 179
  serology, 161, 163
  vectors, 165, 172, 174, 175
Tomato bushy stunt virus, 200
Tomato ringspot virus, 153
  as flat apple cause, 154
  geographical distribution, 162, 164
  grapevine decline and, 154
  natural spread, 165
  physicochemical properties, 159
  as plant death cause, 178
  protein coat polypeptides, 160
  as stem pitting cause, 155
  as union necrosis and decline cause, 154
  vectors, 165, 170, 172, 175
  yellow bud mosaic and, 155
Tomato spotted wilt virus, 202
Tomato top necrosis virus, 153
  geographical distribution, 162
  natural spread, 165
  physicochemical properties, 159
Tomato yellow leaf curl virus, 202
Toscana virus, 98, 100–101
Toxoptera aurantii, as citrus tristeza virus vector, 308, 309, 310
Toxoptera citricidus, as citrus tristeza virus vector, 302, 308, 309, 319, 320
Toxorhynchites, midgut endocrine cells, 36
Transmission
  persistent, of insect-borne viruses, 213–214
  semipersistent, 312–316
    characteristics, 313
    dependence in, 315–316
    foregut in, 193
    helper-assisted, 315
    of insect-borne viruses, 214–215
    specificity, 313–314
    stylets in, 193
    virus localization, 314–315
  transovarial
    of borreliosis, 132–133, 146
    of Lyme disease borrelia, 146
    of rice viruses, 211, 213–214
    by sandflies, 100–101
  venereal, of borreliosis, 133
  vertical
    of borreliosis, 132
    of Lyme disease borrelia, 146
Treponema pallidum, behavior in vector, 131
Trialeurodes vaporariorum, as semipersistent virus vector, 307
Trichispa sericea, as rice yellow mottle virus vector, 229
Trichodorus
  geographical distribution, 171
  onchiostyle, 178
  as tobacco rattle virus vector, 168, 172
  as tobravirus vector, 164, 172
  virus association, 177
  virus retention, 176, 177, 180
  virus transmission mechanism, 178
Trichodorus cylindricus, as tobacco rattle virus vector, 167, 172
Trichodorus hooperi
  geographical distribution, 171
  as tobacco rattle virus vector, 167
Trichodorus pachydermus, as tobacco rattle virus vector, 200
Trichodorus primitivis
  geographical distribution, 171
  as pea early-browning virus vector, 167
  as tobacco rattle virus vector, 167

Trichodorus similis, as tobacco rattle virus vector, 167
Trichodorus viruliferus
as pea early-browning virus vector, 167
as tobacco rattle virus vector, 167
Tsetse fly. *See Glossina morsitans morsitans*
Tungro, 227
outbreaks, 209, 228

Union necrosis and decline, 154
Unkanodes albifascia
as rice black-streaked dwarf virus vector, 216, 217
as rice stripe virus vector, 225
Unkanodes sapporonus
as rice black-streaked dwarf virus vector, 216, 217
as rice stripe virus vector, 225
Uroleucon jaceae, as citrus tristeza virus vector, 308

Vector(s). *See also* under specific viruses; names of specific vectors
vector-virus associations
of nematode-plant viruses, 172–177
of semipersistent viruses, 312–316
Vesiculovirus, vectors, 97, 101
Virus-like particle, in semipersistent transmission, 314–315

Wheat streak mosaic virus, 200
Whitefly. *See also* Aleyrodidae
as plant virus vector, 202
Wilt disease of periwinkle, 244
Wound tumor virus, 198

Xiphinema
geographical distribution, 168–170
virus association, 177
virus retention, 180
virus transmission mechanism, 178
Xiphinema americanum
as cherry rasp-leaf virus vector, 165
geographical distribution, 170
morphometry, 170
as peach rosette mosaic virus vector, 165, 165, 170
as tobacco ringspot virus vector, 170
as tomato black ring virus vector, 165
as tomato ringspot virus vector, 165, 170, 175
Xiphinema americanum sensu lato, 170
Xiphinema americanum sensu stricto, 170
Xiphinema brevicolle, geographical distribution, 170
Xiphinema californicum
as cherry rasp-leaf virus vector, 170
morphometry, 170
as tomato ringspot virus vector, 165, 170
Xiphinema coxi, morphometry, 170
Xiphinema diversicaudatum
as arabis mosaic virus vector, 165, 173, 174, 175, 178–179, 180, 200
geographical distribution, 169–170
morphometry, 170
as strawberry latent ring spot virus vector, 165, 173-174, 175
as strawberry leafroll virus vector, 178–179
virus detection in, 200
virus retention site, 176, 177
Xiphinema index
geographical distribution, 169
as grapevine fanleaf virus vector, 151, 164, 165, 170, 175, 180
Xiphinema italiae, vectors, 165
Xiphinema occidum, as tomato ringspot virus vector, 170
Xiphinema pachtaicum, geographical distribution, 170
Xiphinema rivesi, as tomato ringspot virus vector, 165, 170
Xiphinema utahense, as tomato ringspot virus vector, 170

Xylella, 245
xylem colonization, 247–248
Xylella fastidiosa
as alfalfa dwarf disease pathogen, 257, 258
as almond leaf scorch pathogen, 257, 258
cibarial pump colonization, 249–250, 253
foregut colonization, 253–254
hosts, 244, 246
as leafhopper pathogen, 254–257
needle inoculation, 253
pathogenicity, 246–247
as phony disease of peach causal agent, 244, 245–247
as Pierce's disease of grape causal agent, 243, 243–244, 245–247, 257–258
strain differences, 246–247
vectors, 248–250, 252, 253–254
xylem colonization, 248
Xylem, bacteria association with, 243–266
epidemiology, 257–259
historical background, 243–244
pathogenicity, 245–246
strain differences, 246–247
vectors, 244, 245, 250–257
xylem environment and
physical features, 247–248
xylem chemistry, 248–250
Xylem-feeding insects, 248–250. *See also* Homoptera; Homoptera, as xylem-inhabiting bacteria vectors
phloem-feeding insects versus, 251
as *Xylella fastidiosa* vector, 248–250, 252
cibarial pumps, 249–250, 253
foregut and, 253–254

Yellow bud mosaic disease, 155
Yellow dwarf mycoplasma agent, 213
Yellow orange leaf. *See* Tungro
Yolk, deposition, 34

Zucchini yellow mosaic virus, 324–325